CHEMICAL ENGINEERING METHODS AND TECHNOLOGY

FULLERENES

CHEMISTRY, NATURAL SOURCES AND TECHNOLOGICAL APPLICATIONS

Chemical Engineering Methods and Technology

Additional books in this series can be found on Nova's website under the Series tab.

Additional e-books in this series can be found on Nova's website under the e-book tab.

CHEMICAL ENGINEERING METHODS AND TECHNOLOGY

FULLERENES

CHEMISTRY, NATURAL SOURCES AND TECHNOLOGICAL APPLICATIONS

SHANNON B. ELLIS
EDITOR

nova publishers
New York

Library of Congress Cataloging-in-Publication Data

ISBN: 978-1-63321-385-2

Published by Nova Science Publishers, Inc. † New York

CONTENTS

PREFACE

Fullerenes are molecules composed entirely of carbon atoms, arranged into structures similar to graphite, i.e., stacked graphene sheets of linked hexagonal rings; but they may also contain pentagonal (or sometimes heptagonal) rings. The rings then form nanoparticles of various shapes, such as hollow spheres, ellipsoids, and also cylinders (tubes). Fullerenes are important biologically active molecules with diverse effects on cells and tissues. They are able to oxidize biological molecules and cause serious damage to them, which can lead to cell malfunctioning, genome mutations and cell death. Simultaneously, fullerenes can act as reduction agents. Both the harmful oxidizing effects and the beneficial reducing effects of fullerenes can be exploited for preventing and treating various disorders. This book discusses the chemistry of fullerenes, as well as its natural sources and several technological applications.

Chapter 1 – Fullerenes are spheroidal hollow cage-like carbon nanoparticles with diverse biological activities. They can act either as acceptors or as donors of electrons. Thus, they are able to generate harmful reactive oxygen species which can damage biological molecules and cells. Fullerenes can act as cytotoxic and genotoxic agents, causing oxidative DNA damage, inhibition of detoxificatory and antioxidant enzymes, polyploidy, premature cell senescence, apoptosis and inflammation. However, oxidative damage to cells by fullerenes, further potentiated by irradiation with visible and ultraviolet light, can be utilized for photodynamic therapy against tumors and microbes, including viruses. Simultaneously, fullerenes C_{60} and their derivatives are considered to be the world's most efficient radical scavengers. These molecules are promising for preventing ultraviolet-light-induced skin damage and wrinkle formation, and also for treating inflammatory diseases (e.g., arthritis or allergies), ischemic and hypoxic injuries of myocardium, amyloid-based neurological diseases, alcohol-induced encephalopathy, lead intoxication, or obesity and diabetes-related disorders. The biological response to fullerenes is profoundly influenced by their physical and chemical properties, such as functionalization with various chemical groups (by which the fullerene solubility can be tuned), degree of agglomeration, concentration and use of various solvents. The physicochemical properties of fullerenes and their biological effects may also change in the course of time. In the authors' studies, performed on fresh and aged C_{60} films, the aged films (i.e., one year old) provided better support for the adhesion, spreading, growth, activity of mitochondrial enzymes and viability of human osteoblast-like cells than fresh, i.e., one-week-old C_{60} films. These results can be attributed to changes in the fullerene films during aging, such as fragmentation, oxidation, polymerization and graphitization of fullerenes in an air atmosphere, and thus loss of their reactivity. Also the morphology of fullerene films can influence the cell behavior. On micropatterned fullerene films, i.e., films containing prominences and grooves, the cells adhered and grew preferentially in the grooves, which was more apparent on the fresh films. In addition, fullerenes C_{60} were combined with Ti atoms, because titanium is a metal used for constructing clinically applied bone implants. Hybrid C_{60}/Ti composites provided good substrates for cell adhesion and growth in both fresh and aged forms. This can be explained by the fact that the changes in the fullerene molecules, such as fragmentation, polymerization and oxidation, occurred immediately during C_{60} and Ti co-deposition by interaction of C_{60} molecules and Ti atoms. A favorable result was that the cells on all fullerene C_{60} and C_{60}/Ti films, i.e., fresh, aged, continuous and micropatterned, showed no DNA damage, evaluated by the presence of markers of a DNA damage response, namely markers of DNA double strand breaks in U-2 OS cells (gamma-HA2X and 53BP1). Thus, aged fullerene C_{60} films and fresh and aged C_{60}/Ti provided good substrates for the adhesion, growth and viability of human bone-derived cells, and so they can be used as a bioactive coating for bone implants. In micropatterned form, these films can be used for inducing regionally selective cell adhesion and directed growth for tissue engineering and advanced biotechnologies, such as genomics and proteomics.

Chapter 2 – Professor Kroto in 1985, fullerenes have been studied in depth because of their unique structure and potential as new nanomaterials. This is due to their appealing photo-, electro-chemical, and physical properties, which make them exploitable in different fields. Their electron double bonds allow pristine fullerenes, which are intrinsically hydrophobic, to become readily derivatized and water soluble through the addition of various functional groups. Hence, fullerene derivatives are being investigated for their use in biomedical, cosmetic and industrial applications, ranging from drug-delivery systems and anti-aging formulations to electrical components. Moreover, some of these fullerene derivatives play an important role in the controllable synthesis of new materials and biologically active compounds.

A rapid growth in the development, production, and use of these compounds is being predicted over the next decade, so their dispersion in the environment is to be expected. Moreover, their unique colloidal properties in water can affect their behavior in the environment. For instance, although C_{60} is practically insoluble in water, it can spontaneously form stable aqueous colloidal suspensions containing nanocrystalline aggregates with effective aqueous phase concentrations many orders of magnitude above that corresponding to the aqueous solubility of molecular fullerenes. Consequently, the ability of these carbon nanomaterials to form aggregates could change their bioavailability, leading to additional toxicological concerns, although at the moment data regarding their toxicity is limited. The lack of specific analytical methods for their determination presents a significant limitation when attempting to evaluate the risk these materials present to human health and the environment.

Although significant advances have been made in the analysis of fullerenes in the past few years, there is still an increasing need to develop effective, sensitive, and reliable analytical methods for their determination. Reversed-phase liquid chromatography (RP-LC), using toluene-acetonitrile or toluene-methanol mixtures as mobile phases, has been the method of choice for their analysis. Although ultraviolet (UV) detection is being used for their detection, liquid chromatography coupled to mass spectrometry (LC-MS) is the most commonly proposed technique nowadays for the determination and characterization of this family of compounds in complex matrices. Nowadays, a number of LC-MS applications for the analysis of this family of compounds in environmental waters, biological fluids and cosmetics are available. For most of these applications, both electrospray ionization (ESI) and atmospheric pressure chemical ionization (APCI) have been proposed, but recently atmospheric pressure photoionization (APPI) has appeared as one of the best options for their ionization due to their unique photoionization properties.

This chapter will review the state-of-the-art of liquid chromatography techniques for the analysis of fullerenes. LC and LC-MS conditions, UV-detection systems, ionization sources, and MS analyzers frequently used, as well as strategies for the unequivocal confirmation and quantitative analysis of fullerenes will be discussed by means of relevant applications. Coverage of all kind of applications is beyond the scope of the present contribution, so the authors will focus on the most relevant application fields such as bio-analytical, cosmetic and environmental applications.

Chapter 3 – The present chapter includes the detailed study of the physicochemical properties, structure, and antioxidant effect of both polystyrene films and polystyrene/ fullerene composite films.

The glass transition temperatures, T_g, and specific heat capacities, C_P, in range of 293-423 K have been determined for the films by DSC technique. The plasticization of the polymer is observed in thermal properties of the films under influence of small fullerene additions. The values of T_g and C_P decrease and thermal coefficient of heat capacity, b, increase as fullerene content increases up to 0.02 wt. %. The effect of interaction between polymer and fullerene molecules on thermal properties becomes evident at higher fullerene content in range from 0.02 to 0.1 wt. %.

Electrical properties of the films were investigated. Based on the results of measurements of capacitance and dielectric loss tangent can be conclude that the fullerene-polystyrene composites studied are nonpolar. The authors revealed dependence of the dielectric constant on the film composition is nonmonotonic with a minimum if there is 0.035 wt. % of C_{60}. It is found that the doping polystyrene with small amounts of fullerenes results neither an appreciable increase in the static conductivity nor occurrence of the dielectric relaxation processes.

The structural characteristics of polystyrene films filled with fullerenes were researched by X-ray diffraction technique. It was revealed there is the intermediate-range order in these materials. Insertion of fullerenes into polymer matrix increases a correlation length. It is established by small-angle X-ray scattering that composites containing fullerenes in the amount 0.01 wt. % are characterized by the greatest mobility of the segments in the modified polymer chains.

Solvatochromic effect in the absorption spectra of fullerene in weakly polar solvents and in polymer-containing solution was revealed by electron spectroscopy. It is suggested that there is intermolecular interactions of macromolecules of polystyrene and fullerene in studied solutions and polystyrene/fullerene composite films.

Kinetics of adsorption of Cu(II), Zn(II), and Cd(II) ions by both original polystyrene and polystyrene films filled with fullerenes were studied. It was found that modification of polystyrene increases the adsorption capacity of polymer material.

Effect of fullerene-containing polystyrene nanocomposites on free-radical processes in blood serum has been investigated in vitro. The parameters of lipid peroxidation in native serum after adding nanomaterials were determined by chemiluminescent analysis and spectrophotometry. It was revealed that fullerene-containing nanocomposites can manifest antioxidant activity in blood serum.

Chapter 4 – Since the discovery of a fullerene C_{60}, almost thirty years have passed. During this time, a great number of down-to-earth approaches to the synthesis of fullerene derivatives that constitute a new class of organic and organometallic compounds have been described in the worldwide literature. For structure elucidation of new fullerene adducts, a number of state-of-the-art physicochemical means is widely used, among which ^{13}C NMR possesses the most informative value. Using NMR spectra, regioselectivity and the type of addition to a fullerene core, as well as the symmetry set point can be ascertained with high probability. As the literature shows, these characteristics are determined by a number, intensity and position of the resonances in the ^{13}C NMR spectra. As a rule, no exact and precision assignments of the signals to the individual fullerene carbon atom are achieved restricting only by their enumeration. From a different standpoint, the articles that describe how these problems have been solved require special attention. INADEQUATE NMR technique, for which samples with high concentration (200 mg and more) or labeled with the ^{13}C isotope are necessary was successfully applied in this regard. In this chapter, the results of the examination of the relationship between the ^{13}C NMR chemical shift and structure of fullerenes including their higher derivatives are summarized and discussed. These compounds are mono-adducts from [2+n] cycloaddition (where n=1-4), as well as those obtained by diverse radical and nucleophilic addition reactions. It is also an attempt to review a worldwide and own experience in the use of the quantum chemistry approaches to calculate and ascribe the signals in the ^{13}C NMR spectra of fullerene derivatives. The authors believe that detailed knowledge of the fullerene carbon shielding regularities will create additive rules for exact and unequivocal prediction of the ^{13}C NMR chemical shifts, as well as electronic, optical, and magnetic features of fullerene derivatives.

Chapter 5 – In this review, the research results on the application of organic and inorganic azides in fullerene chemistry are summarized and systematized.

The published data on the chemistry of intermolecular cycloaddition of organic azides to fullerenes C_{60} in terms of thermal and photochemical reactions to give triazolinofullerenes, 5,6-azahomofullerenes, 5,6- and 6,6-aziridinofullerenes are presented in detail.

The first reperesentatives of azidofullerenes were synthesized by the reactions between fullerenes and inorganic azides, namely HN_3, IN_3, BrN_3, metal azides and organometallic azides.

The mechanisms for the formation of new nitrogen-containing fullerene derivatives are discussed.

The bibliography includes 118 references.

Chapter 6 – Molecular switch is a widespread concept in molecular machinery, which is currently a rapidly developed field of nanotechnology. It consists in the ability of a molecular system to exist in two or more stable states, differing by any physical property. Fullerene derivatives, due to the unique structural and electronic properties, have become very attractive for switchable molecular devices. The authors have classified the synthesized and tested fullerene-based switches to recognize what novel fullerene derivatives may be useful for this application.

Previously, the authors have examined polarizability of fullerene derivatives and found that all of them demonstrate remarkably high mean polarizabilities (See: D. Sh. Sabirov, Polarizability of C_{60}/C_{70} Fullerene [2+1]- and [1+1]-Adducts: A DFT-Prognosis // Density

Functional Theory: Principles, Applications and Analysis. Editors: J. Morin and J. M. Pelletier. New York: Nova Publishers, 2013. P. 147–170). This property can be a landmark for design of fullerene-based molecular systems with adjustable properties.

In the present paper, the authors have considered the fullerene-based nanostructures, which can act in two states with different polarizability, *i.e.*, can be potentially useful for application as molecular switch:

1. pairs of isomeric 6.6-closed and 5.6-open fullerene adducts;
2. fullerene dimers with flexible bridge between the cages;
3. reversibly dissociating singly bonded fullerene dimers.

Relations of the authors' theoretical findings to the recent experimental works on molecular switch have been analyzed. It makes them hope that their theoretical propositions will be stimulating for the fullerene-based molecular machinery.

Chapter 7 – The Fullerenes, are organic compounds of high interest, formed by a structure very special of hydrophobic character, presented a only class of spherical molecules of 60 and 70 carbons by general, present a hybridized sp^2 with a conjugated system, this give high aromaticity to organized structure organic, and excellent electrical properties, each fullerene is a closed network of fused pentagons and hexagons, spherical fullerene are called buckyballs and the cylindrical are called buckytubes, which may be single or multiple wall. All these structural features make them special, are considered highly stable molecules, ideally involved in highlighting different reactions like: addition, substitution, involving radicals with metal complexes, polymerization, among others, which provide them a covalent surface modification. Such molecules can be obtained via natural or synthetic. Present important applications in different areas highlighting electronics, medical, automotive, computer, biomedical, among others. This chapter aims to present his obtaining, reactivity and applications of fullerenes.

Chapter 8 – The first natural source of fullerenes appears to be giant systems, where extreme thermodynamic properties may provide an appropriate medium for the birth of these interesting molecules. It may be difficult to find these conditions on the surface of a planet; however, the stars are very interesting bodies with particular characteristics that make them a suitable laboratory for this purpose. The aura of a star comprises a very large region (millions of kilometers), where there are clouds of gases and dust that provide appropriate conditions for these reactions. These conditions can lead to the formation of some light elements (such as atoms) and more complex compounds such as fullerenes and other organic molecules, depending on the proximity of a specific region of the aura to the bulk, as well as the quality of the irradiation of the star. Recent studies have demonstrated that the structures of the compounds found in stars are similar to certain organic substances that occur on Earth. This document aims to show how the stars may constitute the principal natural source for fullerenes and related molecules.

Chapter 9 – The unique physical, chemical and biological properties of fullerenes have opened an intensive research on their possible applications in nanomedicine in the last few decades. The highly unsaturated structure of the hydrophobic pristine fullerene (C_{60}) is suitable for: 1) functionalization of molecule with different functional groups and production of water-soluble derivatives, such as polyhydroxyfullerenes (fullerols or fullerenols) and carboxyfullerenes; 2) conjugation with cyclodextrins, poly(2-oxazolines), isostearic acid, polyethyleneglycol or amino acids; 3) encapsulation into carriers with hydrosoluble surface such as liposomes and nanoparticles of polyvinylpyrrolidone (PVP). Such strategies enable dispersion in aquous vehicles, prevent formation of fullerene aggregates and may increase fullerene capacity to interact with biological environment. Fullerenes as well as their derivatives and C_{60} based composites, demonstrated to have antioxidant activity *in vitro* and *in vivo*. Therefore, they are known as "free-radical sponges". Several studies have demonstrated that fullerene induce suppression of human skin cell injuries caused by ultraviolet (UV) light irradiation and/or peroxides, brain cells damage by peroxide, and oxidative stress induced damage of lung. Thus, they are promissing agents for prophylaxis or therapy of free radical-related skin diseases and other serious disorders such as Alzheimer`s disease, Parkinson`s disease, neutrophilic lung inflammation. PVP-entrapped C_{60} submicron particles (Radical Sponge) are already employed as an ingredient of the „anti-aging" cosmetic products on the current market. Fullerenes are potentially useful for acne treatment, due to their properties as free radical scavengers. On the other hand, fullerenes may become reactive oxygen species (ROS) generators, under visible or UV light exposure. The cytotoxicity of fullerenes is perspective for photo- and radio-therapy of cancers and as antimicrobial agents. The role of fullerenes as antioxidants or prooxidants is dependent on applied dose/ concentration, molecule/particle size, surface morphology, chemical modification/conjugation /encapsulation, and the presence of light or ionizing radiation. HIV-1 protease inhibition is a new application that has been proposed recently for fullerenes and ~~its~~ their derivatives. The future of fullerene family, including novel nanomaterials, such as PEGylated fullerene/iron oxide nanocomposites and pH-activated glycol chitosan/fullerene nanogels, is focused on development of carriers suitable for targeted drug delivery and/or diagnostic strategies. This chapter highlights both benefits and potential health risks of fullerens as active agents of nanopharmaceuticals and cosmetic products.

Chapter 10 – A fullerene is any molecule composed entirely of carbon, in the form of a hollow sphere, ellipsoid, tube, and many other shapes. In this paper were considered works on formation and the synthesis of fullerenes in flames in the following sections, such as, the structure and properties of fullerenes; applications of fullerenes; methods for identification of fullerenes; methods for producing fullerenes; formation and synthesis of fullerenes in flame; synthesis of fullerenes in flames with applied electrical field; mechanism of fullerenes formation. Also was presented modified scheme for soot and fullerenes formation process in flames.

It has been found that if the peripheral zone of the benzene flame is heated by some external source, such as a laser beam, which not just burns the soot but also creates the same conditions as in the middle of the flame, the concentration of fullerenes increases.

The increase in the yield of fullerenes in the case where a ring electrode is positioned above the peripheral part of the reaction zone of a flame is due to the glow-discharge conditions providing an effective synthesis of fullerenes.

Thus, it has been established that the yield of fullerenes increases under the action of a glow discharge in the case where an electrode (a needle or a ring) is positioned directly above a flame, and a maximum yield of fullerenes is attained with the use of a ring electrode positioned above the central region of the flame front. The maximum yield of the fullerene C_{60} was β=15% of the soot formed.

Chapter 11 – The measure of the sphericality of a fullerene cage C_n with n carbon atoms is obtained by computing the topological invariant $\rho = 2W / n\underline{w}$, being W the topological Wiener index of the fullerene molecular graph F and $\underline{w}$ the contribution to W coming from the minimal vertices. Such an originally model is applied to the infinite families of tubular-like 1D fullerenes with exactly $50 + 10n$, $60 + 12n$, $10n$ and $12n$ carbon atoms, correlating molecular sphericality and Hückel energy according to a peculiar parabolic behavior. Formally, the upper limit $1 \leq \rho < 4/3$ is also derived.

In: Fullerenes
Editor: Shannon B. Ellis

ISBN: 978-1-63321-385-2

Chapter 1

INTERACTION OF FULLERENES AND FULLERENE-METAL COMPOSITES WITH CELLS

***Lucie Bacakova*[1*], *Ivana Kopova*[1], *Jiri Vacik*[2] and *Vasily Lavrentiev*[2]**

[1]Institute of Physiology, Academy of Sciences of the Czech Republic, Videnska, Prague, Czech Republic

[2]Nuclear Physics Institute, Academy of Sciences of the Czech Republic, Rez near Prague, Czech Republic

ABSTRACT

Fullerenes are spheroidal hollow cage-like carbon nanoparticles with diverse biological activities. They can act either as acceptors or as donors of electrons. Thus, they are able to generate harmful reactive oxygen species which can damage biological molecules and cells. Fullerenes can act as cytotoxic and genotoxic agents, causing oxidative DNA damage, inhibition of detoxificatory and antioxidant enzymes, polyploidy, premature cell senescence, apoptosis and inflammation. However, oxidative damage to cells by fullerenes, further potentiated by irradiation with visible and ultraviolet light, can be utilized for photodynamic therapy against tumors and microbes, including viruses. Simultaneously, fullerenes C_{60} and their derivatives are considered to be the world's most efficient radical scavengers. These molecules are promising for preventing ultraviolet-light-induced skin damage and wrinkle formation, and also for treating inflammatory diseases (e.g., arthritis or allergies), ischemic and hypoxic injuries of myocardium, amyloid-based neurological diseases, alcohol-induced encephalopathy, lead intoxication, or obesity and diabetes-related disorders. The biological response to fullerenes is profoundly influenced by their physical and chemical properties, such as functionalization with various chemical groups (by which the fullerene solubility can be tuned), degree of agglomeration, concentration and use of various solvents. The physicochemical properties of fullerenes and their biological effects may also change in the course of time. In our studies, performed on fresh and aged C_{60} films, the aged films (i.e., one year old) provided better support for the adhesion, spreading, growth, activity of

[*] Corresponding author: E-mail: lucy@biomed.cas.cz.

mitochondrial enzymes and viability of human osteoblast-like cells than fresh, i.e., one-week-old C_{60} films. These results can be attributed to changes in the fullerene films during aging, such as fragmentation, oxidation, polymerization and graphitization of fullerenes in an air atmosphere, and thus loss of their reactivity. Also the morphology of fullerene films can influence the cell behavior. On micropatterned fullerene films, i.e., films containing prominences and grooves, the cells adhered and grew preferentially in the grooves, which was more apparent on the fresh films. In addition, fullerenes C_{60} were combined with Ti atoms, because titanium is a metal used for constructing clinically applied bone implants. Hybrid C_{60}/Ti composites provided good substrates for cell adhesion and growth in both fresh and aged forms. This can be explained by the fact that the changes in the fullerene molecules, such as fragmentation, polymerization and oxidation, occurred immediately during C_{60} and Ti co-deposition by interaction of C_{60} molecules and Ti atoms. A favorable result was that the cells on all fullerene C_{60} and C_{60}/Ti films, i.e., fresh, aged, continuous and micropatterned, showed no DNA damage, evaluated by the presence of markers of a DNA damage response, namely markers of DNA double strand breaks in U-2 OS cells (gamma-HA2X and 53BP1). Thus, aged fullerene C_{60} films and fresh and aged C_{60}/Ti provided good substrates for the adhesion, growth and viability of human bone-derived cells, and so they can be used as a bioactive coating for bone implants. In micropatterned form, these films can be used for inducing regionally selective cell adhesion and directed growth for tissue engineering and advanced biotechnologies, such as genomics and proteomics.

1. Introduction

Fullerenes are molecules composed entirely of carbon atoms, arranged into structures similar to graphite, i.e., stacked graphene sheets of linked hexagonal rings; but they may also contain pentagonal (or sometimes heptagonal) rings. The rings then form nanoparticles of various shapes, such as hollow spheres, ellipsoids, and also cylinders (tubes). From this point of view, carbon nanotubes, usually considered as an independent type of carbon allotropes, belong to the family of fullerenes [1].

The number of carbon atoms within a fullerene molecule ranges from 20 to several hundreds or thousands (Figure 1). The fullerenes of the C_{60} series can have 240, 540, 960, 1500, 2160 and even more carbon atoms [2]. The most symmetrical, and therefore the most stable fullerene is C_{60}, followed by C_{70}, with the others being much less stable (for a review, see [3, 4]).

Fullerenes were discovered in 1985 by Sir Harold W. Kroto, Richard E. Smalley and Robert F. Curl, Jr. Using a laser to vaporize graphite rods in an atmosphere of helium gas, these authors prepared cage-like molecules composed of 60 carbon atoms (C_{60}) joined together by single and double bonds to form a hollow sphere with 12 pentagonal and 20 hexagonal faces [5].

In 1996, the trio was awarded the Nobel Prize for their pioneering efforts. The C_{60} molecule was named buckminsterfullerene after the American architect R. Buckminster Fuller, whose geodesic dome is constructed on the same structural principles. Another name for the C_{60} molecule (and in general for all spherical fullerenes) is buckyball, due to its resemblance to a soccer ball ([5], http://www.britannica.com/EBchecked/topic/221916/fullerene).

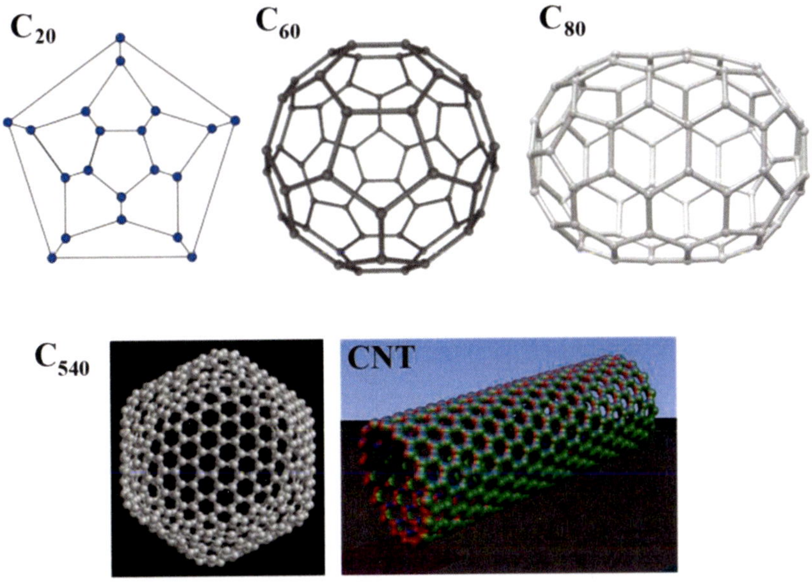

Figure 1. Fullerenes with 20, 60, 80 and 540 carbon atoms (C_{20}, C_{60}, C_{80} and C_{540}, respectively), and a carbon nanotube (CNT). (http://en.wikipedia.org/wiki/Fullerine; http://www.mocpages.com/moc.php/93031).

Although the first discovered fullerene molecule was prepared artificially, fullerenes also occur in nature, e.g., in soot [6], in minerals known as shungites [7], and also in materials of extraterrestrial origin, e.g., meteorites [8]. It is believed that fullerenes are generally widespread in the Universe, e.g., in galactic planetary nebulae [9] or in circumstellar and interstellar regions [10]. Fullerenes from outer space (which can also contain hydrogen atoms) have been even considered as "seeds for life on Earth" [11]. Fullerenes C_{60} to C_{200}, containing helium and argon of extraterrestrial origin, have been found in the Permian-Triassic boundary [12].

Fullerenes are important biologically active molecules with diverse effects on cells and tissues. These molecules are reactive due to the presence of double bonds and bending of sp^2-hybridized carbon atoms, which produces angle strain. Fullerenes can act either as acceptors or as donors of electrons [13]. Thus, they are able to oxidize biological molecules and to cause serious damage to them, which can lead to cell malfunctioning, genome mutations and cell death. Simultaneously, fullerenes can act as reduction agents, and C_{60} is even considered to be the world's most efficient radical scavenger (for a review, see [3, 4]). Both the harmful oxidizing effects and the beneficial reducing effects of fullerenes can be exploited for preventing and treating various disorders. For example, oxidizing effects, further potentiated with irradiation with visible or ultraviolet (UV) light, are the principle of photodynamic therapy against a wide spectrum of tumor cells [14-16], bacteria and fungi [17], and also viruses, e.g., viruses causing influenza [18], hepatitis C [19] and AIDS [20]. The reducing effects of fullerenes are promising for skin care, e.g., for preventing ultraviolet-light-induced skin damage [21] and wrinkle formation [22], and also in treating inflammatory diseases, such as arthritis [23] and allergic disorders [24], ischemic and hypoxic injuries of myocardium [25], amyloid-based neurological diseases [26], alcohol-induced encephalopathy [27], intoxication with heavy metals [28] or obesity and diabetes-related disorders [29, 30].

Fullerenes can also be conjugated with various atoms, chemical functional groups and biomolecules. These substances are usually attached to the fullerene surface, but atoms and small molecules can also be encapsulated inside the cage-like fullerene molecules, which are then referred to as endohedral fullerenes [31]. These modifications to fullerenes open great potential for these molecules to be applied in bioimaging [32], biosensing [33] and delivery of drugs [34] or genes [35]. After therapeutic use of fullerenes, these nanoparticles can be excreted from the organism through the urinary tract and enterons [36].

Fullerenes conjugated with other molecules or atoms also are of great importance for industrial applications, such as hydrogen storage [37], removal of wastes by photocatalytic effect [38] or construction of photovoltaic cells, e.g. by hybridization of fullerenes with other electron-accepting or electron-donating molecules [39]. Fullerene-porphyrin systems mimicked photoinduced energy and electron transfer processes in photosynthesis [40].

The majority of research on fullerenes as biomaterials has been made on fullerenes C_{60}, and thus this review is concentrated mainly on these molecules. The second most promising fullerene molecule for biomedical applications is C_{70}. This molecule has been tested for similar applications as C_{60}, i.e., photodynamic therapy against tumors [41], antiviral activity [42], antioxidant effects [43] and anti-inflammatory and antiallergenic effects [44, 45]. Fullerenes C_{70} were found to be even more efficient photosenzitizers and antioxidants than their C_{60} counterparts [41, 43]. Higher fullerenes (C_{80}, C_{82}) can be used for encapsulating metals, such as lanthanum or gadolinium, and then as reducing agents [46, 47], contrast agents for magnetic resonance imaging, or antitumor agents [48]. Lower fullerenes, namely C_{20}, were used for construction of molecular machines (in the form of so-called flagellum inside a nanotorus formed by carbon atoms [49]) or tested as drug delivery agents [50].

Broad knowledge of the behavior and effects of fullerenes in biological environments has been acquired mainly using fullerenes in the form of free nanoparticles, i.e., using fullerenes in water-based suspensions or in the form of a dust. Less is known about the interaction of cells with fullerenes in condensed form, e.g., in the form of thin layers deposited on a substrate. Our research, which forms a considerable part of this review, has therefore been focused on the adhesion, growth, viability, metabolic activity and potential damage of cells on pure C_{60} films and films of composites of C_{60} with titanium, i.e., a biocompatible metal widely used for constructing implants in the human body.

2. Fullerenes in the Form of Free Nanoparticles

2.1. Negative Effects of Fullerenes

Adverse effects after the administration of fullerenes *in vivo* and *in vitro* have been reported in many studies. Fullerenes have been administered in water in order to study their effects on aquatic animals, and have been introduced into organisms perorally or parenterally, or suspended in cell culture media.

2.1.1. Adverse Effects In Vivo

The toxicity of fullerenes has been proved on aquatic animals, such as zebrafish and invertebrates. In zebrafish, hydroxylated fullerenes $C_{60}(OH)_{24}$ caused significant changes in the expression of genes related to circadian rhythm, cell kinase activity, intracellular trafficking and immune response [51]. In *Dafnia magna*, fullerenes C_{60} increased the heart rate and moving activity, e.g., hopping frequency [52]. Marine mussels *Mytilus galloprovincialis*, exposed to fullerenes C_{60} in a water suspension, changed the lysosomal and oxidative stress biomarkers in their digestive gland [53].

Adverse effects of fullerenes have also been observed in mammals. Fullerenes C_{60} in phosphate buffered saline administered intraperitoneally into mice caused an inflammatory reaction, based on the increased expression of inflammatory genes in peritoneal macrophages, increased serum cytokine levels of IL-1 and IL-6 and increased cell distribution in blood lymphocytes [54]. A similar increase in the inflammatory response was obtained after intrapulmonary instillation of fullerenes C_{60} into mice, manifested by an increase in pro-inflammatory cytokines such as interleukin-1 (IL-1), tumor necrosis factor-alpha (TNF-alpha) and IL-6, in Th1 cytokines such as IL-12 and interferon (IFNr), and in immunoglobulins E (IgE), i.e., immunoglobulins associated with an allergic reaction. In addition, gene expression of the major histocompatibility complex (MHC) class II (H2-Eb1) molecule was stronger than that of the MHC class I (H2-T23), and an increase in T cell distribution was also observed during the experiment period. In addition, cell infiltration and expression of tissue damage-related genes in lung tissue induced an arrest in BAL cells (i.e., cells obtained by bronchoalveolal lavage) in sub-G_1 and G_1 phases of the cell cycle [55]. Fullerenes C_{60} also behaved neurotoxically. Although they probably were not able to cross the blood-brain barrier, intracerebral injection of C_{60} in rats affected neurotransmission in the brain, and led to alterations in the metabolism of monoamines, such as serotonin and dopamine, which was followed by changes in locomotor activity [56].

2.1.2. Adverse Effects at Cellular Level In Vitro

Fullerenes caused alterations in the cellular genome and its expression, damage to the cell membrane, mitochondria, cytoskeleton and other cell components, and also to the cellular ion metabolism. For example, fullerenes C_{60} in a colloidal aqueous suspension were found to be genotoxic, i.e., causing oxidative damage to the DNA of human hepatocarcinoma HepG2 cells [57]. C_{60} also induced apoptosis in human leukemia Jurkat cells, manifested by DNA fragmentation and caspase-3 activation [58]. Bismethanophosphonate fullerenes (BMPF) down-regulated the expression of matrix-related genes in cultured human cells derived from intervertebral disc, namely the expression of genes for matrix proteins aggrecan, type I collagen and type II collagen. BMPFs also down-regulated the expression of IL-1Ra, i.e., an antagonist of the receptor for interleukin-1, an important inflammatory mediator. At the same time, BMPFs up-regulated the expression of matrix metalloproteinase 3, i.e., an enzyme degrading the extracellular matrix. These effects could induce degeneration of the intervertebral disc *in vivo*, and thus compromise fullerene-based treatment of disc diseases [59].

As for damage to the cell membrane, fullerenes caused peroxidation of membrane lipids and phosphorylation of membrane proteins in monkey fibroblast COS-7 cells [60], and also infiltration, weakening and rupture of the membrane of erythrocytes [61, 62]. They also hindered the function of potassium channels in the cell membrane [63] and altered the flux of calcium ions through the cell membranes, which might initiate signaling events leading to cell apoptosis [64].

Other adverse effects of fullerenes involve depolarization of mitochondria [65], release of cytochrome c from mitochondria to the cytoplasm [64], inhibition of cell proliferation [66, 67], and inhibition of microtubule polymerization due to the formation of hydrogen bonding between the nanoparticle and the tubulin heterodimer [68].

The adverse effects of fullerenes have mostly been attributed to the oxidative stress generated by these molecules [66, 67], which can be further potentiated by irradiating the cells with visible and ultraviolet light [58, 60, 64, 69]. This mechanism is promising for photodynamic therapy of infectious and tumor diseases, as mentioned above. On the other hand, fullerenes C_{60} caused oxidative stress, mitochondrial depolarization and caspase activation in mouse B16 melanoma cells *in vitro*, leading to their apoptotic and necrotic death, but potentiated the growth of melanoma *in vivo* in mice, which was probably due to nitric oxide-dependent suppression of the anticancer immune response [65]. In addition, mechanisms independent of oxidative stress can participate in the adverse effects of fullerenes, such as cell cycle block in the G_2/M phase and subsequent inhibition of cell proliferation [70] or G_0/G_1 cell cycle arrest and cellular senescence [71].

2.2. Positive Effects of Fullerenes

In contrast to the negative effects of fullerenes on cells mentioned above, a similar number of studies have reported neutral or even positive effects of fullerenes on the viability and functioning of cells *in vitro* and *in vivo*. The positive effects are generally attributed to the radical scavenging and reducing effects of fullerenes, and involve protective effects against irradiation, DNA damage, skin damage, pathologic immune reaction, cardioprotective and neuroprotective effects and potential for the treatment of metabolic diseases (for a review, see [72-74]).

2.2.1. Protective Effects against Irradiation

Polyhydroxylated fullerenes (also referred to as fullerenols $C_{60}(OH)_n$), i.e., the major water-soluble fullerene derivatives, are potent free radical scavengers and antioxidants in biological systems. These compounds, particularly $C_{60}(OH)_{32-34}•7H_2O$ (HH-F) and $C_{60}(OH)_{44}•8H_2O$ (SHH-F), protected human skin keratinocytes HaCaT against UV-induced damage [75]. In another study on the antioxidant and radiation-protective effects of fullerenes, where human leukemic K562 cells were used as a cell model, fullerenol $C_{60}(OH)_{24}$ nanoparticles rendered these cells more resistant to ionizing X-ray irradiation by inducing significant overexpression of anti-apoptotic Bcl-2 and Bcl-xL genes and cytoprotective genes encoding glutathione S-transferase A4 (GSTA4), Mn-containing superoxide dismutase (MnSOD), nitric oxide synthase (NOS), catalase (CAT) and heme-oxygenase-1 (HO-1) [76]. Exposure of cultured human epithelial A549 cells to C_{60} nanoparticles led to an increase of intracellular reactive oxygen species, but at the same time, the glutathione reductase activity was activated to generate more glutathione to maintain cellular oxidation-reduction equilibrium [77]. Conjugation of glutathione or cysteine with C_{60} enhanced their protective effects on hydrogen peroxide-induced apoptosis in rat pheochromocytoma PC12 cells [78, 79].

Fullerenes can also protect cells against gamma-irradiation. A fullerene derivative, dendro [C_{60}] fullerene-1 (DF-1), protected highly radiosensitive cells, i.e., human lymphocytes and rat intestinal crypt cells against γ-radiation-induced DNA damage and cell death, as measured by inhibition of micronucleus formation. At the same time, DF-1 reduced the levels of reactive oxygen species in the crypt cells. The ability of DF-1 to protect against the cytotoxic effects of radiation was comparable to that of amifostine, a thiol that scavenges oxygen-derived free radicals, but with significant undesirable side effects, such as vasodilation and hypotension [80].

2.2.2. No DNA Damage

Fullerenes C_{60} showed no positive mutagenic response of cells, as proved by a bacterial reverse mutation test on *Salmonella typhimurium* and *Escherichia coli*, by a chromosomal aberration test with cultured Chinese hamster CHL/IU cells under dark conditions and visible light irradiation, and a bone marrow micronucleus test [81]. After intratracheal instillation in rats, fullerene C_{60} nanoparticles did not cause DNA damage in lung cells (evaluated by comet assays) even at doses which evoked an inflammatory response in the lungs [82]. Also in our earlier study on human osteoblast-like U-2 OS cells, fullerenes C_{60} suspended in the culture medium did not affects the adhesion, growth and viability of cells and did not induce DNA damage, evaluated by markers of a DNA damage response, namely phosphorylated histone H2AX and p53 binding protein (53BP1) [83].

2.2.3. Protective Effects on Skin

Fullerenes have also proved to be protective agents against skin injury. Fullerenes C_{60} dissolved in squalane, prepared from olive oil, showed defensive effects against 2,4-nonadienal-induced cell injury in human HaCaT keratinocytes and wrinkle formation in a three-dimensional (3D) human skin tissue model *in vitro* [21]. Moreover, the protective effects of this fullerene-based formulation, designated as LipoFullerene (LF-SQ), against wrinkle formation was proved in clinical tests on Japanese women [22]. The biological safety of LipoFullerene had been proved earlier by reverse mutation tests on four histidine-demanding strains of *Salmonella typhimurium* and a tryptophan-demanding strain of *Escherichia coli*, and by a permeability test in a human skin biopsy built in a diffusion chamber. These tests revealed no mutagenic effects of LipoFullerene and its permeation only into the epidermis, but not into the dermis and dermal veins [84]. Another fullerene formulation, LiposomeFullerene (Lpsm-Flln), i.e., fullerenes C_{60} incorporated in the phospholipid membrane vehicle, protected HaCaT keratinocytes against injury induced by ultraviolet light irradiation and subsequent intracellular reactive oxygen species production [85, 86]. In a 3D human tissue model *in vitro*, these C_{60}-containing liposomes repressed the breakdown of collagen type I/IV, DNA strand cleavage and pycnosis/karyorrhexis of the cell nuclei. Like LF-SQ, Lpsm-Flln was found to be present only in the epidermis without penetrating into the dermis and dermal veins, and is thus without a risk of systemic cytotoxicity [87].

2.2.4. Anti-Inflammatory Effects

Derivatized water-soluble fullerene C_{60}, considered as a strong free-radical scavenger, acted as a protective agent against osteoarthritis *in vivo*. In rabbits with osteoarthritis, treatment with water-soluble C_{60} significantly reduced articular cartilage degeneration [88]. In adjuvant-induced arthritic rats, intra-articular treatment with C_{60} reduced synovitis, decreased the number of osteoclasts and alleviated bone resorption and destruction in the joints [23, 89].

At the cellular level, examined *in vitro*, C_{60} inhibited the catabolic stress-induced production of matrix-degrading enzymes (matrix metalloproteinases 1, 3, and 13), down-regulation of matrix production, apoptosis and premature senescence in human chondrocytes [88]. Fullerenes C_{60} significantly suppressed the TNF-alpha-induced production of proinflammatory cytokines in synovial fibroblasts, synovial-infiltrating lymphocytes and macrophages *in vitro* [23]. At the same time, C_{60} significantly inhibited the responses of osteoclast precursor cells to a cytokine referred to as a receptor activator of nuclear factor kappa-B ligand (RANKL), which induces osteoclast differentiation [23, 89]. Conjugation of C_{60} fullerenes with dexamethason, i.e., a widely used anti-inflammatory drug with serious side effects, attenuated the pro-apoptotic action of this drug on thymocytes [90]. Fullerenes also acted as potent antiallergenic agents, suppressing antigen-driven type I hypersensitivity. After preincubation with C_{60} fullerenes, human mast cells and peripheral blood basophils exhibited significant inhibition of IgE-dependent mediator release and IgE-induced elevation of reactive oxygen species in the cytoplasm. Furthermore, in experiments performed *in vivo* in mice, fullerenes C_{60} prevented the release of histamine and a drop in core body temperature using a mast cell-dependent model of anaphylaxis [91].

2.2.5. Immunomodulatory Effects

On the other hand, fullerenes can activate the innate immunity and thereby inhibit the growth of tumors. For example, conjugation of tuftsin, i.e., a naturally occurring immunostimulating tetrapeptide (Thr-Lys-Pro-Arg) with antimicrobial and anticancer potential with fullerenes C_{60} enhanced the phagocytosis, chemotaxis activities and the expression of major histocompatibility complex class II molecules in murine peritoneal macrophages *in vitro* [92]. Thus, fullerenes can be considered as promising immunomodulatory drugs and vaccine adjuvants.

2.2.6. Cardioprotective Effects

Other important and therapeutically promising properties of fullerenes are their antiischemic and cardioprotective effects. Fullerene-based low toxic nanocationite particles (porphyrin adducts of cyclohexyl fullerene-C_{60}), designed for targeted delivery of the paramagnetic magnesium-stable isotope to the heart muscle, promoted overproduction of ATP in myocardium and thus had a positive influence on the heart cell energy metabolism, which may help to prevent and treat local myocardial hypoxic disorders [93]. In addition, the C_{60} fullerene-porphyrin adduct protected the heart muscle from doxorubicin-induced damage [25].

2.2.7. Neuroprotective Effects

Fullerenes are also promising neuroprotective agents. Fullerenes prevented degeneration of pyramidal neurons, manifested by RNA degradation and a decrease in proteosynthesis, induced in rats by intrahippocampal microinjections of amyloid-beta-25-35 ($A\beta_{25-35}$), i.e., an important peptide participating in the pathogenesis of Alzheimer's disease [26]. The mechanism of this action is explained by the radical scavenging properties of fullerenes, which are able to add multiple radicals per molecule [94]. PEG-C_{60}-3, a C_{60} fullerene derivative incorporating poly(ethylene glycol), and its pentoxifylline-bearing hybrid (PTX-C_{60}-2), decreased the reactive oxygen species and maintained the mitochondrial membrane potential in mouse neuroblast Neuro-2A cells exposed to $A\beta_{25\text{-}35}$ [94]. Sodium fullerenolate $Na_4[C_{60}(OH)_{\sim 30}]$ (NaFL), a water-soluble polyhydroxylated C_{60} fullerene derivative, destroyed amyloid fibrils of the $A\beta_{1\text{-}42}$ peptide in the brain in mice after intraperitoneal drug injection [95]. Another promising neuroprotective fullerene derivative is the C_{60}-poly(2-ethyl-2-oxazoline) complex, which was uptaken by catecholaminergic CATH.a neuronal cells *in vitro,* and attenuated the increase in intra-neuronal superoxide induced by angiotensin II stimulation [96]. Aqueous solutions of hydrated C_{60} fullerenes (C_{60}HyFn), added to drinking water, protected the central nervous system from damage caused by oxidative stress, prevented the loss of astrocytes, and their marker, i.e., glial fibrillary acidic proteins, significantly improves the behavioral response and the eliminated emotional deficits induced by chronic alcohol uptake in rats [27]. Similarly, C_{60}-methionine derivative protected human SH-SY5Y neuroblastoma cells *in vitro* against cytotoxic damage by lead [28]. Last but not least, $C_{60}(C(COOH)_2)_2$ nanoparticles selectively entered oxidation-damaged cerebral microvessel endothelial cells (rather than undamaged cells) and protected them from apoptosis [97].

2.2.8. Treatment of Metabolic Diseases

Fullerenes have potential for treating obesity and diabetes-related disorders. Squalane-dissolved fullerene C_{60} and highly hydroxylated fullerenes C_{60} also inhibited differentiation of mouse OP9 and 3T3-L1 preadipocytes into adipocytes *in vitro*. The underlying mechanisms involved the inhibition of intracellular accumulation of lipid droplets and reactive oxygen species, and also the localization of fullerene derivatives at the cytoskeleton, which controls the transport of lipids [29, 98, 99]. Administration of hydrated C_{60} fullerenes (C_{60}HyFn) in the drinking water of male rats with diabetes induced by streptozotocin treatment significantly reduced diabetes-induced oxidative stress and associated complications, such as testicular dysfunction and spermatogenic disruption, and thus restored the reproductive functions in these rats [30].

2.3. Factors Modulating the Fullerene-Cell Interaction

It is clearly apparent that fullerenes exert a dual effect on cells, tissues and organisms. This knowledge leads to a cardinal question: how to induce a desired action of fullerenes, i.e., killing or protecting the cells? This question is still relatively far from being solved, although it is known that the adverse or positive effects of fullerenes can be adjusted by concentration, aggregation, functionalization, association with various carriers and the use of various solvents. For example, increased concentration and aggregation of fullerenes C_{60} had a negative impact on the viability of Chinese hamster ovary cells and Madin-Darby canine kidney cells [100]. On the other hand, the hydroxylated fullerene $C_{60}(OH)_{25}$ was observed to intensify the H_2O_2-induced damaging effect on mouse peritoneal macrophage plasma membranes at a concentration of 0.05 mg/ml, and to reduce it at a higher concentration of 0.5 mg/ml [101]. The cytotoxicity of fullerenes C_{60} and C_{70} was decreased by their solubilization by polyhydroxylation, but was increased by their solubilization in tetrahydrofuran [102]. Complexes of fullerenes C_{60} with poly(vinyl pyrrolidone) protected test cells from UV-radiation damage, while complexes of C_{60} with gamma-cyclodextrin increased the fullerene phototoxicity [103]. An alternative way to modulate the cytotoxicity of fullerenes can be to organize them into compact forms. This could diminish the release of free fullerene particles from the material and their penetration into cells.

3. Fullerenes in Compact Form

3.1. Common Characteristics

In order to exert cytotoxic effects, fullerenes should enter into cells, i.e., they should be dispersed in the surrounding environments of cells. This dispersion could be prevented by organizing fullerenes into compact forms, which can be three- or two-dimensional, i.e., bulk materials or thin films firmly attached to various substrates (Figure 2).

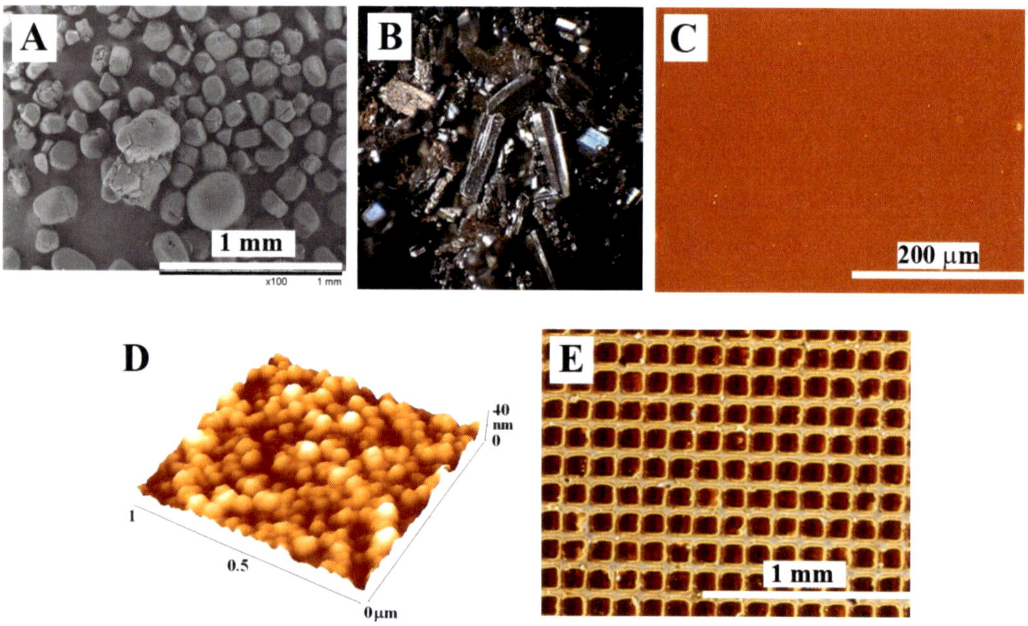

Figure 2. Fullerenes in the form of fullerite (A), C_{60} fullerenes in crystalline form (B), C_{60} fullerenes as thin continuous films (C, D) or micropatterned films (E) deposited on glass substrates. A, B: http://en.wikipedia.org/wiki/Fullerite#Fullerite; C-E: Grausova et al. 2009 [124]. Scanning electron microscope image (A), optical microscope images (B, C, E), AFM image (D).

Fullerites are a solid-state manifestation of fullerenes and related compounds. By high-pressure and high-temperature processing, fullerites can be further converted into a nanocrystalline form of diamond [104]. Nanocrystalline diamond films have repeatedly been shown to be excellent substrates for the adhesion, growth and osteogenic differentiation of cells, and are promising for coating bone implants [105, 106]; for a review see [3, 4]. However, fullerites have been used almost exclusively for technical applications, where the extraordinary electronic, magnetic, thermodynamic, optic and acoustic properties of fullerites are needed [107-110]. Reports on the behavior of fullerites in biological systems are rare, but their effects seem to be beneficial. For example, the antioxidant capacity of blood serum was significantly decreased by fullerene soot but not by fullerite [111]. As for fullerenes deposited on various substrates in the form of thin films, there have also been relatively few reports on their behavior in biological systems and on their interaction with cells. Fullerene C_{60} derivatives bearing thiol groups were adsorbed on Au_{111}-coated mica surfaces. These fullerene monolayers protected human undifferentiated neuroblastoma SH-SY5Y cells in cultures on these substrates toward L-glutamate induced excitotoxicity [112]. Fullerene-containing films were also constructed by incorporating C_{60} into synthetic lipid bilayer membranes [113]. Fullerenes C_{60} were preferentially incorporated into the hydrophobic region of these membranes and increased its specific conductivity. Upon interaction of fullerenes C_{60} with cellular plasma membrane, pores or other types of local defects were formed, and permitted the passage of water molecules across the cell membrane, resulting in the externalization of phosphatidylserine from the inner to the outer plasma membrane leaflet [114]. However, most of the research on so-called thin fullerene films has in fact been made on films based on carbon nanotubes. Pure carbon nanotube films [115] or composite films of carbon nanotubes and graphene [116], poly(L-lactide) [117], poly(lactic-*co*-glycolic acid) [118], polylactide and caprolactone [119], polypyrrole [120] or hydroxyapatite [121] provided good supports for the adhesion, growth, differentiation and functioning of bone marrow mesenchymal stem cells [118], pluripotent stem cells [115], fibroblasts [116], osteoblasts [117, 119, 121 and neuronal cells [120]. Our studies were therefore focused on thin fullerene C_{60} films deposited in the form of continuous and micropatterned films on glass substrates. Both types of films provided good support for the adhesion, growth and viability of human osteoblast-like MG 63 cells. While the cell colonization of continuous fullerene films was homogeneous, micropatterned surfaces, i.e., films with prominences and grooves, promoted preferential growth of cells in grooves among the prominences [122-124]. We have also found that not only the morphology, but also the aging of the films can influence the cell behavior. This can be explained by changes in the reactivity of fullerenes in time due to the spontaneous oxidation and polymerization of fullerenes in an air atmosphere. We therefore tested the dependence between the age of fullerene films (one week and one year) and the proliferation, viability, metabolic activity, morphology and potential DNA damage to human osteoblast-like cells of the MG-63 and U-2 OS lines [83].

3.2. Comparison of Fresh and Aged Fullerene C_{60} Films of Various Morphology

C_{60} fullerene films were synthesized in the Molecular Beam Epitaxy (MBE) systems (at NPI ASCR Rez), and were deposited on glass coverslips (Menzel Glaser, Braunschweig, Germany; diameter 12 mm) either as continuous films or through a metallic mesh with regular rectangular openings (100 μm × 150 μm) as a micropatterned array of prominences and grooves. The samples were stored in an air atmosphere at room temperature in a dark and dry place, and were evaluated either one week after deposition (fresh samples) or after one year (aged samples) [83]. As was demonstrated in our earlier studies, these films were mechanically resistant to manipulation associated with cell cultivation and cell processing, such as washing, swabbing with cotton, fixation and staining [2, 122]. For studies of cell adhesion, proliferation, viability, metabolic activity and morphology, the samples were seeded with human osteoblast-like MG-63 cells. For studies of potential DNA damage, human osteoblast-like U-2 OS cells were used instead of MG-63 cells, because these cells are p53-deficient. The number of initially adhered MG-63 cells on day 1 after seeding on fresh (i.e., one-week-old) continuous and micropatterned fullerene films was significantly lower than on the aged (i.e., one-year-old) C_{60} layers, and also than on the reference microscopic glass coverslips (Figure 3).

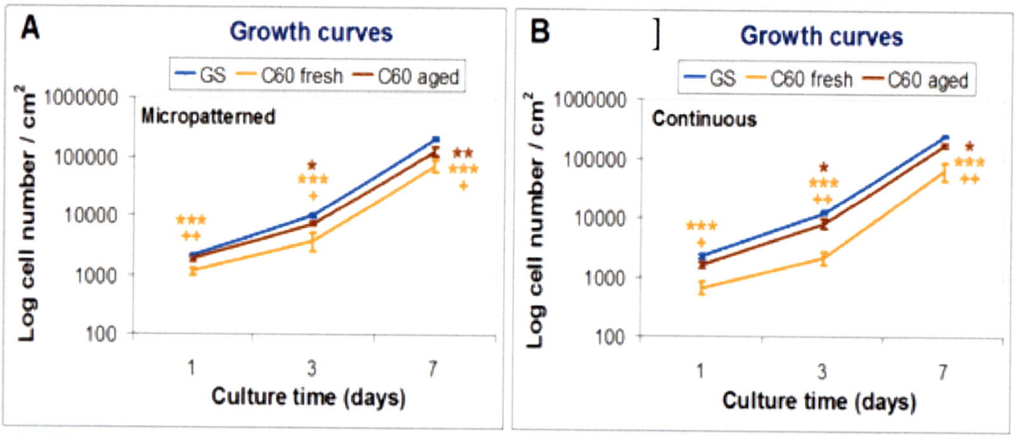

* significant difference from GS; + significant difference from C_{60} aged. $P \leq 0.001$ (***/ +++); $P \leq 0.01$ (**/ ++); $P \leq 0.05$ (*/ +) [83].

Figure 3. Growth curves of human osteoblast-like MG-63 cells on micropatterned (A) or continuous (B) fresh and aged fullerene films. GS: microscopic glass coverslips, reference material. Mean ± SEM from 3 experiments, ANOVA, Student-Newman-Keuls method.

After 3 and 7 days of cultivation, cells growing on both types of fresh fullerene films also reached significantly lower population densities than those on the aged layers and control glass coverslips. In accordance with this, the metabolic activity of the cells, measured by the XTT test, was also lower on the fresh films than on the aged films and on the microscopic glass coverslips. In addition, the cells cultivated on the fresh C_{60} films were poorly spread with a rounded morphology (Figure 4). The trypan blue exclusion test, based on the permeability of the membrane of non-viable cells, revealed that on day 7 after seeding, the cell populations on the fresh fullerene films contained a lower percentage of viable cells than the populations on the aged C_{60} films and on the microscopic glass coverslips.

On the other hand, the cells on the aged fullerene films were well-spread, i.e., polygonal or spindle-shaped, similarly as on the glass coverslips (Figure 4). The number of initially adhered cells (day 1 after seeding) on the aged fullerene films was similar as on glass, but on days 3 and 7, the number was significantly lower on the fullerene films (Figure 3). This result was in accordance with lower metabolic activity of the cells, measured by the XTT test, on both fresh and aged fullerene films than on glass coverslips. These findings can be explained by higher hydrophobicity of the fullerene films (water drop contact angles from 92.7° to 99.3°) than of the ethanol-cleaned microscopic glass coverslips (contact angle ~60°). Nevertheless, the adhesion, growth and morphology of the cells were better on the aged fullerene films than on the fresh films. This can be attributed to changes in the fullerene films during aging, e.g., fragmentation, oxidation, polymerization and graphitization of fullerenes in an air atmosphere, as revealed by Raman spectroscopy and XPS [83].

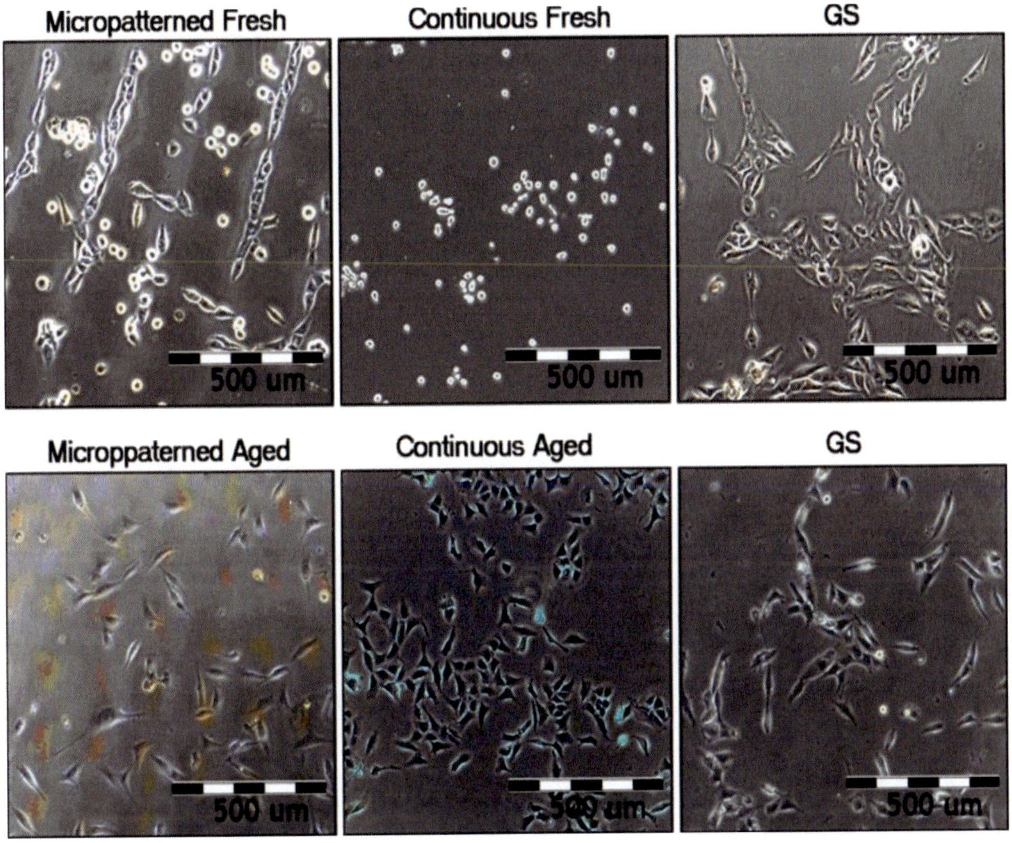

Figure 4. Morphology and preferential growth of human osteoblast-like MG-63 cells on micropatterned or continuous fresh and aged fullerene films. Native cultures on day 3 after seeding. GS: microscopic glass coverslips, reference material [83].

Although the fullerene C_{60} films, particularly the fresh ones, attenuated the cell adhesion, spreading, metabolic activity and viability, the cells on these films did not show DNA damage, evaluated by the presence of markers of a DNA damage response (DDR), namely markers of DNA double strand breaks in U-2 OS cells. These markers included gamma-H2AX (phosphorylated histone H2AX, a marker of early DDR) and 53BP1 (p53 binding protein).

Flow cytometry revealed no increase in the percentage of cells with enhanced phosphorylation of histone H2AX cultured either on fresh or aged fullerene films in comparison with the reference glass coverslips. The visualization of both DDR markers by immunofluorescence staining also revealed no increased recruitment and formation of either gamma-H2AX or 53BP1 foci (Figure 5).

Thus, it can be concluded that fullerene C_{60} films in their aged form can be considered as promising materials in bone tissue engineering, namely for potential bioactive coating of bone implants.

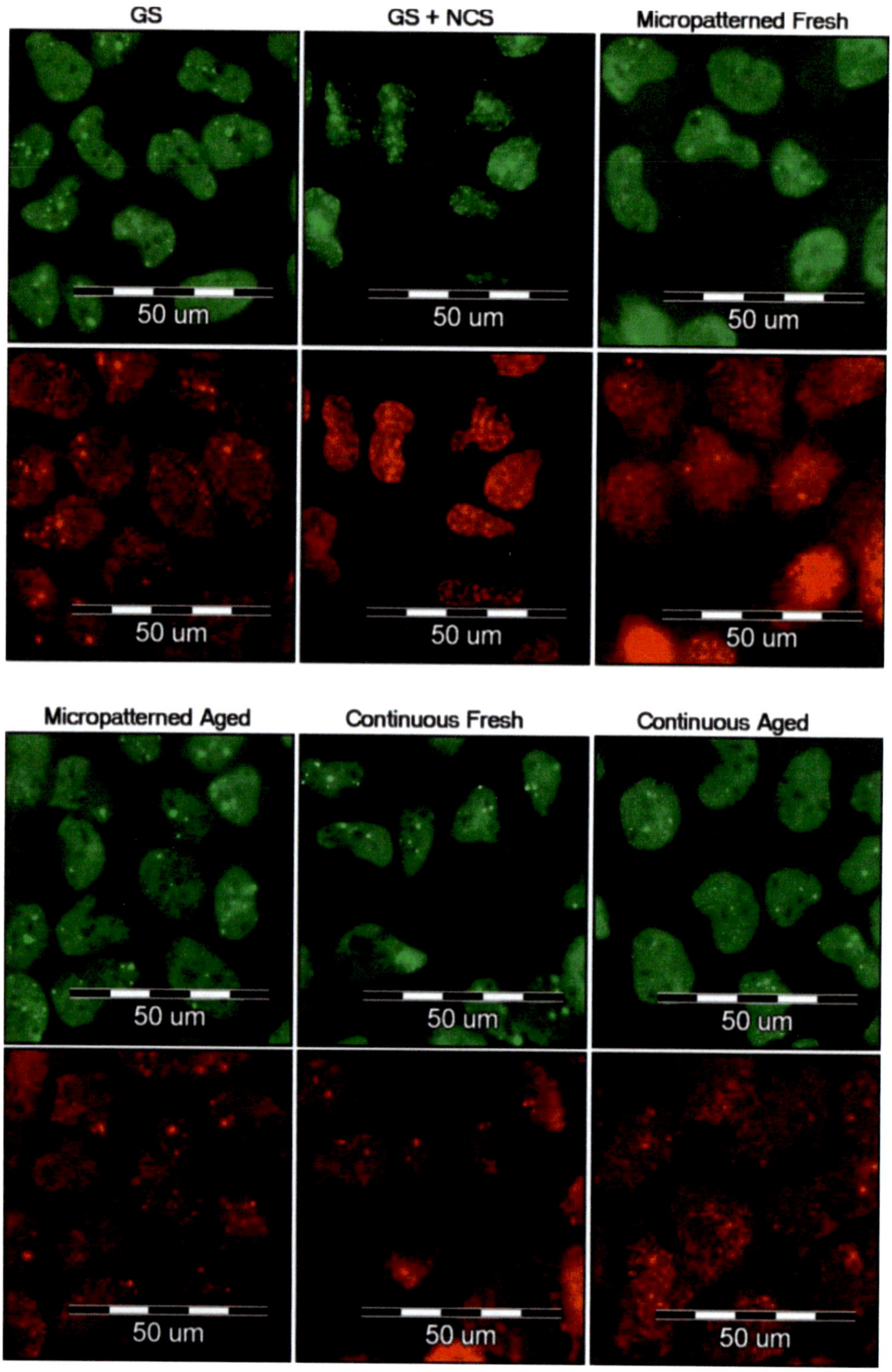

Figure 5. Immunofluorescence staining of markers of the DNA damage response: 53BP1 (green) and gamma-H2AX (red) in human osteoblast-like U-2 OS cells on micropatterned or continuous fresh and aged fullerene films after 7 days of cultivation. GS: microscopic glass coverslips, reference material; GS+NCS: positive control to DNA damage response, induced by 1 hour incubation of U-2 OS in neocarzinostatin (NCS; 700 ng/mL) [83].

3.3. Comparison of Continuous and Micropatterned Fullerene C_{60} Films of Various Age

Cells cultivated on fresh micropatterned films adhered and grew preferentially in grooves among the prominences of C_{60} (Figure 4). This can be explained by a synergistic action of certain physical and chemical properties of the fullerene bulges that are less appropriate for cell adhesion, e.g., their hydrophobicity, a relatively steep rise, as well as the tendency of spherical ball-like fullerene C_{60} molecules to diffuse out of the prominence. However, on aged micropatterned C_{60} films, the preferential adhesion and growth of cells in grooves among the prominences almost disappeared (Figure 4). This could be mainly due to a decrease in the height of the prominences after diffusion of the fullerenes (Figure 6) and also due to other changes in the fullerenes during aging, as mentioned above, such as their fragmentation, oxidation, polymerization and graphitization.

The preferential growth of cells in grooves among the prominences could be utilized in biotechnologies requiring regionally-selective cell adhesion and directed growth, such as tissue engineering and construction of patterned surfaces for genomics and proteomics.

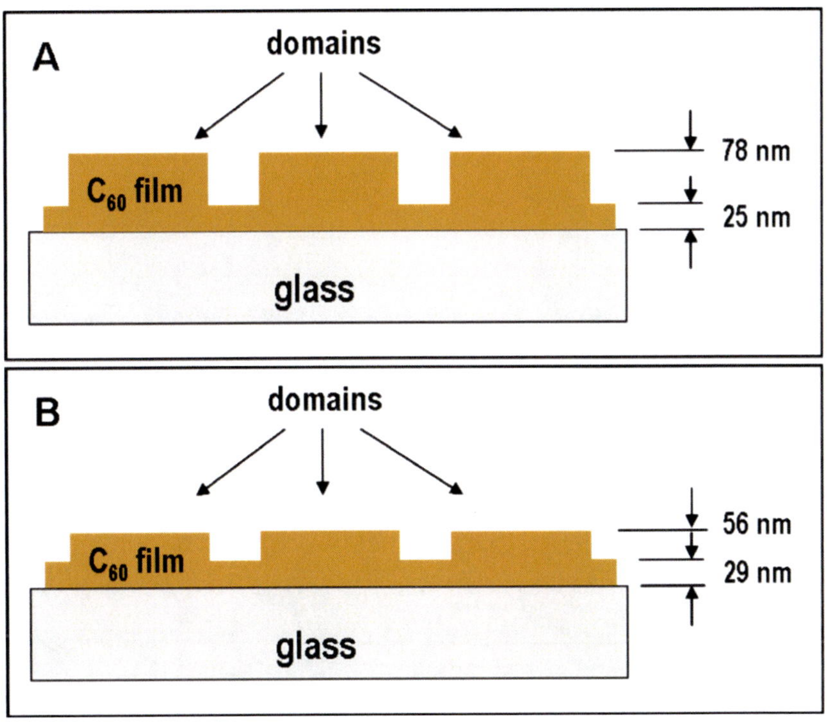

Figure 6. Height of the prominences and the grooves of the micropatterned fresh (A) and aged (B) C_{60} films. Note that fullerenes C_{60} are present both in prominences and in grooves of the micropatterned films [83].

In addition, the differences in cell numbers and morphology between fresh and aged fullerene films, i.e., the lower cell numbers and the lack of cell spreading on fresh films, were more pronounced on the continuous films than on the micropatterned films (Figs. 3, 4). These results could be attributed to the microscale surface roughness of the micropatterned fullerene films, hierarchically combined with a nanostructure which could, at least to a certain degree, compensate the relatively high hydrophobicity of the fullerene films, i.e., a property not suitable for cell adhesion and growth.

4. FULLERENE-METAL COMPOSITES

4.1. Common Characteristics

Composites of fullerenes and metals or related compounds can be divided into four main groups: (1) exohedral fullerenes, in which the metal components are attached to the exterior of the fullerenes, (3) fullerene-containing multicomposites, in which metals and other compounds are localized in the exterior of the fullerenes (2) composites in which fullerenes are attached to metallic substrates in the form of particles or films, and (4) endohedral fullerenes, in which metals are encapsulated inside the fullerene molecule (Figure 7). All these types of composites are promising materials for a wide range of industrial and biomedical applications.

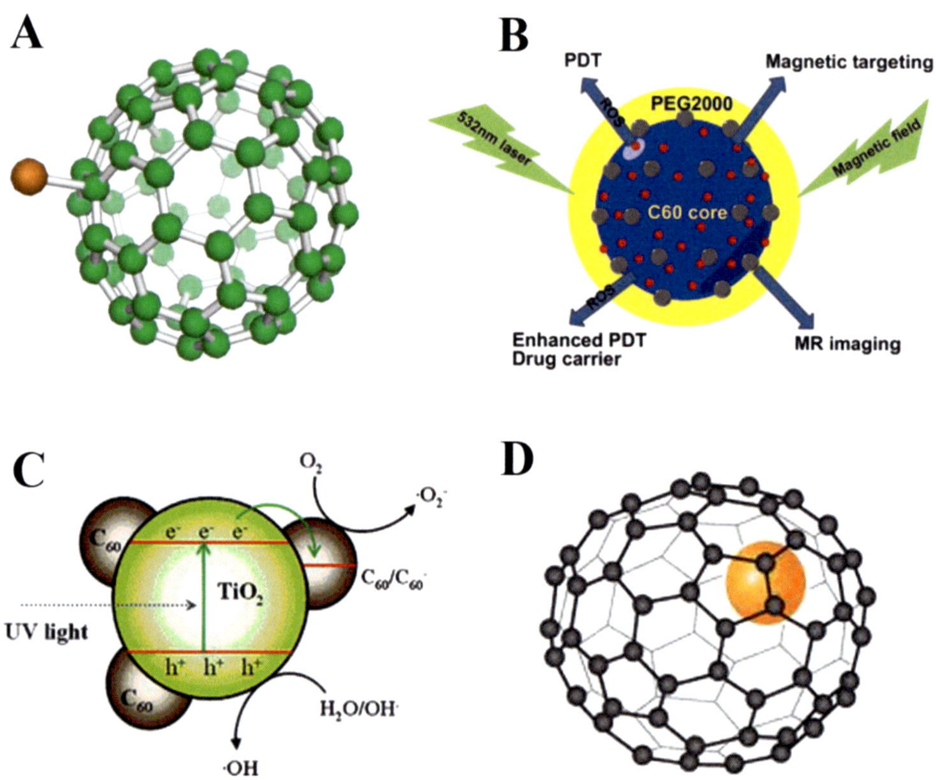

Figure 7. Schemes of an exohedral fullerene (A), multicomposites containing fullerene (blue) and metal (gray) components (B), fullerenes attached to a metal-oxide particle (C), endohedral fullerene (D). A, D: http://eng.thesaurus.rusnano.com/wiki/article1933; B: Shi et al. 2013 [32]; C: Yu et al. 2011 [135].

Ad (1). Osmium complexes attached to the exterior of fullerene C_{60} exhibited a significant photocatalytic effect, which can be used for removing non-biodegradable wastes [38]. Fullerenes C_{60} coated with cobalt [37] or doped with sodium [125] or lithium [126] exhibited high hydrogen storage ability. A similar effect has been reported when Ti atoms are clustered on the C_{60} surface [127]. Fullerenes C_{60} adsorbed with silver acted as good antimicrobial agents by denaturation of disulfide bonds, which prevented the spread of the HIV-1 virus [128]. Fullerenes C_{60} derivatized with neodymium demonstrated great potential in killing both normal and drug-resistant cancer cells [129].

Ad (2). As for multicomposites of fullerenes, metals and other compounds, composite poly(3-hexylthiophene)/copper phthalocyanine/fullerene C_{60} films have recently been developed for photovoltaics [130]. Nanocomposites of fullerene, iron oxide, poly(ethylene glycol) and also hematoporphyrin monomethyl ether have great potential for photodynamic therapy, targeted drug delivery and magnetic resonance imaging [32]. Magnesium-25 carrying porphyrin-fullerene nanoparticles had protective effects against diabetic neuropathy in streptozotocin-treated rats [131]. A composite of fullerene C_{60}, ferrocene and thiolated chitosan has been used for constructing an electrochemical immunosensor for bacteria detection [132]. Another promising way to apply this material is for immobilizing enzymes, e.g., glucose oxidase, on magnetite-fullerene nanocomposites [133].

Ad (3). As for fullerenes attached to metallic and related substrates, a fullerene-C_{60}-modified gold electrode was more effective in sensing prednisolone than indium tin oxide modified with gold nanoparticles [134]. Modification of TiO_2 and CdS-TiO_2 particles with fullerenes C_{60} enhanced their photocatalytic and sonocatalytic activities [135, 136]. Supramolecular composites containing fullerenes C_{60} immobilized on nanosilica particles are promising for photodynamic therapy against tumor cells [137].

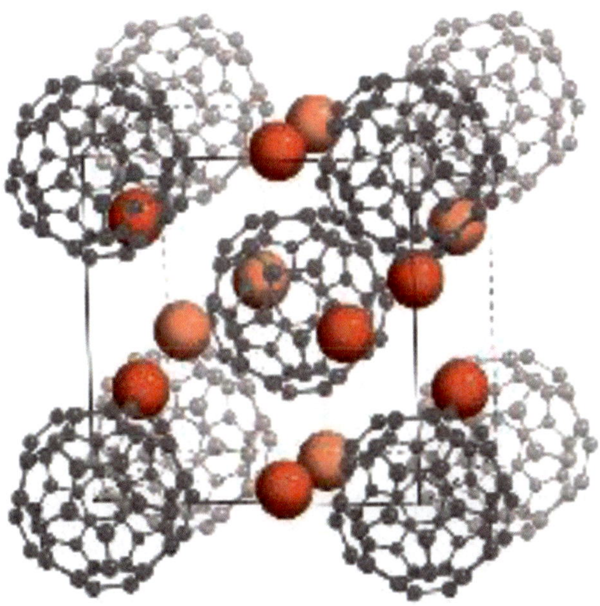

Figure 8. Crystalline structure of the Cs_3C_{60} fulleride (http://eng.thesaurus.rusnano.com/upload/iblock/f15/fulleride.jpg).

Ad (4). Endohedral fullerenes can trap atoms and small molecules to generate endohedral complexes applicable as superconductors, drug carriers, molecular reactors, and ferroelectric materials [31]. A photoelectrochemical solar cell composed of supramolecular nanoclusters of lithium-encapsulating fullerene ($Li^{+}@C_{60}$) and zinc sulphonated meso-tetraphenylporphyrin exhibited significantly better photoelectrochemical performance than the reference system containing only a single component [138]. Fullerenes with encapsulated lanthanum ($La_2@C_{80}$ and $La@C_{82}$) acted as electron donors and thus as reducing agents [46, 47]. Fullerenes encapsulating palladium can be applied for constructing biosensors, e.g., for glucose [139] and bacteria [140]. Gadolinium-containing endohedral fullerenes are pursued as contrast agents for magnetic resonance imaging for diagnostics of various diseases [35, 141]. At the same time, these fullerenes, namely polyhydroxylated metallofullerenes $Gd@C_{82}(OH)_x$ inhibited the growth of tumors in mice [48].

A special issue are fullerides, i.e., metallic salts formed from fullerenes (Figure 8), characterized by high superconductivity [142]. The metallic components of fullerides include alkali metals, such as lithium, sodium, potassium, rubidium and cesium [143], and rare earth metals, such as samarium, ytterbium and europium [144]. To the best of our knowledge, fullerides have been applied only for technical purposes, while no applications in biotechnologies and medicine have yet been established.

4.2. Composite C_{60}/Ti Films with Various Ti concentrations

Titanium is a biocompatible metal widely used for constructing stomatological implants, and in the form of alloys, such as Ti-6Al-4V or newly-developed beta-titanium alloys, and also for orthopedic implants, such as load-bearing joint replacements [145-147] (Harcuba et al. 2012, Strasky et al. 2013, Jirka et al. 2013, for a review, see [148] Vandrovcova et al. 2011). We therefore chose this metal for preparing binary fullerene C60/transitional metal thin films for potential biomedical application. Similar (related) materials, i.e., amorphous hydrogenated carbon or hydrocarbon plasma polymers enriched with Ti, proved to be good supports for the adhesion, growth and phenotypic maturation of human osteoblast-like MG-63 cells, vascular endothelial CPAE cells and primary rat aortic smooth muscle cells in our earlier studies [149, 150] (Bacakova et al. 2001, Grinevich et al. 2009). In our studies, we therefore investigated the adhesion, growth, metabolic activity, viability and potential DNA damage of human bone-derived cells in cultures on C60/Ti composite films with three Ti concentrations. A combination of fullerenes with Ti was also expected to further strengthen the films and to make them more resistant to the release of fullerene nanoparticles.

The C_{60}/Ti composite films were prepared by co-deposition of C_{60} and Ti onto microscopic glass coverslips (Menzel Glaser, Germany, diameter 12 mm) in a MBE system using the Knudsen cell and the e^{-}-gun for vaporization of C_{60} and Ti (Figure 9). Three C_{60}/Ti systems with different phase ratios were fabricated, i.e., with low concentration (25%, i.e., 25 Ti atoms and 75 C_{60} molecules), medium concentration (45%) and high concentration (70%) of Ti atoms in the composite.

The composites were synthesized with a micropatterned morphology by deposition through a contact mask (a metallic mesh) producing rectangular C_{60}/Ti prominences on the substrates with an average size of 128 μm per 98 μm (12,500 $μm^2$) and with 50 μm spacing. However, as was revealed by Raman spectroscopy and AFM, these spaces (grooves) also contained the C_{60}/Ti composites. The total thickness of the systems varied from several tens of nm (in the grooves among the prominences) to about 300 nm (on the tops of the prominences).

The samples then were stored for 1-2 weeks (fresh samples) or for 1 year (aged samples) in an air atmosphere at room temperature in a dark and dry place.

Raman spectroscopy revealed fragmentation of the fullerene molecules in both fresh and aged fullerene films, probably due to the destruction of fullerene molecules by Ti atoms during co-deposition. Other features were polymerization, hybridization and oxidation of the fullerene allotropes. Like in pure C_{60} films, measurements of the static water drop contact angle revealed relatively high hydrophobicity of the C_{60}/Ti films (contact angle in the range from 90.0° to 98.4°). AFM revealed decreased height of the prominences and increased height of the grooves in the aged films, most probably due to the diffusion of fullerenes from the prominences to the grooves [151].

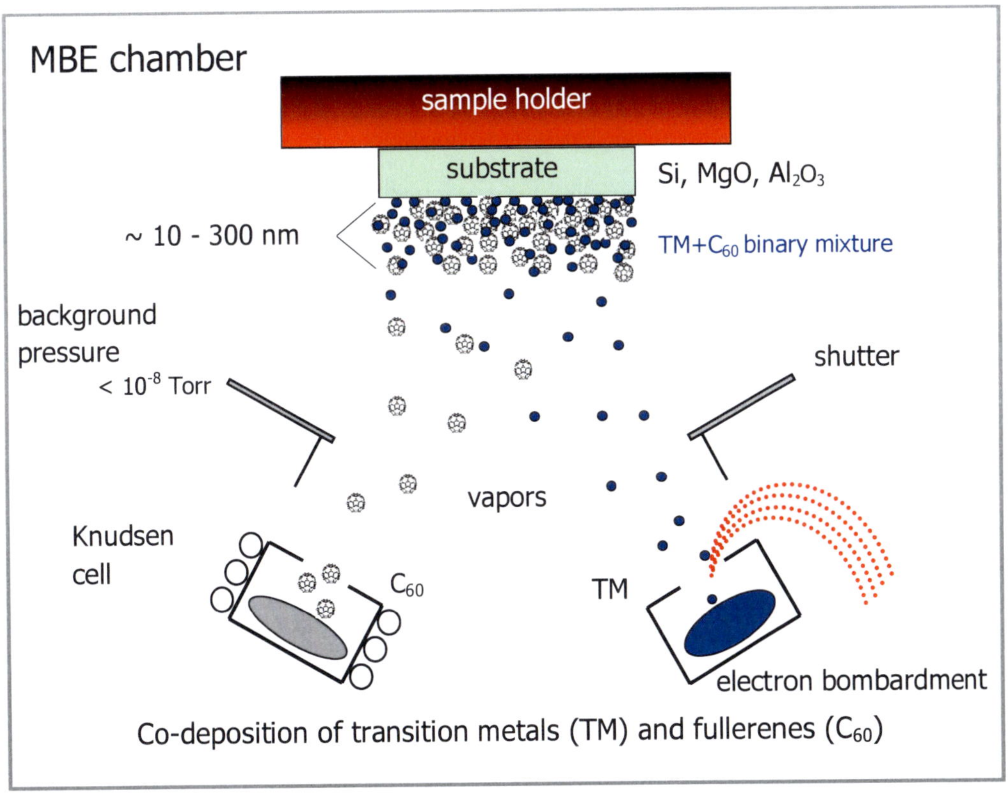

Figure 9. A scheme of preparation of hybrid fullerene C_{60}/metal composites. Deposition rates: DR(M) = DR(C_{60}) ~ 5 Å/s. Temperatures during deposition: RT [151].

The samples were then seeded with human osteoblast-like MG-63 or U-2 OS cells. The number of initially adhered MG-63 cells on fresh (i.e., one week old) C_{60}/Ti composites was on an average lower than the values for the reference microscopic glass coverslips. The decrease in cell numbers was positively correlated with increasing Ti concentration; however, these differences were not statistically significant. Similar results were obtained on day 3 after seeding. On day 7 after seeding, the slight differences in cell number had fully disappeared

(Figure 10A). On aged C_{60}/Ti films, the cell numbers on all tested samples were almost the same in all culture intervals (Figure 10B).

The cell morphology was also similar on both fresh and aged C_{60}/Ti composites of all Ti concentrations. The cells were generally well-spread, polygonal or spindle-shaped. Preferential growth in grooves among the prominences was also slightly apparent, particularly on fresh composites with low and high Ti concentration (Figure 11).

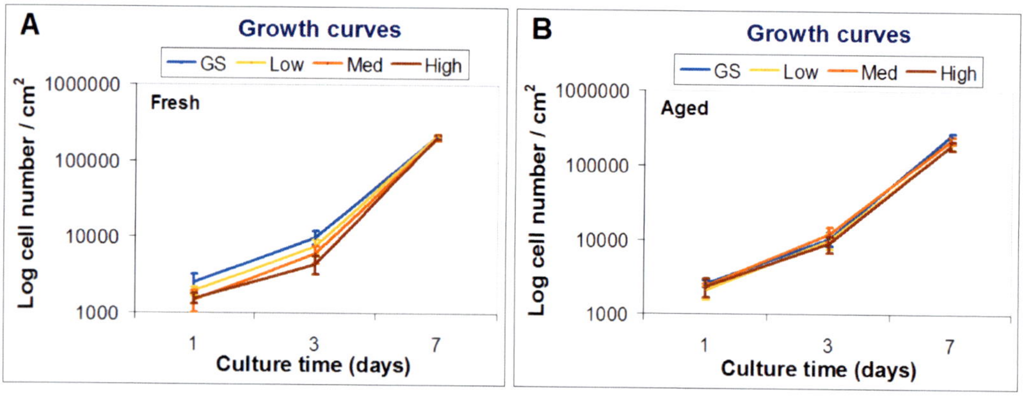

Figure 10. Growth curves of human osteoblast-like MG-63 cells on fresh (A) or aged (B) C_{60}/Ti composites with various Ti concentrations (Low: 25%, Medium: 45%, High: 70%). GS: microscopic glass coverslips, a reference material. Mean ± SEM from 3 samples, ANOVA, Student-Newman-Keuls method. No significant differences among the experimental groups were found [151].

As for the other parameters of cell behavior on the C_{60}/Ti composites, the cell metabolic activity, measured by the XTT test and by cell viability, measured by the trypan blue exclusion test, did not differ significantly among the tested groups of samples. The cells on all tested samples also showed no markers of a DNA damage response (DDR). After 3 and 7 days of cultivation on various Ti concentrations of fresh and aged C_{60}/Ti composites, the level of gamma-H2AX phosphorylation was analyzed by flow cytometry.

The results show no increase in the percentage of cells with enhanced phosphorylation of histone H2AX, cultured on layers with various Ti concentrations in comparison with the reference glass coverslips. Moreover, there was no effect of the age of the C_{60}/Ti composites on DDR [151].

Thus, it can be concluded that both fresh and aged C_{60}/Ti composites are suitable substrates for the adhesion and growth of human bone cells. The difference between pure fullerene films and C_{60}/Ti composites may lie in the fact that, in the composites, the changes in the fullerene molecules, such as fragmentation, polymerization and oxidation, occur not only by aging of the material, but immediately during C_{60} and Ti co-deposition by the interaction of C_{60} molecules and Ti atoms. In other words, the C_{60}/Ti composites exhibited similar biocompatibility as the mix of amorphous carbon and titanium, but different biocompatibility than fullerene C_{60} solids.

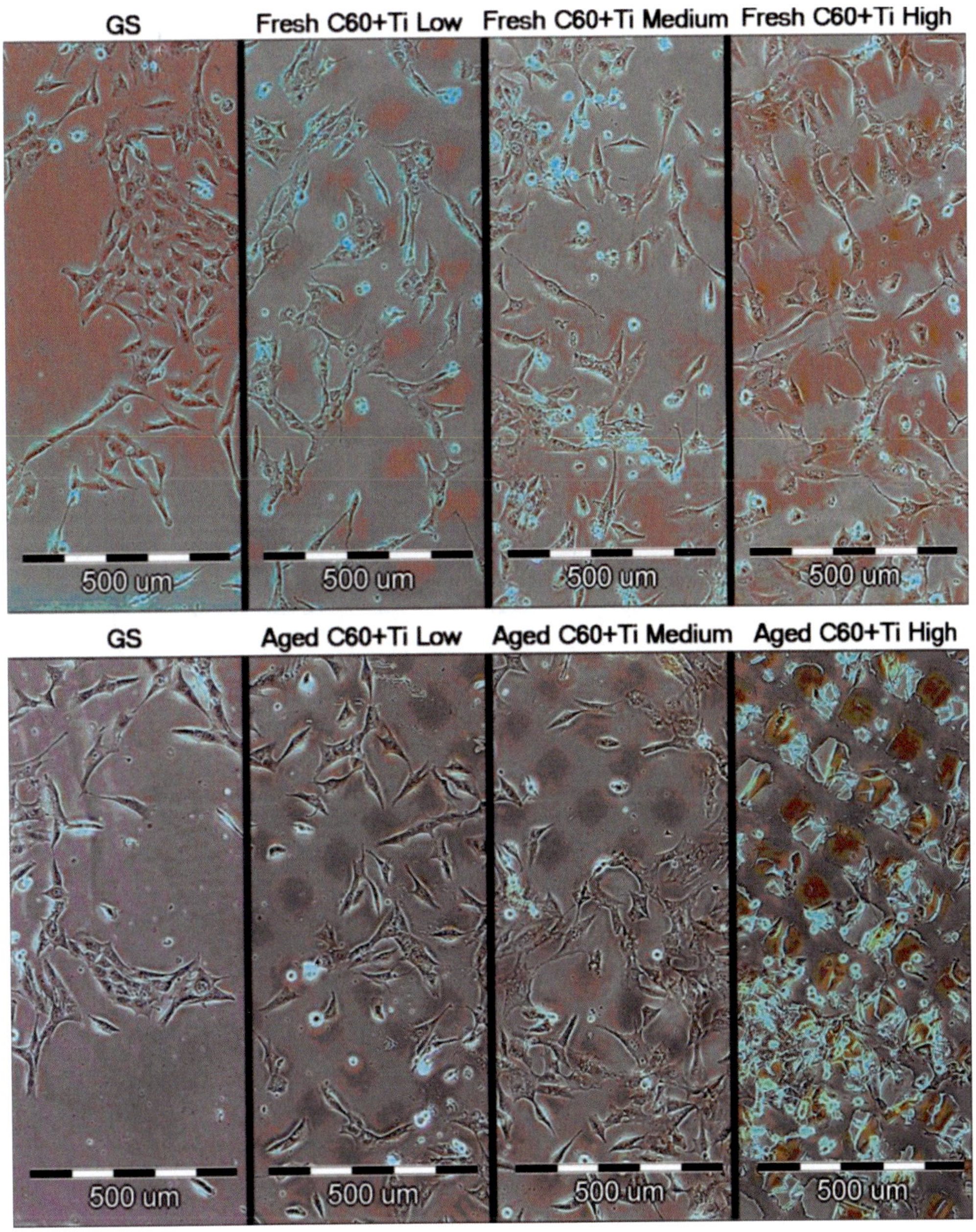

Figure 11. Morphology of human osteoblast-like MG-63 cells on day 3 after seeding on 1-week-old and 1-year-old C_{60}/Ti composites with various concentrations of Ti (Low: 25%, Medium: 45%, High: 70%). GS: control microscopic glass coverslip [151].

CONCLUSION

Fullerenes are carbon nanoparticles with dual effects on cells, i.e., damaging and protective. This dual effect is based on the oxidative or reducing effects of fullerenes, i.e., acceptance or donation of electrons. The oxidative damage caused by fullerenes includes damage to important cell components, such as the cell membrane, cytoskeleton, mitochondria and particularly the cell genome, and can be further potentiated by irradiation with visible and ultraviolet light. However, even this serious damage to cells by fullerenes has a beneficial application – in photodynamic therapy against tumor cells and infectious agents. Positive reducing effects of fullerenes include quenching of harmful radicals, protective effects against various types of irradiation, protective effects on skin, anti-inflammatory, antiallergenic and immunomodulatory effects, cardioprotective and neuroprotective effects and the potential of fullerenes for treating obesity and diabetes-related disorders. The positive or harmful action of fullerenes on cells can be modulated by concentrating them, aggregating them, derivatizing them, and by using various solvents or irradiation. However, this issue needs further investigation. An alternative approach is the use of fullerenes in a compact form, which could reduce the release of free fullerene molecules and their penetration into cells, i.e., as a thin film deposited on appropriate substrates. Pure C_{60} films and composite C_{60}/Ti films provided good substrates for the adhesion, growth, viability and metabolic activity of human osteoblast-like cells without DNA damage, especially in their aged and less reactive form. In their micropatterned form, these films promoted regionally-selective adhesion and growth of cells. Thus, fullerene C_{60}-based films are promising for use in coating bone implants, for tissue engineering and for other advanced biotechnologies, such as genomics and proteomics.

ACKNOWLEDGMENTS

Our studies included in this chapter were supported by the Grant Agency of the Czech Republic (grants No. P107/11/1856 and P108/12/1168). Robin Healey (Czech Technical University in Prague) and Mrs. Sherryl Ann Vacik (P&A Consultants, Angel Fire, New Mexico, U.S.A.) are gratefully acknowledged for their language revision of the text.

REFERENCES

[1] Yakobson, B; Smalley, RE. Fullerene Nanotubes: $C_{1,000,000}$ and Beyond. *Am. Sci.*, 1997, *85*, 324-337.

[2] Janner, A. Alternative approaches to onion-like icosahedral fullerenes. *Acta Crystallogr. A Found. Adv.*, 2014, *70*, 168-180.

[3] Bacakova, L; Grausova, L; Vandrovcova, M; Vacik, J; Frazcek, A; Blazewicz, S; Kromka, A; Vanecek, M; Nesladek, M; Svorcik, V; Kopecek, M. Carbon nanoparticles as substrates for cell adhesion and growth. In: Simone Luca Lombardi, Editor. Nanoparticles: New Research. Hauppauge, New York, USA; Nova Science Publishers, Inc., 2008, 39-107. ISBN 978-1-60456-704-5

[4] Bacakova, L; Grausova, L; Vacik, J; Kromka, A; Biederman, H; Choukourov, A; Stary, V. Nanocomposite and nanostructured carbon-based films as growth substrates for bone cells. In: Boreddy Reddy, Editor. Advances in Diverse Industrial Applications of Nanocomposites. Intech; Open Access Publisher; 2011, 399-435. ISBN 978-953-307-202-209.

[5] Kroto, HW; Heath, JR; O'Brien, SC; Curl RF. "C60: Buckminsterfullerene". *Nature*, 1985, *318*, 162-163.

[6] Buseck, PR; Tsipursky, SJ; Hettich, R. Fullerenes from the geological environment. *Science*, 1992, *257*, 215-217.

[7] Mossman, D; Eigendorf, G; Tokaryk, D; Gauthier-Lafaye, F; Guckert, KD; Melezhik, V; Farrow, CEG. Testing for fullerenes in geologic materials: Oklo carbonaceous substances, Karelian shungites, Sudbury Black Tuff. *Geology*, 2003, *31,* 255-258.

[8] Beckcr, L; Porcda, RJ; Bunch, TE. Fullerenes: An extraterrestrial carbon carrier phase for noble gases. *Proc. Natl. Acad. Sci. USA*, 2000, 2979-2983.

[9] Cami, J; Bernard-Salas, J.; Peeters, E.; Malek, S.E. "Detection of C60 and C70 in a Young Planetary Nebula". *Science,* 2009, *329*, 1180–1182.

[10] Dunk, PW; Adjizian, J-J; Kaiser, NK; Quinn, JP; Blakney, GT; Ewels, C; Marshall, AG; Kroto, HW. Metallofullerene and fullerene formation from condensing carbon gas under conditions of stellar outflows and implication to stardust. *Proc. Natl. Acad. Sci. USA*, 2013, 18081-18086.

[11] Atkinson, N. Buckyballs could be plentiful in the Universe. Universe Today. Retrieved 2010-10-28. http://www.universetoday.com/76732/buckyballs-could-be-plentiful-in-the-universe/

[12] Becker, L; Poreda, RJ; Hunt, AG; Bunch, TE; Rampino, M. Impact event at the Permian-Triassic boundary: evidence from extraterrestrial noble gases in fullerenes. *Science*, 2001, *291*, 1530-1533.

[13] Guildi, DM. Fullerenes; electron accepting building blocks in photoactive donor-bridge-acceptor dyads. http://www.photobiology.com/reviews/6/

[14] Liu, J; Tabata Y. Photodynamic therapy of fullerene modified with pullulan on hepatoma cells. *J. Drug Target.*, 2010, *18*, 602-610.

[15] Hu, Z; Zhang, C; Huang, Y; Sun, S; Guan, W; Yao, Y. Photodynamic anticancer activities of water-soluble C(60) derivatives and their biological consequences in a HeLa cell line. *Chem. Biol. Interact.*, 2012, *195,* 86-94.

[16] Asada, R; Liao, F; Saitoh, Y; Miwa, N. Photodynamic anti-cancer effects of fullerene [C60]-PEG complex on fibrosarcomas preferentially over normal fibroblasts in terms of fullerene uptake and cytotoxicity. *Mol. Cell. Biochem.*, 2014, *390*, 175-184.

[17] Milanesio, ME; Spesia, MB; Cormick, MP; Durantini, EN. Mechanistic studies on the photodynamic effect induced by a dicationic fullerene C60 derivative on Escherichia coli and Candida albicans cells. *Photodiagnosis Photodyn. Ther.*, 2013, *10,* 320-327.

[18] Shoji, M; Takahashi, E; Hatakeyama, D; Iwai, Y; Morita, Y; Shirayama, R; Echigo, N; Kido, H; Nakamura, S; Mashino, T; Okutani, T; Kuzuhara, T. Anti-influenza activity of c60 fullerene derivatives. *PloS One*, 2013, *8*, e66337.

[19] Xiao, SL; Wang, Q; Yu, F; Peng, YY; Yang, M; Sollogoub, M; Sinaÿ, P; Zhang, YM; Zhang, LH; Zhou, DM. Conjugation of cyclodextrin with fullerene as a new class of HCV entry inhibitors. *Bioorg. Med. Chem.*, 2012, *20,* 5616-5622.

[20] Toshima, K. Target-selective degradation of proteins and oligosaccharides by light-activated hybrid molecules for molecular-targeted photodynamic therapy. *Future Med. Chem.,* 2009, *1*, 1113-1124.

[21] Kato, S; Kikuchi, R; Aoshima, H; Saitoh, Y; Miwa, N. Defensive effects of fullerene-C60/liposome complex against UVA-induced intracellular reactive oxygen species generation and cell death in human skin keratinocytes HaCaT, associated with

intracellular uptake and extracellular excretion of fullerene-C60. *J. Photochem. Photobiol. B.*, 2010, *98*, 144-151.
[22] Kato, S; Aoshima, H; Saitoh, Y; Miwa, N. Defensive effects of fullerene-C60 dissolved in squalane against the 2,4-nonadienal-induced cell injury in human skin keratinocytes HaCaT and wrinkle formation in 3D-human skin tissue model. *J. Biomed. Nanotechnol.*, 2010, *6*, 52-58.
[23] Yudoh, K; Karasawa, R; Masuko, K; Kato, T. Water-soluble fullerene (C60) inhibits the development of arthritis in the rat model of arthritis. *Int. J. Nanomedicine,* 2009, *4*, 217-225.
[24] Ryan, JJ; Bateman, HR; Stover, A; Gomez, G; Norton, SK; Zhao, W; Schwartz, LB; Lenk, R; Kepley, CL. Fullerene nanomaterials inhibit the allergic response. *J. Immunol.*, 2007, *179*, 665-672.
[25] Shetab Boushehri, SV; Ostad, SN; Sarkar, S; Kuznetsov, DA; Buchachenko, AL; Orlova, MA; Minaii, B; Kebriaeezadeh, A; Rezayat, SM. The C60-fullerene porphyrin adducts for prevention of the doxorubicin-induced acute cardiotoxicity in rat myocardial cells. *Acta Med. Iran.*, 2010, 48, 342-350.
[26] Makarova, EG; Gordon, RY; Podolski, IY. Fullerene C60 prevents neurotoxicity induced by intrahippocampal microinjection of amyloid-beta peptide. *J. Nanosci. Nanotechnol.*, 2012, *12*, 119-126.
[27] Tykhomyrov, AAI; Nedzvetsky, VS; Klochkov, VK; Andrievsky, GV. Nanostructures of hydrated C60 fullerene (C60HyFn) protect rat brain against alcohol impact and attenuate behavioral impairments of alcoholized animals. *Toxicology,* 2008, *246*, 158-165.
[28] Chen, T; Li, YY; Zhang, JL; Xu, B; Lin, Y; Wang, CX; Guan, WC; Wang, YJ; Xu, SQ. Protective effect of C(60)-methionine derivate on lead-exposed human SH-SY5Y neuroblastoma cells. *J. Appl. Toxicol.*, 2011, *31,* 255-261.
[29] Xiao, L; Aoshima, H; Saitoh, Y; Miwa, N. Highly hydroxylated fullerene localizes at the cytoskeleton and inhibits oxidative stress in adipocytes and a subcutaneous adipose-tissue equivalent. *Free Radic. Biol. Med.*, 2011, *51*, 1376-1389.
[30] Bal, R; Türk, G; Tuzcu, M; Yilmaz, O; Ozercan, I; Kuloglu, T; Gür, S; Nedzvetsky, VS; Tykhomyrov, AA; Andrievsky, GV; Baydas, G; Naziroglu, M. Protective effects of nanostructures of hydrated C(60) fullerene on reproductive function in streptozotocin-diabetic male rats. *Toxicology,* 2011; *282*, 69-81.
[31] Hu, YH; Ruckenstein, E. Endohedral chemistry of C60-based fullerene cages. *J. Am. Chem. Soc.*, 2005; 127, 11277-11282.
[32] Shi, J; Yu, X; Wang, L; Liu, Y; Gao, J; Zhang, J; Ma, R; Liu, R; Zhang, Z. PEGylated fullerene/iron oxide nanocomposites for photodynamic therapy, targeted drug delivery and MR imaging. *Biomaterials,* 2013, *34,* 9666-9677.
[33] Wei, L; Lei, Y; Fu, H; Yao, J. Fullerene hollow microspheres prepared by bubble-templates as sensitive and selective electrocatalytic sensor for biomolecules. *ACS Appl. Mater. Interfaces,* 2012, *4*, 1594-1600.
[34] Blazkova, I; Viet Nguyen, H; Kominkova, M; Konecna, R; Chudobova, D; Krejcova, L; Kopel, P; Hynek, D; Zitka, O; Beklova, M; Adam, V; Kizek, R. Fullerene as a transporter for doxorubicin investigated by analytical methods and in vivo imaging. *Electrophoresis,* 2014, *35,* 1040-1049.

[35] Sitharaman, B; Zakharian, TY; Saraf, A; Misra, P; Ashcroft, J; Pan, S; Pham, QP; Mikos, AG; Wilson, LJ; Engler, DA. Water-soluble fullerene (C60) derivatives as nonviral gene-delivery vectors. *Mol. Pharm.*, 2008, *5*, 567-578.

[36] Lin, CM; Lu, TY. C60 fullerene derivatized nanoparticles and their application to therapeutics. *Recent Pat. Nanotechnol.*, 2012, *6*, 105-113.

[37] Bao, D; Gao, P; Shen, X; Chang, C; Wang, L; Wang, Y; Chen, Y; Zhou, X; Sun, S; Li, G; Yang, P. Mechanical ball-milling preparation of fullerene/cobalt core/shell nanocomposites with high electrochemical hydrogen storage ability. *ACS Appl. Mater. Interfaces*, 2014, *6*, 2902-2909.

[38] Chamberlain, TW; Champness, NR; Schröder, M; Khlobystov, AN. A piggyback ride for transition metals: encapsulation of exohedral metallofullerenes in carbon nanotubes. *Chemistry*, 2011, *17*, 668-674.

[39] Ito, O; D'Souza, F. Recent advances in photoinduced electron transfer processes of fullerene-based molecular assemblies and nanocomposites. *Molecules*, 2012, *17*, 5816-5835.

[40] Imahori, H. Porphyrin-fullerene linked systems as artificial photosynthetic mimics. *Org. Biomol. Chem.*, 2004, *2*, 1425-1433.

[41] Liu, Q; Guan, M; Xu, L; Shu, C; Jin, C; Zheng, J; Fang, X; Yang, Y; Wang, C. Structural effect and mechanism of C70-carboxyfullerenes as efficient sensitizers against cancer cells. *Small*, 2012, *8*, 2070-2077.

[42] Kornev, AB; Peregudov; AS, Martynenko; VM, Balzarini, J; Hoorelbeke, B; Troshin, PA. Synthesis and antiviral activity of highly water-soluble polycarboxylic derivatives of [70]fullerene. *Chem. Commun. (Camb.)*, 2011, *47*, 8298-8300.

[43] Liu, Q; Zheng, J; Guan, M; Fang, X; Wang, C; Shu, C. Protective effect of C70-carboxyfullerene against oxidative-induced stress on postmitotic muscle cells. *ACS Appl. Mater. Interfaces*, 2013, *5*, 4328-4333.

[44] Norton, SK; Wijesinghe, DS; Dellinger, A; Sturgill, J; Zhou, Z; Barbour, S; Chalfant, C; Conrad, DH; Kepley, CL. Epoxyeicosatrienoic acids are involved in the C(70) fullerene derivative-induced control of allergic asthma. *J. Allergy Clin. Immunol.*, 2012, *130*, 761-769.

[45] Norton, SK; Dellinger, A; Zhou, Z; Lenk, R; Macfarland, D; Vonakis, B; Conrad, D; Kepley, CL. A new class of human mast cell and peripheral blood basophil stabilizers that differentially control allergic mediator release. *Clin. Transl. Sci.*, 2010, *3*, 158-169.

[46] Takano, Y; Obuchi, S; Mizorogi, N; García, R; Herranz, MÁ; Rudolf, M; Guldi, DM; Martín, N; Nagase, S; Akasaka, T. An endohedral metallofullerene as a pure electron donor: intramolecular electron transfer in donor-acceptor conjugates of La2@C80 and 11,11,12,12-tetracyano-9,10-anthra-p-quinodimethane (TCAQ). *J. Am. Chem. Soc.*, 2012, *134*, 19401-19408.

[47] Tsuchiya, T; Rudolf, M; Wolfrum, S; Radhakrishnan, SG; Aoyama, R; Yokosawa, Y; Oshima, A; Akasaka, T; Nagase, S; Guldi, DM. Coordinative interactions between porphyrins and C60, La@C82, and La2@C80. *Chemistry*, 2013, *19*, 558-565.

[48] Zhou, G; Li, Y; Liu, Y; Ge, C; Li, W; Sun, B; Li, B; Gao, Y; Chen, C. Subcellular distribution of polyhydroxylated metallofullerene Gd@C82(OH)22 in different tissues of tumor-bearing mice. *J. Nanosci. Nanotechnol.*, 2010, *10*, 8597-8602.

[49] Neto, AM; Santos, JC; Santos, EC. Molecular dynamics study of the flagella inside the carbon torus. *J. Nanosci. Nanotechnol.*, 2011, *11*, 1230-1236.

[50] Shityakov, S; Förster, C. Multidrug resistance protein P-gp interaction with nanoparticles (fullerenes and carbon nanotube) to assess their drug delivery potential: a theoretical molecular docking study. *Int. J. Comput. Biol. Drug Des.* 2013, *6*, 343-357.

[51] Jovanović, B; Ji, T; Palić, D. Gene expression of zebrafish embryos exposed to titanium dioxide nanoparticles and hydroxylated fullerenes. *Ecotoxicol. Environ. Saf.*, 2011, *74*, 1518-1525.

[52] Lovern, SB; Strickler, JR; Klaper, R. Behavioral and physiological changes in Daphnia magna when exposed to nanoparticle suspensions (titanium dioxide, nano-C60, and C60HxC70Hx). *Environ. Sci. Technol.*, 2007, *41*, 4465-4470.

[53] Canesi, L; Fabbri, R; Gallo, G; Vallotto, D; Marcomini, A; Pojana, G. Biomarkers in Mytilus galloprovincialis exposed to suspensions of selected nanoparticles (Nano carbon black, C60 fullerene, Nano-TiO2, Nano-SiO2). *Aquat. Toxicol.*, 2010, *100*, 168-177.

[54] Park, EJ; Roh, J; Kim, Y; Park, K. Induction of inflammatory responses by carbon fullerene (C60) in cultured RAW264.7 cells and in intraperitoneally injected mice. *Toxicol. Res.*, 2010, *26*, 267-273.

[55] Park, EJ; Kim, H; Kim, Y; Yi, J; Choi, K; Park, K. Carbon fullerenes (C60s) can induce inflammatory responses in the lung of mice. *Toxicol. Appl. Pharmacol.*, 2010, *244*, 226-233.

[56] Yamada, T; Jung, DY; Sawada, R; Matsuoka, A; Nakaoka, R; Tsuchiya, T. Effects intracerebral microinjection and intraperitoneal injection of [60]fullerene on brain functions differ in rats. *J. Nanosci. Nanotechnol.*, 2008, *8*, 3973-3980.

[57] Matsuda, S; Matsui, S; Shimizu, Y; Matsuda, T. Genotoxicity of colloidal fullerene C_{60}. *Environ. Sci. Technol.*, 2011, *45*, 4133-4138.

[58] Palyvoda, KO; Grynyuk, II; Prylutska, SV; Samoylenko, AA;, Drobot, LB; Matyshevska, OP. Apoptosis photoinduction by C60 fullerene in human leukemic T cells. *Ukr. Biokhim. Zh.*, 2010, *82*, 121-127.

[59] Yang, X; Wan, Y; Qiao, X; Arlet, V; Li, X. Transcriptional alteration of matrix-related gene expression in cultured human disc cells by nanoparticles of a bismethanophosphonate fullerene. *Cell Biol. Int.*, 2010, *34*, 837-844.

[60] Vileno, B; Jeney, S; Sienkiewicz, A; Marcoux, PR; Miller, LM; Forró, L. Evidence of lipid peroxidation and protein phosphorylation in cells upon oxidative stress photo-generated by fullerols. *Biophys. Chem.*, 2010, *152*, 164-169.

[61] Trpkovic, A; Todorovic-Markovic, B; Kleut, D; Misirkic, M; Janjetovic, K; Vucicevic, L; Pantovic, A; Jovanovic, S; Dramicanin, M; Markovic, Z; Trajkovic, V. Oxidative stress-mediated hemolytic activity of solvent exchange-prepared fullerene (C60) nanoparticles. *Nanotechnology*, 2010, *21*, 375102.

[62] Zhang, X; Zhang, Y; Zheng, Y; Wang, B. Mechanical characteristics of human red blood cell membrane change due to C60 nanoparticle infiltration. *Phys. Chem. Chem. Phys.*, 2013, *15*, 2473-2481.

[63] Kraszewski, SI; Tarek, M; Treptow, W; Ramseyer, C. Affinity of C60 neat fullerenes with membrane proteins: a computational study on potassium channels. *ACS Nano*, 2010, *4*, 4158-4164.

[64] Grebinyk, SM; Palyvoda, KO; Prylutska, SV; Grynyuk, II; Samoylenko, AA; Drobot, LB; Matyshevska, OP. Photoactivated fullerene C60 induces store-operated Ca2 entry and cytochrome c release in Jurkat cells. *Ukr. Biokhim. Zh.*, 2012, *84*, 58-63.

[65] Zogovic, NS; Nikolic, NS; Vranjes-Djuric, SD; Harhaji, LM; Vucicevic, LM; Janjetovic, KD; Misirkic, MS; Todorovic-Markovic, BM; Markovic, ZM; Milonjic, SK; Trajkovic, VS. Opposite effects of nanocrystalline fullerene (C(60)) on tumour cell growth in vitro and in vivo and a possible role of immunosupression in the cancer-promoting activity of C(60). *Biomaterials,* 2009, *30*, 6940-6946.

[66] Horie, M; Nishio, K; Kato, H; Shinohara, N; Nakamura, A; Fujita, K; Kinugasa, S; Endoh, S; Yamamoto, K; Yamamoto, O; Niki, E; Yoshida, Y; Iwahashi, H. In vitro evaluation of cellular responses induced by stable fullerene C60 medium dispersion. *J. Biochem.*, 2010, *148*, 289-298.

[67] Horie, M; Nishio, K; Kato, H; Shinohara, N; Nakamura, A; Fujita, K; Kinugasa, S; Endoh, S; Yoshida, Y; Hagihara, Y; Iwahashi H. In vitro evaluation of cellular influences induced by stable fullerene C_{70} medium dispersion: induction of cellular oxidative stress. *Chemosphere*, 2013, *93*, 1182-1188.

[68] Ratnikova, TA; Govindan, PN; Salonen, E; Ke, PC. In vitro polymerization of microtubules with a fullerene derivative. *ACS Nano,* 2011, *5*, 6306-6314.

[69] Zhao, B; Bilski, PJ; He, YY; Feng, L; Chignell, CF. Photo-induced reactive oxygen species generation by different water-soluble fullerenes (C) and their cytotoxicity in human keratinocytes. *Photochem. Photobiol.*, 2008, *84*, 1215-1223.

[70] Harhaji, L; Isakovic, A; Raicevic, N; Markovic, Z; Todorovic-Markovic, B; Nikolic, N; Vranjes-Djuric, S; Markovic, I; Trajkovic, V. Multiple mechanisms underlying the anticancer action of nanocrystalline fullerene. *Eur. J. Pharmacol.*, 2007, *568*, 89-98.

[71] Gao, J; Wang, HL; Shreve, A; Iyer, R. Fullerene derivatives induce premature senescence: a new toxicity paradigm or novel biomedical applications. *Toxicol. Appl. Pharmacol.*, 2010, *244*, 130-143.

[72] Lens, M. Recent progresses in application of fullerenes in cosmetics. *Recent Pat. Biotechnol.*, 2011, *5*, 67-73.

[73] Injac, R; Prijatelj, M; Strukelj, B. Fullerenol nanoparticles: toxicity and antioxidant activity. *Methods Mol. Biol.*, 2013, *1028*, 75-100.

[74] Yang, X; Ebrahimi, A; Li, J; Cui, Q. Fullerene-biomolecule conjugates and their biomedicinal applications. *Int. J. Nanomedicine*, 2014, *9*, 77-92.

[75] Saitoh, Y; Miyanishi, A; Mizuno, H; Kato, S; Aoshima, H; Kokubo, K; Miwa, N. Super-highly hydroxylated fullerene derivative protects human keratinocytes from UV-induced cell injuries together with the decreases in intracellular ROS generation and DNA damages. *J. Photochem. Photobiol. B.*, 2011, *102*, 69-76.

[76] Stankov, K; Borisev, I; Kojic, V; Rutonjski, L; Bogdanovic, G; Djordjevic, A. Modification of antioxidative and antiapoptotic genes expression in irradiated K562 cells upon fullerenol C60(OH)24 nanoparticle treatment. *J. Nanosci. Nanotechnol.*, 2013, *13*, 105-113.

[77] Wang, F; Jin, C; Liang, H; Tang, Y; Zhang, H; Yang, Y. Effects of fullerene C60 nanoparticles on A549 cells. *Environ. Toxicol. Pharmacol.*, 2014, *37*, 656-661.

[78] Hu, Z;, Liu, S;, Wei, Y;, Tong, E; Cao, F; Guan, W. Synthesis of glutathione C60 derivative and its protective effect on hydrogen peroxide-induced apoptosis in rat pheochromocytoma cells. *Neurosci. Lett.*, 2007, *429*, 81-86.

[79] Hu, Z; Guan, W; Wang, W; Huang, L; Xing, H;, Zhu, Z. Protective effect of a novel cystine C(60) derivative on hydrogen peroxide-induced apoptosis in rat pheochromocytoma PC12 cells. *Chem. Biol. Interact.*, 2007, *167*, 135-144.

[80] Theriot, CA; Casey, RC; Moore, VC; Mitchell, L; Reynolds, JO; Burgoyne, M; Partha, R; Huff, JL; Conyers, JL; Jeevarajan, A; Wu, H. Dendro[C(60)]fullerene DF-1 provides radioprotection to radiosensitive mammalian cells. *Radiat. Environ. Biophys.*, 2010, *49*, 437-445.

[81] Shinohara, N;, Matsumoto, K; Endoh, S; Maru, J; Nakanishi, J. In vitro and in vivo genotoxicity tests on fullerene C60 nanoparticles. *Toxicol. Lett.*, 2009, *191*, 289-296.

[82] Ema, M; Tanaka, J; Kobayashi, N; Naya, M; Endoh, S; Maru, J; Hosoi, M; Nagai, M, Nakajima M, Hayashi M, Nakanishi J. Genotoxicity evaluation of fullerene C_{60} nanoparticles in a comet assay using lung cells of intratracheally instilled rats. *Regul. Toxicol. Pharmacol.*, 2012, *62*, 419-424.

[83] Kopova, I; Bacakova, L; Lavrentiev, V; Vacik, J: Growth and potential damage of human bone-derived cells on fresh and aged fullerene C_{60} films. *Int. J. Mol. Sci.*, 2013, *14*, 9182-9204.

[84] Kato, S; Aoshima, H; Saitoh, Y; Miwa, N. Biological safety of LipoFullerene composed of squalane and fullerene-C60 upon mutagenesis, photocytotoxicity, and permeability into the human skin tissue. *Basic Clin. Pharmacol. Toxicol.*, 2009, *104*, 483-487.

[85] Kato, S; Kikuchi, R; Aoshima, H; Saitoh, Y; Miwa, N. Defensive effects of fullerene-C60/liposome complex against UVA-induced intracellular reactive oxygen species generation and cell death in human skin keratinocytes HaCaT, associated with intracellular uptake and extracellular excretion of fullerene-C60. *J. Photochem. Photobiol. B.*, 2010, *98*, 144-1451.

[86] Kato, S; Aoshima, H; Saitoh, Y; Miwa, N. Fullerene-C60 incorporated in liposome exerts persistent hydroxyl radical-scavenging activity and cytoprotection in UVA/B-irradiated keratinocytes. *J. Nanosci. Nanotechnol.*, 2011, *11*, 3814-3823.

[87] Kato, S; Aoshima, H; Saitoh, Y; Miwa, N. Fullerene-C60/liposome complex: Defensive effects against UVA-induced damages in skin structure, nucleus and collagen type I/IV fibrils, and the permeability into human skin tissue. *J. Photochem. Photobiol. B.*, 2010, *98*, 99-105.

[88] Yudoh, K; Shishido, K; Murayama, H; Yano, M; Matsubayashi, K; Takada, H; Nakamura, H; Masuko, K; Kato, T; Nishioka, K. Water-soluble C60 fullerene prevents degeneration of articular cartilage in osteoarthritis via down-regulation of chondrocyte catabolic activity and inhibition of cartilage degeneration during disease development. *Arthritis Rheum.*, 2007, *56*, 3307-3318.

[89] Yudoh, K; Karasawa, R; Masuko, K; Kato, T. Water-soluble fullerene (C60) inhibits the osteoclast differentiation and bone destruction in arthritis. *Int. J. Nanomedicine,* 2009, *4*, 233-239.

[90] Zhang, Y; Wang, L; Sun, Y; Zhu, Y; Zhong, Z; Shi, J; Fan, C; Huang, Q. Conjugation of dexamethasone to C60 for the design of an anti-inflammatory nanomedicine with reduced cellular apoptosis. *ACS Appl. Mater. Interfaces*, 2013, *5*, 5291-5297.

[91] Ryan, JJ; Bateman, HR; Stover, A; Gomez, G; Norton, SK; Zhao, W; Schwartz, LB; Lenk, R;, Kepley, CL. Fullerene nanomaterials inhibit the allergic response. *J. Immunol.*, 2007, *179*, 665-672.

[92] Xu, Y; Zhu, J; Xiang, K; Li, Y; Sun, R; Ma, J; Sun, H; Liu, Y. Synthesis and immunomodulatory activity of [60]fullerene-tuftsin conjugates. *Biomaterials,* 2011, *32*, 9940-9949.

[93] Rezayat, SM; Boushehri, SV; Salmanian, B; Omidvari, AH; Tarighat, S; Esmaeili, S; Sarkar, S; Amirshahi, N; Alyautdin, RN; Orlova, MA; Trushkov, IV; Buchachenko, AL; Liu, KC; Kuznetsov, DA. The porphyrin-fullerene nanoparticles to promote the ATP overproduction in myocardium: 25Mg2+-magnetic isotope effect. *Eur. J. Med. Chem.*, 2009, *44*, 1554-1569.

[94] Lee, CM; Huang, ST; Huang, SH; Lin, HW; Tsai, HP; Wu, JY; Lin, CM; Chen, CT. C60 fullerene-pentoxifylline dyad nanoparticles enhance autophagy to avoid cytotoxic effects caused by the β-amyloid peptide. *Nanomedicine,* 2011, *7*, 107-114.

[95] Bobylev, AG; Kornev, AB; Bobyleva, LG; Shpagina, MD; Fadeeva, IS; Fadeev, RS; Deryabin, DG; Balzarini, J; Troshin, PA; Podlubnaya, ZA. Fullerenolates: metallated polyhydroxylated fullerenes with potent anti-amyloid activity. *Org. Biomol. Chem.*, 2011, *9*, 5714-5719.

[96] Tong, J; Zimmerman, MC; Li, S; Yi, X; Luxenhofer, R; Jordan, R; Kabanov, AV. Neuronal uptake and intracellular superoxide scavenging of a fullerene (C60)-poly(2-oxazoline)s nanoformulation. *Biomaterials*, 2011, *32*, 3654-3665.

[97] Lao, F; Chen, L; Li, W; Ge, C; Qu, Y; Sun, Q; Zhao, Y; Han, D; Chen, C. Fullerene nanoparticles selectively enter oxidation-damaged cerebral microvessel endothelial cells and inhibit JNK-related apoptosis. *ACS Nano*, 2009, *3*, 3358-3368.

[98] Saitoh, Y; Xiao, L; Mizuno, H; Kato, S; Aoshima, H; Taira, H; Kokubo, K; Miwa N. Novel polyhydroxylated fullerene suppresses intracellular oxidative stress together with repression of intracellular lipid accumulation during the differentiation of OP9 preadipocytes into adipocytes. *Free Radic. Res.*, 2010, *44*, 1072-1081.

[99] Xiao, L; Aoshima, H; Saitoh, Y; Miwa, N. The effect of squalane-dissolved fullerene-C60 on adipogenesis-accompanied oxidative stress and macrophage activation in a preadipocyte-monocyte co-culture system. *Biomaterials,* 2010, *31*, 5976-5985.

[100] Han, B; Karim, MN. Cytotoxicity of aggregated fullerene C60 particles on CHO and MDCK cells. *Scanning*, 2008, *30*, 213-220.

[101] Pirutin, SK; Turovetskiĭ, VB; Kedrov, AV; Kudriashov, IuB; Shaĭtan, KV; Rubin, AB. Effect of hydroxylated fullerene C60(OH)25 on macrophage plasma membrane integrity. *Radiats. Biol. Radioecol.*, 2012, *52*, 252-256.

[102] Harhaji, L; Isakovic, A; Vucicevic, L; Janjetovic, K; Misirkic, M; Markovic, Z; Todorovic-Markovic, B; Nikolic, N; Vranjes-Djuric, S; Nikolic, Z; Trajkovic, V. Modulation of tumor necrosis factor-mediated cell death by fullerenes. *Pharm. Res.*, 2008, *25*, 1365-1376.

[103] Eropkin, MIu; Piotrovskiĭ, LB; Eropkina, EM; Dumpis, MA; Litasova, EV; Kiselev, OI. Effect of polymer carrier origin and physical state on fullerene C60 phototoxicity in vitro. *Eksp. Klin. Farmakol.*, 2011, *74*, 28-31.

[104] Blank, V; Popov, M; Pivovarov, G; Lvova, N; Gogolinsky, K; Reshetov, V. Ultrahard and superhard phases of fullerite C60: Comparison with diamond on hardness and wear. *Diam. Rel. Mater.*, 1998, *7*, 427, 1998.

[105] Grausova, L; Bacakova, L; Kromka, A; Potocky, S; Vanecek, M; Nesladek, M; Lisa, V. Nanodiamond as promising material for bone tissue engineering. *J. Nanosci. Nanotechnol.,* 2009, *9*, 3524-3534.

[106] Grausova, L; Kromka, A; Burdikova, Z; Eckhardt, A; Rezek, B; Vacik, J; Haenen, K; Lisa, V; Bacakova, L. Enhanced growth and osteogenic differentiation of human osteoblast-like cells on boron-doped nanocrystalline diamond thin films. *PLoS One*, 2011, *6*, e20943.

[107] Ramirez, AP; Haddon, RC; Zhou, O; Fleming, RM; Zhang, J; McClure, SM; Smalley, RE. Magnetic susceptibility of molecular carbon: nanotubes and fullerite. *Science*, 1994, *265*, 84-86.

[108] Zubov, VI; Zubov, IV. Thermal vacancy effects on the thermodynamic properties and stability of fullerites. *J. Phys. Chem. B*, 2005, *109*, 14627-14631.

[109] Zhao, J; Feng, M; Yang, J; Petek, H. The superatom states of fullerenes and their hybridization into the nearly free electron bands of fullerites. *ACS Nano*, 2009, *3*, 853-864.

[110] Duda, JC; Hopkins, PE; Shen, Y; Gupta, MC. Exceptionally low thermal conductivities of films of the fullerene derivative PCBM. *Phys. Rev. Lett.*, 2013, *110*, 015902.

[111] Rogers, EJ; Hsieh, SF; Organti, N; Schmidt, D; Bello, D. A high throughput in vitro analytical approach to screen for oxidative stress potential exerted by nanomaterials using a biologically relevant matrix: human blood serum. *Toxicol. In Vitro*, 2008, *22*, 1639-1647.

[112] Giust, D; Albasanz, JL; Martín, M; Marega, R; Delforge, A; Bonifazi, D. [60]Fullerene-based monolayers as neuroprotective biocompatible hybrid materials. *Chem. Commun. (Camb)*, 2011, *47*, 10617-10619.

[113] Nakanishi, T; Morita, M; Murakami, H; Sagara, T; Nakashima, N. Structure and electrochemistry of self-organized fullerene-lipid bilayer films. *Chemistry*, 2002, *8*, 1641-1648.

[114] Prylutska, S; Bilyy, R; Overchuk, M; Bychko, A; Andreichenko, K; Stoika, R; Rybalchenko, V; Prylutskyy, Y; Tsierkezos, NG; Ritter, U. Water-soluble pristine fullerenes C60 increase the specific conductivity and capacity of lipid model membrane and form the channels in cellular plasma membrane. *J. Biomed. Nanotechnol.*, 2012, *8*, 522-527.

[115] Holy, J; Perkins, E; Yu, X. Adhesion, proliferation and differentiation of pluripotent stem cells on multi-walled carbon nanotubes. *IET Nanobiotechnol.*, 2011, *5*, 41-46.

[116] Ryoo, SR; Kim, YK; Kim, MH; Min, DH. Behaviors of NIH-3T3 fibroblasts on graphene/carbon nanotubes: proliferation, focal adhesion, and gene transfection studies. *ACS Nano*, 2010, *4*, 6587-6598.

[117] Armentano, I; Marinucci, L; Dottori, M; Balloni, S; Fortunati, E; Pennacchi, M; Becchetti, E; Locci, P; Kenny, JM. Novel poly(L-lactide) PLLA/SWNTs nanocomposites for biomedical applications: material characterization and biocompatibility evaluation. *J. Biomater. Sci. Polym. Ed.*, 2011, *22*, 541-556.

[118] Lin, C; Wang, Y; Lai, Y; Yang, W; Jiao, F; Zhang, H; Ye, S; Zhang, Q. Incorporation of carboxylation multiwalled carbon nanotubes into biodegradable poly(lactic-co-glycolic acid) for bone tissue engineering. *Colloids Surf. B Biointerfaces*, 2011, *83*, 367-375.

[119] Lahiri, D; Rouzaud, F; Namin, S; Keshri, AK; Valdés, JJ; Kos, L; Tsoukias, N; Agarwal, A. Carbon nanotube reinforced polylactide-caprolactone copolymer: mechanical strengthening and interaction with human osteoblasts in vitro. *ACS Appl. Mater. Interfaces*, 2009, *1*, 2470-2476.

[120] Lu, Y; Li, T; Zhao, X; Li, M; Cao, Y; Yang, H; Duan, YY. Electrodeposited polypyrrole/carbon nanotubes composite films electrodes for neural interfaces. *Biomaterials*, 2010, *31*, 5169-5181.
[121] Lobo, AO; Corat, MA; Ramos, SC; Matsushima, JT; Granato, AE; Pacheco-Soares, C; Corat, EJ. Fast preparation of hydroxyapatite/superhydrophilic vertically aligned multiwalled carbon nanotube composites for bioactive application. *Langmuir,* 2010, *26*, 18308-18314.
[122] Bacakova, L; Grausova, L; Vacik, J; Fraczek, A; Blazewicz, S; Kromka, A; Vanecek, M; Svorcik, V. Improved adhesion and growth of human osteoblast-like MG 63 cells on biomaterials modified with carbon nanoparticles. *Diam. Relat. Mater.,* 2007, *16*, 2133-2140.
[123] Grausova, L; Vacik, J; Bilkova, P; Vorlicek, V; Svorcik, V; Soukup, D;, Bacakova, M; Lisa, V; Bacakova, L. Regionally-selective adhesion and growth of human osteoblast-like MG 63 cells on micropatterned fullerene C(60) layers. *J. Optoelectronics Adv. Mater.*, 2008, *10*, 2071-2076.
[124] Grausova, L; Vacik, J; Vorlicek, V; Svorcik, V; Slepicka, P; Bilkova, P; Vandrovcova, M; Lisa, V; Bacakova, L. Fullerene C(60) films of continuous and micropatterned morphology as substrates for adhesion and growth of bone cells. *Diam. Relat. Mater.*, 2009, *18*, 578-586.
[125] Chandrakumar, KR; Ghosh, SK. Alkali-metal-induced enhancement of hydrogen adsorption in C60 fullerene: an ab Initio study. *Nano Lett.*, 2008, *8*, 13-19.
[126] Teprovich, JA, Jr; Wellons, MS; Lascola, R; Hwang, SJ; Ward, PA; Compton, RN; Zidan, R. Synthesis and characterization of a lithium-doped fullerane (Li(x)-C60-H(y)) for reversible hydrogen storage. *Nano Lett.*, 2012, *12*, 582-589.
[127] Sun, Q; Wang, Q; Jena, P; Kawazoe, Y. Clustering of Ti on a C60 surface and its effect on hydrogen storage. *J. Am. Chem. Soc.,* 2005, *127*, 14582-14583.
[128] Azizi, K; Sohrabinia, A. A DFT study on the interaction between adsorbed silver on C_{60} and disulfide bond. *J. Mol. Graph. Model.*, 2012, *38*, 354-359.
[129] Wei, P; Zhang, L; Lu, Y; Man, N; Wen, L. C60(Nd) nanoparticles enhance chemotherapeutic susceptibility of cancer cells by modulation of autophagy. *Nanotechnology,* 2010, *21*, 495101.
[130] Derouiche, H; Mohamed, AB. Thermal annealing effect on poly(3-hexylthiophene): fullerene:copper-phthalocyanine ternary photoactive layer. *ScientificWorld Journal,* 2013, *2013*, 914981.
[131] Hosseini, A;, Sharifzadeh, M; Rezayat, SM; Hassanzadeh, G; Hassani, S; Baeeri, M; Shetab-Bushehri, V; Kuznetsov, DA; Abdollahi, M. Benefit of magnesium-25 carrying porphyrin-fullerene nanoparticles in experimental diabetic neuropathy. *Int. J. Nanomedicine,* 2010, *5*, 517-523.
[132] Li, Y; Fang, L; Cheng, P; Deng, J; Jiang, L; Huang, H; Zheng, J. An electrochemical immunosensor for sensitive detection of Escherichia coli O157:H7 using C60 based biocompatible platform and enzyme functionalized Pt nanochains tracing tag. *Biosens. Bioelectron.,* 2013, *49*, 485-491.
[133] Kalska-Szostko, B; Rogowska, M. Preparation of magnetite-fullerene nanocomposite with enzyme immobilization. *J. Nanosci. Nanotechnol.*, 2012, *12*, 6907-6912.

[134] Goyal, RN; Oyama, M; Bachheti, N; Singh, SP. Fullerene C60 modified gold electrode and nanogold modified indium tin oxide electrode for prednisolone determination. *Bioelectrochemistry,* 2009, 74, 272-277.

[135] Yu, J; Ma, T; Liu, G; Cheng, B. Enhanced photocatalytic activity of bimodal mesoporous titania powders by C60 modification. *Dalton Trans.*, 2011, *40*, 6635-6644.

[136] Meng, ZD; Zhu, L; Choi, JG; Park, CY; Oh, WC. Sonocatalytic degradation of Rhodamine B in the presence of C60 and CdS coupled TiO2 particles. *Ultrason. Sonochem.,* 2012, *19*, 143-150.

[137] Scharff, P; Ritter, U; Matyshevska, OP; Prylutska, SV; Grynyuk, II; Golub, AA; Prylutskyy, YI; Burlaka, AP. Therapeutic reactive oxygen generation. *Tumori*, 2008, *94*, 278-283.

[138] Ohkubo, K; Kawashima, Y; Sakai, H; Hasobe, T; Fukuzumi, S. Enhanced photoelectrochemical performance of composite photovoltaic cells of Li(+)@C60-sulphonated porphyrin supramolecular nanoclusters. *Chem. Commun. (Camb),* 2013, *49*, 4474-4476.

[139] Zhong, X; Yuan, R; Chai, Y. In situ spontaneous reduction synthesis of spherical Pd@Cys-C60 nanoparticles and its application in nonenzymatic glucose biosensors. *Chem. Commun. (Camb),* 2012, *48*, 597-599.

[140] Wang, H; Bai, L; Chai, Y; Yuan, R. Synthesis of multi-fullerenes encapsulated palladium nanocage, and its application in electrochemiluminescence immunosensors for the detection of streptococcus suis serotype 2. *Small*, 2014, *10*, 1857-1865.

[141] Tóth, E; Bolskar, RD; Borel, AV; González, G; Helm, L; Merbach, AE; Sitharaman, B; Wilson, LJ. Water-soluble gadofullerenes: toward high-relaxivity, pH-responsive MRI contrast agents. *J. Am. Chem. Soc.,* 2005, *127*, 799-805.

[142] Kamarás, K; Klupp, G. Metallicity in fullerides. *Dalton Trans.*, 2014, *43*, 7366-7378.

[143] Yao, M; Pischedda, V; San Miguel, A. Raman study of the electron-phonon interaction in light alkali metal intercalated metallic fullerides. *J. Phys. Condens. Matter*, 2011, *23*, 115701.

[144] Prassides, K; Takabayashi, Y; Nakagawa, T. Mixed valency in rare-earth fullerides. *Philos. Trans. A Math. Phys. Eng. Sci.,* 2008, *366*, 151-161.

[145] Harcuba, P; Bacakova, L; Strasky, J; Bacakova, M, Novotna, K; Janecek, M: Surface treatment by electric discharge machining of Ti-6Al-4V alloy for potential application in orthopaedics. *J. Mech. Behav. Biomed. Mater.*, 2012, *7*, 96-105.

[146] Strasky, J; Havlikova, J; Bacakova, L; Harcuba, P; Mhaede, M; Janecek, M. Characterization of electric discharge machining, subsequent etching and shot-peening as a surface treatment for orthopedic implants. *Appl. Surf. Sci.*, 2013, *213*, 73-78.

[147] Jirka, I; Vandrovcova, M; Frank, O; Tolde, Z; Plsek, J; Luxbacher, T; Bacakova, L; Stary V. On the role of Nb-related sites of an oxidized β-TiNb alloy surface in its interaction with osteoblast-like MG-63 cells. *Mater. Sci. Eng. C Mater. Biol. Appl.*, 2013, *33*, 1636-1645.

[148] Vandrovcova, M; Bacakova, L. Adhesion, growth and differentiation of osteoblasts on surface-modified materials developed for bone implants. *Physiol. Res.*, 2011, *60*, 403-417.

[149] Bacakova, L; Stary, V; Kofronova, O; Lisa, V. Polishing and coating carbon fibre-reinforced carbon composites with a carbon-titanium layer enhances adhesion and

growth of osteoblast-like MG63 cells and vascular smooth muscle cells in vitro. *J. Biomed. Mater. Res.*, 2001, *54*, 567-578.

[150] Grinevich, A; Bacakova, L; Choukourov, A; Boldyryeva, H; Pihosh, Y; Slavinska, D; Noskova, L; Skuciova, M; Lisa, V; Biederman, H: Nanocomposite Ti/hydrocarbon plasma polymer films from reactive magnetron sputtering as growth support for osteoblast-like and endothelial cells. *J. Biomed. Mat. Res. A,* 2009, *88A*, 952-966.

[151] Kopova, I; Bacakova, L; Lavrentiev, V; Vacik, J. Growth and potential damage of human bone-derived cells cultured on fresh and aged C60/Ti films. *PloS One*, submitted.

In: Fullerenes
Editor: Shannon B. Ellis

ISBN: 978-1-63321-385-2

Chapter 2

LIQUID CHROMATOGRAPHY IN THE ANALYSIS OF FULLERENES

Alina Astefanei, Oscar Núñez* [*] *and Maria Teresa Galceran
Department of Analytical Chemistry, University of Barcelona,
Martí i Franquès, Barcelona, Spain

ABSTRACT

Since the existence of the remarkably stable carbon cluster C_{60} was first reported by Professor Kroto in 1985, fullerenes have been studied in depth because of their unique structure and potential as new nanomaterials. This is due to their appealing photo-, electro-chemical, and physical properties, which make them exploitable in different fields. Their electron double bonds allow pristine fullerenes, which are intrinsically hydrophobic, to become readily derivatized and water soluble through the addition of various functional groups. Hence, fullerene derivatives are being investigated for their use in biomedical, cosmetic and industrial applications, ranging from drug-delivery systems and anti-aging formulations to electrical components. Moreover, some of these fullerene derivatives play an important role in the controllable synthesis of new materials and biologically active compounds.

A rapid growth in the development, production, and use of these compounds is being predicted over the next decade, so their dispersion in the environment is to be expected. Moreover, their unique colloidal properties in water can affect their behavior in the environment. For instance, although C_{60} is practically insoluble in water, it can spontaneously form stable aqueous colloidal suspensions containing nanocrystalline aggregates with effective aqueous phase concentrations many orders of magnitude above that corresponding to the aqueous solubility of molecular fullerenes. Consequently, the ability of these carbon nanomaterials to form aggregates could change their bioavailability, leading to additional toxicological concerns, although at the moment data regarding their toxicity is limited. The lack of specific analytical methods for their determination presents a significant limitation when attempting to evaluate the risk these materials present to human health and the environment.

[*] E-mail address: oscar.nunez@ub.edu.

Although significant advances have been made in the analysis of fullerenes in the past few years, there is still an increasing need to develop effective, sensitive, and reliable analytical methods for their determination. Reversed-phase liquid chromatography (RP-LC), using toluene-acetonitrile or toluene-methanol mixtures as mobile phases, has been the method of choice for their analysis. Although ultraviolet (UV) detection is being used for their detection, liquid chromatography coupled to mass spectrometry (LC-MS) is the most commonly proposed technique nowadays for the determination and characterization of this family of compounds in complex matrices. Nowadays, a number of LC-MS applications for the analysis of this family of compounds in environmental waters, biological fluids and cosmetics are available. For most of these applications, both electrospray ionization (ESI) and atmospheric pressure chemical ionization (APCI) have been proposed, but recently atmospheric pressure photoionization (APPI) has appeared as one of the best options for their ionization due to their unique photoionization properties.

This chapter will review the state-of-the-art of liquid chromatography techniques for the analysis of fullerenes. LC and LC-MS conditions, UV-detection systems, ionization sources, and MS analyzers frequently used, as well as strategies for the unequivocal confirmation and quantitative analysis of fullerenes will be discussed by means of relevant applications. Coverage of all kind of applications is beyond the scope of the present contribution, so we will focus on the most relevant application fields such as bio-analytical, cosmetic and environmental applications.

INTRODUCTION

The nanotechnology era was born with the first synthesis of carbon-based nanomaterials (carbon nanoparticles, CNPs), C_{60} fullerene (Buckminster fullerene, or Bucky ball), by Kroto *et al.* [1] in the 1980s. The discovery of fullerenes by the laser vaporization technique resulted in the award of Chemistry Nobel Prize to Curl, Kroto, and Smalley in 1996. These compounds are molecules arranged into a spherical shape resembling a geodesic dome, and there are multiple spherical configurations involving different numbers of carbon atoms (*e.g.* C_{60}, C_{70}, and C_{80}). In general their structure is composed of closed carbon cages formed by 12 pentagons and a number of hexagons that increases with the size of the fullerene (2n/2-10, n = number of C atoms). As an example, Figure 1 depicts the structure of the most common fullerene (C_{60}).

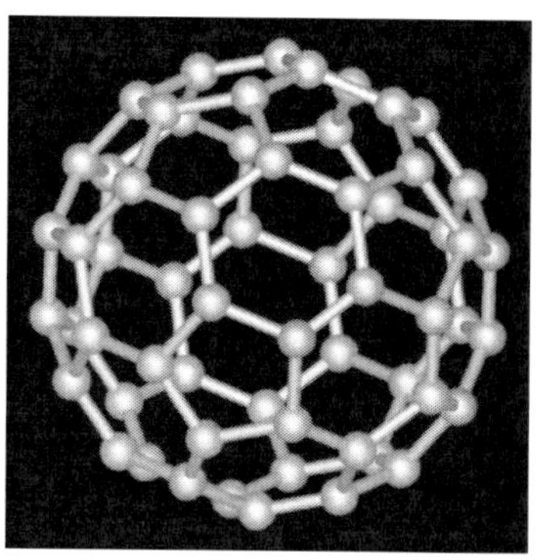

Figure 1. Structure of C_{60} fullerene. Permission granted to copy, distribute and/or modify under the terms of the GNU Free Documentation Lisence, Version 1.2 or any later version published by the Free Software Foundation.

Fullerenes are the most prominent class of CNPs in terms of research investigations and current applications in consumer goods by virtue of their exceptional, characteristic, and varied properties. An increase on human and environmental exposure to these compounds is expected because of their industrial scale production (currently exceeding several tones/year) [2]. However, their occurrence in the environment can be due to both natural and anthropogenic causes [3]. Regarding natural origin, it has been reported that some fullerenes have interstellar origin, being brought to earth by comets or asteroids [4,5]. Additionally, it is also believed that they have been generated from polycyclic aromatic hydrocarbons derived from algal matter by metamorphosis at temperatures between 300 and 500 °C in the presence of elemental sulfur [6] or by natural combustion processes.

Nowadays, fullerenes are used in a wide range of applications such as in optical and electronic devices (polymer additives, solar cells, photovoltaic devices, electro-optical devices) [7-12], in commercial cosmetic products [13-15], as well as in biomedicine (antiviral, anticancer, and antioxidant agents, in drug delivery systems, or as gene carriers) [16-21]. Currently, surface-functionalized fullerenes are produced in larger quantities than empty fullerenes [22-24] in an attempt to create more biologically compatible forms [23-25]. It should be taken into account that underivatized fullerenes are extremely insoluble in water [26] and will likely exist on the water surface or at the water-sediment interface until physical mixing occurs within the aquatic environment. For instance, the solubility of C_{60} in water is 8 ng/L [27] or even lower. However, despite its high hydrophobicity, C_{60} can spontaneously form stable colloidal suspensions consisting of nanocrystalline aggregates providing effective aqueous phase concentrations many orders of magnitude above the one corresponding to its aqueous solubility. In addition, the chemical functionalization of C_{60} fullerene spheres produces a large number of different derivatives that combine the desirable properties of C_{60} with the solubilizing power of the functional groups [28-30].

There is much controversy concerning the toxicity of this family of compounds. Most of the existing reports showed that fullerene compounds are non-toxic and that they even have a wide spectrum of positive and unique biological activity [15,31-33], while other works are suggesting the possibility of some adversary effects. For instance, Seda et al. [34] showed that aqueous suspensions of gallic acid–C_{70} fullerene reduced the fecundity in *Daphnia magna*, and Lovern and Klaper [35] reported that exposure to C_{60} fullerene can induce *Daphnia magna* mortality. Recently it was found that the *in vitro* toxicity of fullerene C_{60} is correlated with its ability to undergo aggregation [36], and, specifically the surface of its aggregated forms may effectively bind with proteins like human serum albumin leading to functional changes of the drug binding sites of the protein [37]. Nevertheless, these compounds are being considered as emerging contaminants, thus, sensitive and reliable methodologies for their analysis are needed.

Chromatographic Separation of Fullerene Compounds

Because of their similar structure and limited solubility in common solvents, together with their tendency to form aggregates depending on the solvent composition, the separation of fullerene compounds is challenging, especially when dealing with higher fullerenes. Liquid chromatography is one of the most effective methods for their separation being investigated since the 1990s [38-42]. Table 1 summarizes LC methods reported in the literature for the analysis of this family of compounds [13,14,17,32,39,40,42-67]. Most of the works are dealing with the analysis of C_{60} fullerene, either alone or together with higher fullerenes (C_{70} to C_{98}), and few works are also studying functionalized fullerenes, most of them based on the C_{60} bulky structure [32,44,45,52,53,57,61,65-67]. Only recently, Kolkman et al. [58] and Carboni et al. [57] reported the determination of a C_{70}-functionalized fullerene ([6,6]-Phenyl-C71-butyric acid methyl ester (C_{70}-PCBM)) with other pristine and C_{60} fullerene derivatives. Reversed-phase LC using C18 columns is the method of choice for the separation of fullerenes either using toluene [59], or toluene:methanol [39,44-46,48,49,52-54] and toluene:acetonitrile [14,17,43,47,51,56] mixtures as mobile phases. Other solvents such as n-alkanes (pentane, hexane, and heptane) [50] and acetonitrile:dichlormethane (50:50 *v/v*) [40] have also been proposed as mobile phases with C18 columns. In most of these works, the use of conventional 3.5-5 μm particle size columns is described, obtaining long analysis times. Only very few papers are proposing the use of columns with sub-2μm totally porous particles [54,55] or core-shell partially porous particles (which are constituted of a fused core and a layer of porous silica coating) [53] for their analysis. A considerably reduction in analysis time (2.5 to 4.3 min) can be achieved with these columns, with higher column efficiency, while maintaining good chromatographic separation. As an example, the separation of five pristine fullerenes (C_{60}, C_{70}, C_{76}, C_{74} and C_{84}) with a core-shell C18 Ascentis Express (Supelco, Bellefonte, PA, USA) column and a sub-2μm C18 Hypersil Gold (Thermo Fisher Scientific, San José, CA, USA) column is shown in Figure 2 [54].

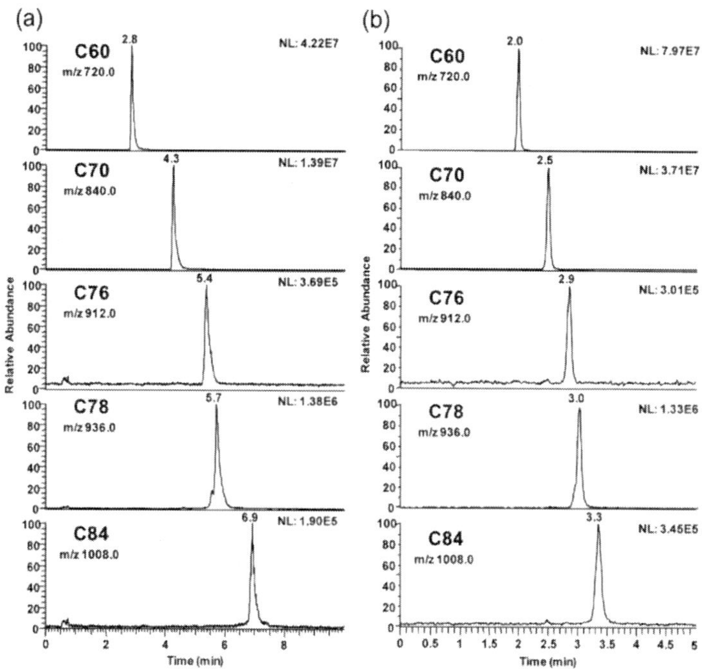

Figure 2. Chromatographic separation of fullerenes (100 μg/L in toluene/methanol 50:50 *v/v*) in (a) a core-shell C18 Ascentis Express (150 x 2.1 mm, 2.7 μm particle size) column, and (b) a sub-2μm C18 Hypersil Gold (150 x 2.1 mm, 1.9 μm particle size) column. Separation: isocratic elution using toluene:methanol 45:55 (*v/v*) as mobile phase. Flow rate: 500 μL/min. Acquisition performed by LC-APPI-MS. Reprinted with permission from reference [54]. Copyright (2012) American Chemical Society.

Table 1. Liquid chromatography methods for the analysis of fullerenes

Fullerenes	Sample	Column	Chromatographic conditions	Detection	Analysis time	Ref.
C_{60}, C_{70}	Soils and sediments	Luna C18 150 x 4.60 mm, 5 µm, 100 Å	Toluene:acetonitrile (60:40, *v/v*) 0.75 mL/min	UV-Vis 334 nm	9 min	[43]
C_{60}, C_{70}, C_{60}-pyrr, PCBM, ThCBM	River sediments and soils	Purospher STAR RP-18 125 x 2 mm, 5 µm	Toluene:methanol (50:50, v/v) 3 mL/min	Mass spectrometry (-) ESI-MS (triple quadrupole)	-	[44]
C_{60}, C_{70}	Cosmetics	Nova-Pak C18 150 x 3.9 mm, 4 µm , 60 Å	Toluene:acetonitrile (55:45, v/v) 1 mL/min	Mass spectrometry (-) APCI-MS	6 min	[14]
C_{60}, C_{70}, C_{60}-pyrr	Wastewater	Purospher STAR C18 125 x 2 mm, 5 µm	Toluene:methanol (50:50, v/v) 0.4 mL/min	Mass spectrometry (-) ESI-MS (QqLIT)	4 min	[45]
C60	Water	Nova-Pak C18 150 x 3.9 mm, 4 µm, 60 Å	Toluene:methanol (50:50, v/v) 1 mL/min 0.2 mL/min	UV-Vis 333 nm Mass spectrometry (-) APCI/ESI-MS (LTQ-FT Orbitrap)	-	[46]
C_{60}, C_{70}	Standard mixture	Gemini C18 150 x 2.0 mm, 5 µm, 110 Å	Toluene:methanol (70:30, v/v) 0.2 mL/min	Mass spectrometry (-) APPI-MS (single quadrupole) Dopant: toluene	5 min	[49]
C_{60}, C_{70}	Standard mixture	Handmade C18, 250 x 4.6 mm, 120 Å	n-Pentane n-Hexane n-Heptane	UV-Vis	15 min 8 min 6 min	[50]
C_{60}	Biological samples	Delta Pak C18 150 x 3.9 mm, 5 µm, 300 Å	Toluene:acetonitrile (60:40, v/v) 1 mL/min	UV-Vis 333 nm	3 min	[17]
C_{60}	Blood and tissues	LiChrocart C18 4.6 x 20 mm, 5 µm, 100 Å	Toluene:acetonitrile (60:40, v/v) Toluene:acetonitrile (55:45, v/v) Toluene:methanol (55:45, v/v) 1 mL/min	UV-Vis 333 nm Mass spectrometry (-) APCI-MS (ion trap)	6 min	[51]

Table 1. (Continued)

Fullerenes	Sample	Column	Chromatographic conditions	Detection	Analysis time	Ref.
C_{60}, C_{70}	Geological materials	Nova-Pak C18 3.9 x 300 mm, 4 µm , 60 Å	Toluene-methanol (40:60, v/v) 2 mL/min	UV-Vis 330 nm	12 min	[39]
C_{60}, C_{70}, C_{60}-pyrr, CPTAE, PCBM, ThCBM	Aerosols	Luna C18 (2)-HST 50 x 2.0 mm, 2.5 µm, 100 Å	Gradient elution: (A) Toluene (B) Methanol 0.3 mL/min	Mass spectrometry (-) ESI-MS (triple quadrupole)	-	[52]
C_{60},C_{70}, PCBM	Water	Ascentis Express C18 50 x 2.1 mm, 2.7 µm, 90 Å	Toluene:methanol (50:50, v/v) 0.2mL/min	Mass spectrometry (-) APPI-MS/MS (ion trap) Dopant: toluene	2.5 min	[53]
PCBM, PCBB, C_{60}-pyrr, C_{60}, C_{70}, C_{76}, C_{78}, C_{84}	Water and sediments	Hypersil GOLD C18 150 x 2.1 mm, 1.9 µm , 175 Å	Gradient elution (A) Toluene (B) Methanol 0.5 mL/min	Mass spectrometry (-) APPI-MS (triple quadrupole) Dopant: toluene	4.3 min	[55]
C_{60}, C_{70}, C_{76}, C_{78},C_{84}	Pond water	Hypersil GOLD C18 150 x 2.1 mm, 1.9 µm , 175 Å	Toluene:methanol (45:55, v/v) 0.5 mL/min	Mass spectrometry (-) APPI-MS (triple quadrupole) Dopant: toluene	3.5 min	[54]
C_{60}, C_{70}, C_{76}, C78, C_{82},C_{84}	Standard mixture	Develosil C30-UG-5 18.2% C 250 x 4.6 mm, 5 µm, 14 Å	Toluene:acetonitrile (80:20, v/v) 1 mL/min	UV-Vis 312 nm	40 min	[56]
C_{60}, C_{70}, PCBB, PCMO, PCBM, bis-PCBM,ThCB, C_{70}-PCBM	Soils	Cosmosil Buckyprep 250 x 4.6 mm, 5 µm, 120 Å	Gradient elution (A)Toluene (B)Acetonitrile 1 mL/mmin	UV-Vis 305 and 332 nm	25 min	[57]
C_{60}, C_{7}0, PCBM, C_{70}-PCBM, bisPCBM, PCBB, ThCBM, PCBO	Water	Cosmosil Buckyprep 250 x 2 mm, 5 µm, 120 Å	Gradient elution (A)Toluene (B)Acetonitrile 0.2 mL/min	Mass spectrometry (-) ESI-MS (hybrid LTQ Orbitrap)	30 min	[58]

Fullerenes	Sample	Column	Chromatographic conditions	Detection	Analysis time	Ref.
C_{60}	Cosmetic products	Cosmosil Buckyprep 250 x 4.6 mm, 5 µm, 120 Å	Toluene (100%)	UV-Vis 285 nm	-	[13]
C_{60}	Wastewater	Cosmosil Buckyprep 250 x 4.6 mm, 5 µm, 120 Å	Toluene (100%) 1 mL/min	UV-Vis 332 nm Mass spectrometry (-) APCI-MS (single quadrupole)	-	[59]
C_{60}	Standard solution	Cosmosil Buckyprep 150 x 4 mm	Toluene:methanol (95:5, v/v) 1mL/min	Mass spectrometry (-) APPI-MS (triple quadrupole) Dopant: toluene	-	[60]
C_{60}, C_{70}, PCBM	Water	Cosmosil Buckyprep-PYE 150 x 4.6 mm, 5 µm, 120 Å	Toluene:acetonitrile (80:20, v/v) 1.3 mL/min	UV-Vis 330 and 333nm	6.5 min	[61]
C_{60}, C_{70}, C_{76}, C_{78},C_{80}, C_{82},C_{84},C_{86},C_{88}, C_{90}	Soot extract	Cosmosil Buckyprep-PYE 150 x 4.6 mm, 5 µm, 110 Å	Gradient elution (A) Dichloromethane (B) Acetonitrile 1mL/min	UV-Vis 350 nm Mass spectrometry (-) APCI-MS (triple quadrupole) Dopant: toluene	37 min	[40]
C_{60}, C_{70}	Standard mixture	Magnesia-zirconia stationary phase modified with 2,4,6-trinitrophenol and pyrenebutytic acid (lab made) 150 x 4.6 mm	Toluene:cyclohexane. (50:50, v/v) 1 mL/min	UV-Vis 410 nm	15 min	[62]
C_{60}, C_{70} and higher fullerenes (C_{76}-C_{84})	Standard mixture	Two charge transfer stationary phases: p-nitobenzoic acid-bonded silica (PNBA), naphtyl acetic acid-bonded silica (NPAA) (lab made)	Toluene:cyclohexane (30:70, v/v)	UV-Vis 410 nm	10 min	[63]

Table 1. (Continued)

Fullerenes	Sample	Column	Chromatographic conditions	Detection	Analysis time	Ref.
C_{60}, C_{70}, C_{76}, C_{78}, C_{82}, C_{84}, C_{86}, C_{88}, C_{90}, C_{92}, C_{94}	Standard mixture	Zn (II) Tetraphenylporphyrin silica stationary phase 4.6 x 250 mm	Carbon disulfide: toluene (45:55, v/v) 2 mL/min	UV-Vis 430 nm Mass Spectrometry MALDI-TOF-MS (337 nm laser wavelength)	8 min	[42]
C_{60}, C_{70} , C_{76},C_{78}, C_{84}, C_{86}	Standard mixture	Hydroxyphenyl-triphenylporphyrin (HPTPP) silica stationary phase 250 x 4.6 mm	Chlorobenzene (100%)	UV-Vis 410 nm	50 min	[64]
Fullerols ($C_{60}(OH)_x(ONa)_y$, x+y=24,y=6-8)	Standard solution	Waters XBridge amide 150 x 4.6 mm, 3.5 μm, 130 Å	Water:acetonitrile (10:90, *v/v*) 0.5 mL/min	UV-Vis 254 nm Mass spectrometry MALDI-TOF-MS (+)ESI-MS (triple quadrupole)	3 min	[65]
Fullerol $C_{60}(OH)_{24}$	Standard solution	Zorbax SB-C18 150 x 4.6 mm, 5 μm, 80 Å	H_2O (100%) 1 mL/min	UV-Vis 280 nm Mass spectrometry (+) ESI-MS (Q-TOF)	3 min	[66]
Chiral C_{60} fullerene derivatives (Hexakis and tris aducts)	Standard mixture	R-(-) TAPA bound to silica gel 250 x 0.25 mm, 5 μm	Acetonitrile:water (75:25, *v/v*) 0.4 μL/min (Hexakis adduct) Toluene n-heptane (68:32, *v/v*) 0.5 μL/min (Tris adduct)	UV-Vis 254 nm 315 nm	170 min 240 min	[67]
Dendritic Methano C_{60} fullerene octadecaacid	Standard solution	Superphere C18, 250 x 4.6 mm, 5μm	Acetonitrile: aqueous buffer (0.1 M, pH 2) (50:50 *v/v*) 0.8 mL/min	UV-Vis 320 nm	10 min	[32]

An interesting result of this work was that in contrast to the general retention behavior observed for small molecules, which usually show slightly lower retention in core-shell columns than in sub-2μm particle size columns [68-70], pristine fullerenes showed higher retention in the core-shell partially porous column (Figure 2a), and especially for larger fullerenes. The authors attributed this behavior to the size and symmetry of pristine fullerenes, which facilitated a higher interaction in the core-shell column, with a smaller pore size (90 Å) than the sub-2 μm column (175 Å). This effect was also observed by Astefanei et al. [55] who recently proposed to use the same column for the separation of three C_{60}-fullerene derivatives (N-methyl fulleropyrrolidine (C_{60}-pyrr); [6,6]-Phenyl C_{61} butyric acid methyl ester (PCBM); and [6,6]-Phenyl C_{61} butyric acid buthyl ester (PCBB)) with the five pristine fullerenes. In this case, only C_{60}-pyrr, the compound with the smallest functional group, showed a slightly higher retention in the core-shell column as happens with pristine fullerenes [54]. The other two C_{60}-fullerene derivatives showed the typical retention behavior observed for small molecules when comparing core-shell columns and columns with totally porous particles (sub-2μm) [68-70].

The same authors also compared the retention of this family of compounds by using different organic solvent compositions as mobile phases in the sub-2μm particle size column [55]. Although the eluotropic strength of the mixture toluene:acetonitrile is higher than that of toluene:methanol, fullerene compounds showed higher retention times when acetonitrile mixtures were used. This behavior was also reported by Jinno et al. [71]. The authors suggest that this phenomenon can be related to the ability of fullerenes to form aggregates in polar solvents and to the fact that increasing polarity seems to lead to the formation of smaller aggregates [72,73]. Thus, fullerene aggregates in toluene:acetonitrile mixtures would be bigger than in toluene:methanol mixtures and, as a result, fullerenes will be more retained when toluene:acetonitrile is used.

Other reversed-phase columns have also been proposed for the separation of fullerenes. For instance, Ohta et al. [56] studied the chromatographic separation of pristine fullerenes (C_{60}, C_{70}, C_{76}, C_{78}, C_{82} and C_{84}) in monomeric type C30, C18 and C8 alkyl bonded stationary phases using toluene:acetonitrile (60:40 *v/v*) as mobile phase. The C8 chains were too short to have an effective interaction with fullerene molecules because they did not have enough depth to catch and hold the bulky molecules within the ligand space available in the column. In contrast, C30 chains resulted to be very effective for the interaction with fullerene molecules because of its very long chain, so a strong retention was achieved. Nevertheless, this stationary phase is not very practical for a fast chromatographic analysis of this family of compounds because very long analysis times (higher than 40 min) were obtained.

The chromatographic separation of this family of compounds has also been reported using other stationary phases (Table 1). A 3-(1-pyrenyl)propyl group bonded silica based column (Cosmosil Buckyprep) specifically designed for fullerene separation is commercially available from Nacalai USA Inc. (San Diego, CA, USA), and it has been proposed by several authors [13,57-60]. This column is able to retain fullerenes with a mobile phase of 100% toluene [13,59], although toluene mixtures with acetonitrile [57,58] or methanol [60] has also been used. This column has been mainly proposed in the literature for the analysis of C_{60} [13,59,60]. Recently, Carboni et al. [57] reported the use of this column for the separation of C_{60} and C_{70} and several functionalized fullerenes (C_{60} and C_{70} exohedral-derivatives) and the chromatogram obtained is shown in Figure 3. The elution order in this column is correlated with: (i) the number and presence of functionalization on the fullerene cage, which increases the solubility of the compounds in the mobile phase and decreases the surface available for the interaction with the pyrenyl-propyl groups of the stationary phase (functionalized fullerenes elute earlier than pristine fullerenes), and (ii) the size of the fullerene cage (C_{60} structures elute earlier than C_{70} ones). The authors attributed the jagged peak shape of the first-eluted compound (bis[60]PCBM) to either the presence of different isomers (60 positions are available on the structure for the attachment of the two functional groups) or by the formation of micelles in the solution.

As depicted in Figure 3, long analysis times are usually obtained with this specific Cosmosil Bulkyprep column (25 min, although the whole method lasted 32 min to allow the system to equilibrate prior to the next analysis), which is disadvantageous in comparison to the short analysis times that can be achieved nowadays with sub-2μm or core-shell column technology as previously commented [53-55]. Nevertheless, this specially designed column for fullerene separation exceeds the injection volume of a standard C18 column by a factor of 35, therefore it is ideal for a preparative-scale separations [74]. A variation of this column, a Cosmosil Bulkyprep-PYE with a 2-(1-pyrenyl)ethyl group bonded silica stationary phase, has also been applied for the separation of C_{60}, C_{70} and PCBM using toluene:acetonitrile (80:20 *v/v*) as mobile phase with analysis times of 6.5 min [61], and for the analysis of pristine fullerenes up to C_{90} in soot extracts but with a long analysis time (37 min) [40].

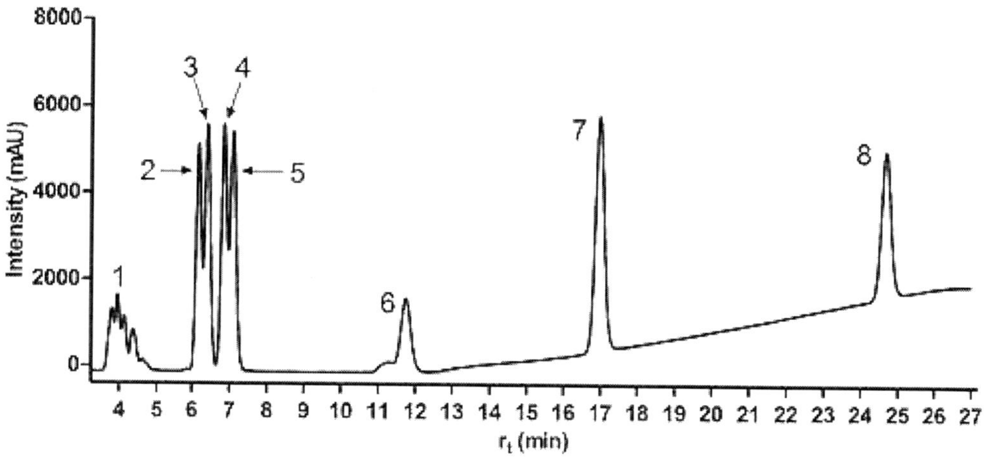

Figure 3. HPLC-UV chromatogram of fullerenes and functionalized fullerenes in a Cosmosil Buckyprep (250 x 4.6 mm, 5μm particle size) column. Separation under gradient elution from toluene:acetonitrile (75:25 *v/v*) to 100% toluene in 6 min, and then isocratic elution at 100% toluene. Peak identification: (1) [6,6]-Bis-Phenyl-C_{61}-butyric acid methyl ester (bis[60]PCBM); (2) [6,6]-Phenyl-C_{61}-butyric acid octyl ester ([60]PCMO); (3) [6,6]-Phenyl-C_{61}-butyric acid butyl ester ([60]PCBB); (4) [6,6]-Phenyl-C_{61}-butyric acid methyl ester ([60]PCBM); (5) [6,6]-Thienyl-C_{61}-butyric acid methyl ester ThCBM); (6) [6,6]-Phenyl-C_{71}-butyric acid methyl ester (C_{70}-PCBM); (7) C_{60}; and (8) C_{70}. Reprinted with permission from reference [57]. Copyright (2014) Elsevier.

The use of charge-transfer stationary phases such as p-nitrobenzoic acid-bonded silica (PNBA) and naphtyl acetic acid-bonded silica (NPAA) using toluene:cyclohexane solution as mobile phase has also been reported for the analysis of pristine fullerenes [63]. Additionally, a magnesia-zirconia stationary phase modified with 2,4,6-trinitrophenol acid [62] and other stationary phases based on phenylporphyrin immobilized silica, such as tetraphenylporphyrin-silica [42] and hydroxyphenyltriphenylporphyrin-silica [64], have also been proposed for the separation of these compounds. These last stationary phases provide very good selectivity and efficiency, being the phenyl group of the immobilized porphyrin structure critical for fullerene separation. For instance, Xiao et al. [42] reported better selectivity using tetraphenylporphyrin stationary phase for C_{60} and C_{70} separation compared to the commercially-available Cosmosil Buckyprep columns. However, their use is limited by the high cost of phenylporphyrin materials.

Lately, there is an increasing interest in developing methods for the analysis of water-soluble fullerenes, compounds with an expanding use in medicine, cosmetics and industry. Among them, the chromatographic separation of fullerenols (or fullerols) is challenging because of their stronger hydroxylation and the associated negative surface charges that make difficult the application of conventional C18 reversed-phase separation approaches. For this reason, some authors have proposed the use of hydrophilic interaction liquid chromatography (HILIC) to solve this problem. For instance, Chao et al. [65] achieved a good efficiency for the analysis of highly water-soluble hydroxylated fullerene derivatives (fullerols $C_{60}(OH_x)(ONa)_y$ with x+y=24, y=6-8) using an amide-type HILIC column (Zorbax SB-C18, 150 x 4.6 mm, 5 µm particle size) under isocratic conditions with water:acetonitrile (10:90 *v/v*) as mobile phase. Figure 4 shows the LC-MS chromatograms of a 60 pg/ml fullerol solution using different MS scan methods. Under this conditions, a peak signal at 3.1 min with a plate number higher than 2,000 was obtained for the fullerol.

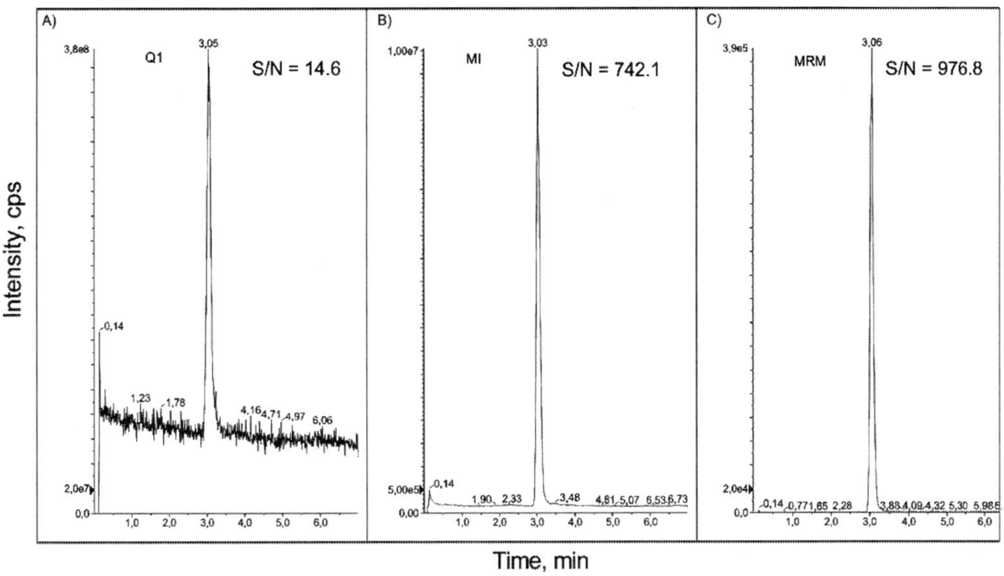

Figure 4. Representative LC-MS chromatograms of a 60 pg/mL fullerol solution using different MS scan methods. (A) Q1 scan of masses between *m/z* 520 and 1150, (B) multiple ion scan using the major peaks of the 74 *m/z* series, and (C) MRM using the transition of the major 74 *m/z* series peaks to *m/z* 73. The signal-to-noise (S/N) ratios for each peak are indicated. Reprinted with permission from reference [65]. Copyright (2011) American Chemical Society.

Working with the same group of compounds, Silion et al. [66] proposed the use of a Zorbax SB-C18 reversed-phase column (150 x 4.6 mm, 5 µm particle size) to study fullerenol $C_{60}(OH)_{24}$ and using water as mobile phase (1 mL/min). Under these unconventional reversed-phase conditions, a very small peak at approximately 1.5 min retention time was observed, corresponding to contaminants from the standard, and a major peak at 2.6 min, corresponding to the studied fullerenol.

Among other water-soluble functionalized fullerenes, the separation of chiral water-soluble fullerene derivatives using a chiral charge-transfer complexing agent (2-(2,4,5,7-tetranitro-9-fluorenylideneaminooxy) propionic acid) has also been reported [67]. The authors obtained high retention factors due to strong π-donor-acceptor interactions with a chiral stationary phase, but with a total analysis time of 3-4 h.

As regards the chromatographic detection most of the works are using UV (see Table 1) because of the high absorptivity coefficients of these compounds (carbon bulky structure with a high number of double bounds) and the simplicity of this detection system. It should be pointed out that many of the published works are only performing studies with standard mixtures (for characterization or column comparison) [32,42,50,56,62-67], but they are not dealing with sample analysis, so detection sensitivity is not a handicap. Moreover, unequivocal identification and confirmation in complex matrices is not achieved with UV-detection. Concluding, liquid chromatography coupled to mass spectrometry (LC-MS) is becoming the technique of choice if the objective is the analysis of fullerenes and functionalized fullerenes.

LIQUID CHROMATOGRAPHY-MASS SPECTROMETRY

Presently, the necessity to control and analyze emerging contaminants, such as fullerene compounds, in very complex matrices, *i.e.*, environmental (sediments, soils, waters), biological (blood, tissues) and cosmetic samples, and the legal requirements for unequivocal identification and confirmation of positive and negative samples [75] drives the need to develop sensitive and reliable analytical methodologies. LC-MS is one of the best techniques to solve these requirements and it is becoming very popular in the analysis of fullerenes by employing different MS analyzers such as single quadrupole (Q) [49,59], triple quadrupole (QqQ) [14,40,44,47,48,52,54,55,60], ion-trap (IT) [51,53], and quadrupole-linear ion-trap (QqLIT) [45] instruments. High-resolution mass spectrometry (HRMS) either using time-of-flight (TOF) [42,65], hybrid quadrupole-time-of-flight (Q-TOF) [66], and very recently hybrid linear trap-Orbitrap (LTQ-Orbitrap) [46,58] analyzers has also been proposed by some authors to analyze fullerenes.

However, the first step in the LC-MS analysis of fullerenes is their ionization. Atmospheric pressure ionization (API) sources (Table 1), electrospray ionization (ESI) [44-46,48,52,58,66] and atmospheric pressure chemical ionization (APCI) [14,40,46,47,51] are the most frequently used for both, fullerenes and fullerene derivatives and, in general, by using negative ionization mode. In all cases, negative-ESI is providing the molecular radical anion $[M]^{-\bullet}$ as the most abundant peak in the MS spectra of fullerenes and fullerene derivatives, but several authors also reported the presence of other radical anions, for instance $[MO]^{-\bullet}$ for pristine fullerenes described as oxidized ion [45,52,58]. As an example, Figure 5 shows the chromatogram of a 100 ng/L standard mixture of C_{60}, C_{70} and C_{60}-pyrr and the corresponding full scan MS spectra [45]. As commented, the molecular radical anions at *m/z* 720 and *m/z* 840 for C_{60} and C_{70}, respectively, were the most abundant signals, but fullerene oxide ions ($[C_{60}O]^{-\bullet}$ at *m/z* 736; $[C_{70}O]^{-\bullet}$ at *m/z* 756) were also observed at low abundances, and the authors proposed their use for confirmation purposes. APCI has been reported only for the ionization of pristine fullerenes and in all cases the molecular radical ion $[M]^{-\bullet}$ was also the most abundant one in the MS spectra, and was monitored for quantification purposes [14,40,46,47,51].

In 2000, Bruins and coworkers developed atmospheric pressure photoionization (APPI) as a complementary API source for LC-MS [76], expanding the application of LC-MS to non-polar compounds and compounds which are difficult to ionize by ESI and/or APCI [77,78]. Since then, APPI has been proposed as an alternative to classical API sources for the LC-MS analysis of fullerene and fullerene derivatives [46,49,53-55,60]. Some authors used C_{60} as a test analyte to study the performance of APPI but were not focused on its analysis, and they reported that fullerene ionization occurred through a solvent-mediated ionization mechanism yielding also the radical molecular ion $[M]^{-\bullet}$ [79-81]. Núñez et al. [54] and more recently Astefanei et al. [55] compared the ionization efficiency of pristine fullerenes (C_{60}, C_{70}, C_{76}, C_{78} and C_{84}) and three C_{60}-fullerene derivatives (PCBM, PCBB and C_{60}-pyrr), respectively, with several API sources: ESI, heated-electrospray ionization (H-ESI, Thermo Fisher Scientific), APCI and APPI, and the signal responses obtained for the $[M]^{-\bullet}$ are shown in Figure 6. As can be seen, APPI provided the highest responses for all compounds. It is well-known that the use of toluene as dopant improves ionization efficiency in LC-APPI-MS [82]. Toluene is usually added post-column into the LC eluent flow, but in most of the LC chromatographic separations of fullerenes it is one of the major components of the mobile phase (Table 1).

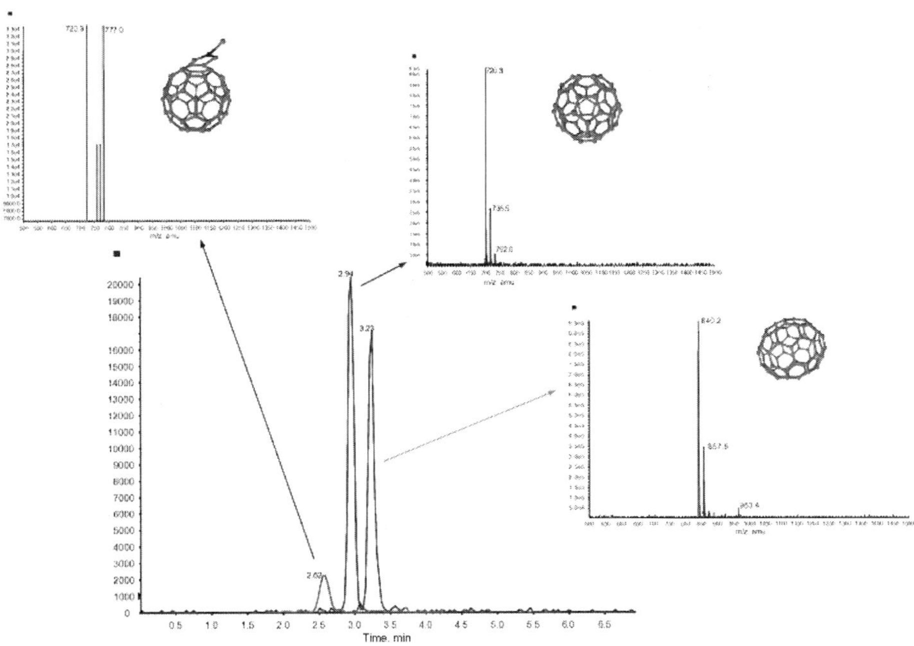

Figure 5. LC-ESI-QqLIT chromatogram of a 100 ng/L standard mixture of C_{60}, C_{70} and C_{60}-pyrr analyzed in negative ionization mode. Full scan MS spectra of each compound, and its chemical structure, is also depicted. Reprinted with permission from reference [45]. Copyright (2010) Elsevier.

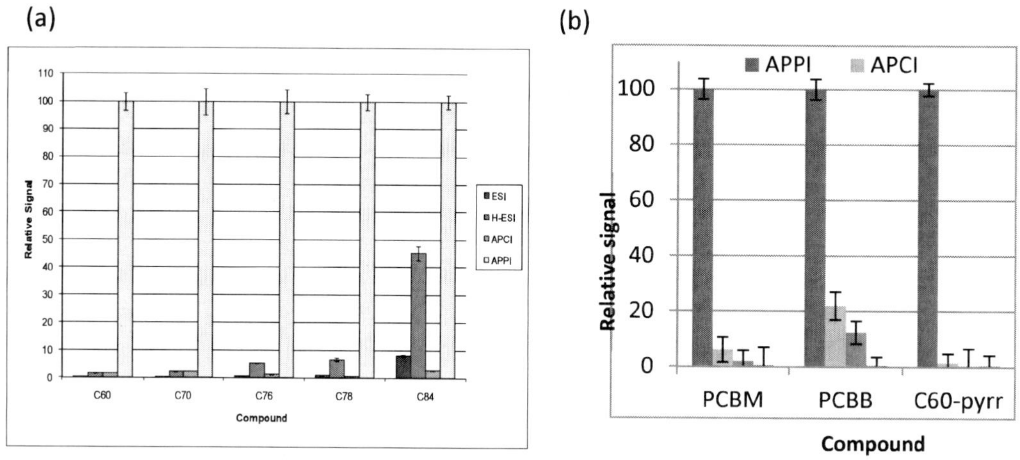

Figure 6. Signal of the $[M]^{-\bullet}$ ion of fullerenes with different API sources by LC-MS. (a) Absolute response values for pristine fullerenes normalized to the signal observed for APPI with each fullerene. Adapted with permission from reference [54]. Copyright (2012) American Chemical Society. (b) Absolute response values for three C_{60}-fullerene derivatives normalized to the signal observed for APPI with each compound.

This fact could explain the higher ionization efficiency observed when using solvent mediated negative-ionization APPI in comparison to the one found with other API sources. Similar results were obtained by Chen et al. [53] when comparing APPI, ESI and APCI ionization efficiency for C_{60}, C_{70} and PCBM. Regarding H-ESI, this source gave better results than ESI for pristine fullerenes, and mainly for larger fullerenes, while APCI showed a bad performance for the ionization of these compounds [54]. In contrast, for C_{60}-fullerene derivatives better results were observed with APCI in comparison to both ESI and H-ESI, which was attributed by the authors to the fact that for these specific compounds electron capture was favored in the APCI plasma [55].

APPI single MS spectra of pristine fullerenes in negative ionization mode reported by Núñez et al. [54] were dominated by the isotopic cluster of the molecular ions *m/z* M, M+1, M+2, and M+3 (see, as an example, the LC-APPI-MS spectra of C_{60} with a triple quadrupole instrument in Figure 7a). A remarkable result observed in this work was that APPI isotope patterns for pristine fullerenes did not match with those obtained with ESI and APCI, nor with the theoretical ones, showing higher than expected relative abundances of *m/z* M+1, M+2, and M+3 [51,83]. These discrepancies increased with the bulky fullerene structure, and for fullerenes with more than 70 carbon atoms the relative abundance of *m/z* M+1 was even higher than *m/z* M. The authors investigated this phenomena by studying the tandem mass spectrometry spectra of the molecular cluster ions *m/z* M and M+1 (Figures 7b and 7c), showing that two isobaric ions, the ^{13}C isotopic contribution ($[C_{59}{}^{13}C]^{-\bullet}$), as expected, and the addition of a hydrogen to one of the fullerene double bonds ($[C_{60}+H]^{-\bullet}$), were contributing to the abundance of *m/z* M+1.

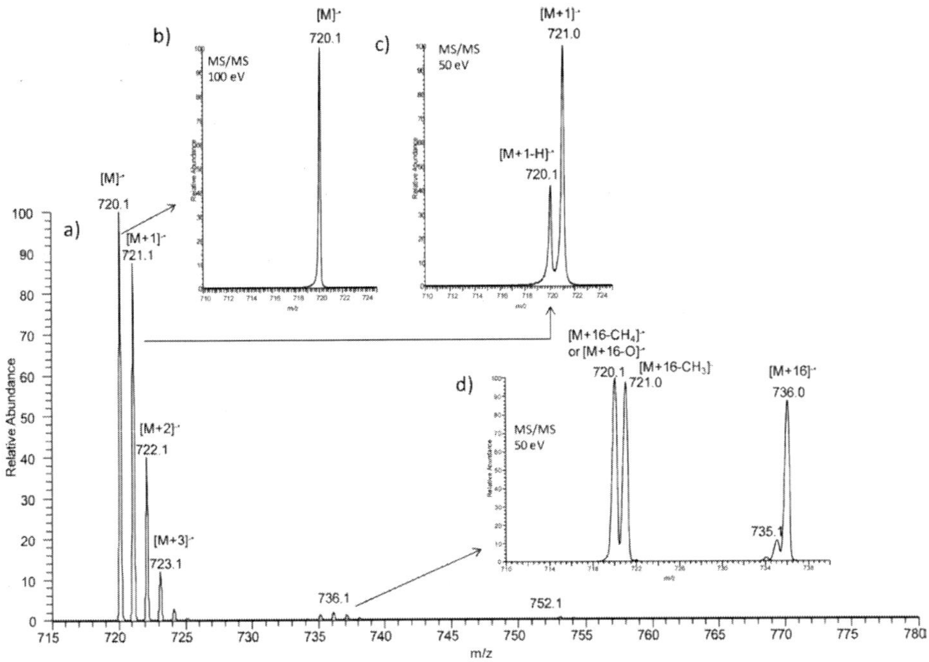

Figure 7. LC-APPI-MS spectra of C_{60} fullerene in a triple quadrupole instrument. (a) Full scan MS spectrum, (B) MS/MS spectrum of the precursor ion at *m/z* 720, (C) MS/MS spectrum of the precursor ion at *m/z* 721, and (D) MS/MS spectrum of the precursor ion at *m/z* 736. Reproduced with permission from reference [54]. Copyright (2012) American Chemical Society.

The identity of this non-expected $[M+H]^{-\bullet}$ ion was confirmed by the authors by HRMS and accurate mass measurements in a Fourier Transform-Ion Cyclotron Resonance (FTICR) instrument (Thermo Fisher Scientific) at a resolving power of 400,000 FWHM (full width at half maximum) [54]. Two ions were present in *m/z* M+1, the most abundant one at *m/z* 721.0038 (mass error -0.115 ppm) due to the ^{13}C contribution and the addition of hydrogen to a double bond at *m/z* 721.0087 (mass error 0.411 ppm). Later on, the same research group demonstrated that the addition of hydrogen to the pristine fullerene double bond in the APPI ion source was related to the fullerene concentration [55], and that both experimental and theoretical isotopic patterns matched at concentrations lower than ~100 μg/L.

Núñez et al. [54] also studied and characterized other cluster ions observed in the single MS spectra of pristine fullerenes at higher *m/z* values (M+16 and M+32). As previously commented, these ions have also been observed when using ESI and APCI and were generally explained by the addition of oxygen to the fullerene bulky structures [45,46,48,49]. For the correct characterization of these ions tandem mass spectrometry experiments in a triple quadrupole instrument were carried out. The authors observed that when the ion at *m/z* M+16 was selected as precursor ion, two product ions shifted in mass by 15 and 16 Da were observed (see Figure 7d for C_{60}). These results revealed that this M+16 signal cannot be explained only due to the addition of oxygen ($[M+O]^{-\bullet}$) as previously described in the literature [49], but can also be explained as $[M+CH_4]^{-\bullet}$ (involving a kind of methylation process), or to the presence of both isobaric ions. In order to correctly describe these signals the authors performed HRMS and accurate mass measurements using an Orbitrap mass analyzer operating at a mass resolving power of 100,000 FWHM, and the results obtained for the signal of M+16 of fullerene C_{60} are shown in Figure 8.

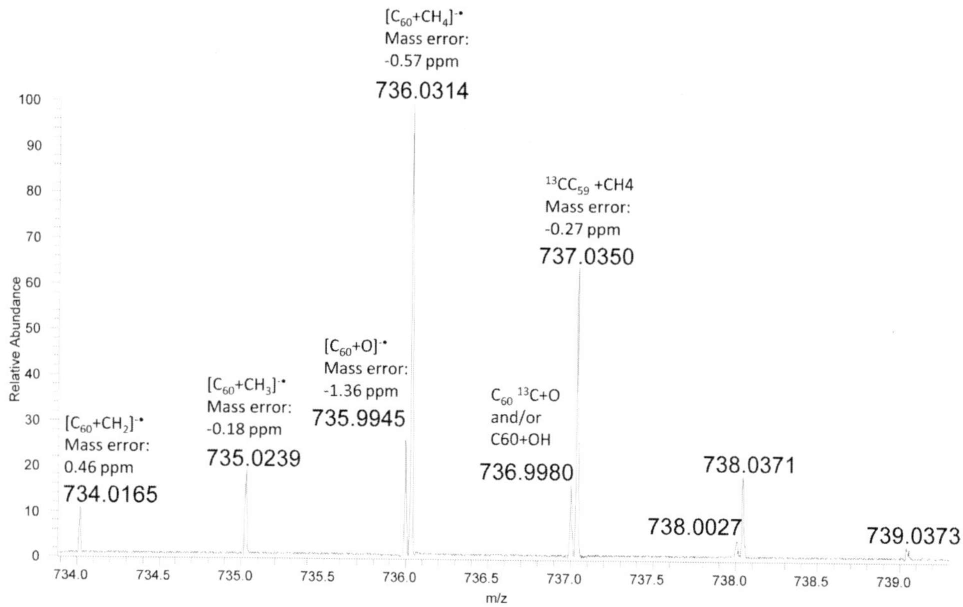

Figure 8. Orbitrap-MS spectra at a resolving power of 100,000 FWHM of C_{60} cluster ion at M+16. Reproduced with permission from reference [54]. Copyright (2012) American Chemical Society.

As can be seen in Figure 8, two ions were observed for M+16, an ion at *m/z* 735.9945 (mass error -1.36 ppm) assigned to $[M+O]^{-\bullet}$ that could be generated via the radical addition of oxygen to yield an epoxy group, and an ion at *m/z* 736.0314 assigned to $[M+CH_4]^{-\bullet}$ and explained by the gas-phase ion molecule reaction of $[M]^{-\bullet}$ with a radical $^{\bullet}CH_3$, and subsequent stabilization by hydrogen addition [84].

Other ionization techniques, such as matrix-assisted laser desorption ionization (MALDI) have also been proposed for the characterization of pristine fullerenes [42] and fullerols [65]. For instance, Xiao et al. [42] used MALDI-TOF-MS for the characterization of six fractions collected from the chromatographic separation of a pyridine extract of crude soot which contained higher fullerenes, and they were able to identify several fullerenes from C_{76} to C_{94}. Recently, Chao et al. [65] characterized fullerols by obtaining the MALDI-TOF-MS spectra. For that purpose, 1 µL of fullerol $C_{60}(OH)_x(ONa)_y$ (which has a composition of x+y=24 with y generally around 6-8) sample was dispersed in a MALDI matrix (2,5-dihydrobenzoic acid) and, after drying, mass spectra were obtained by taking 1000 shots. In the MALDI-MS spectra, the authors found a major peak at *m/z* 720, which corresponds to the molecular mass of C_{60}. In addition, a peak series from *m/z* 720 to 576 with a spacing of 24 *m/z* was found. This has also been observed in earlier studies of fullerenes and has been attributed to sequential C_2 losses [85]. The authors also studied the same fullerol sample with LC-ESI-MS(/MS) and they observed that the ESI-MS analysis of fullerols resulted in a complex spectrum (see as example Figure 9), with only a weak peak corresponding to the protonated molecular mass of fullerol $[C_{60}(OH)_{24}+H]^+$ at *m/z* 1129. In contrast to MALDI-TOF spectra, no C_2 series were found, but a decreasing series spaced by 74 *m/z* units from *m/z* 1129 to 537 ($[M\text{-}n74+H]^+$ with n=1-8) was observed, that the authors tentatively explained by subsequent neutral losses of $C_2H_2O_3$.

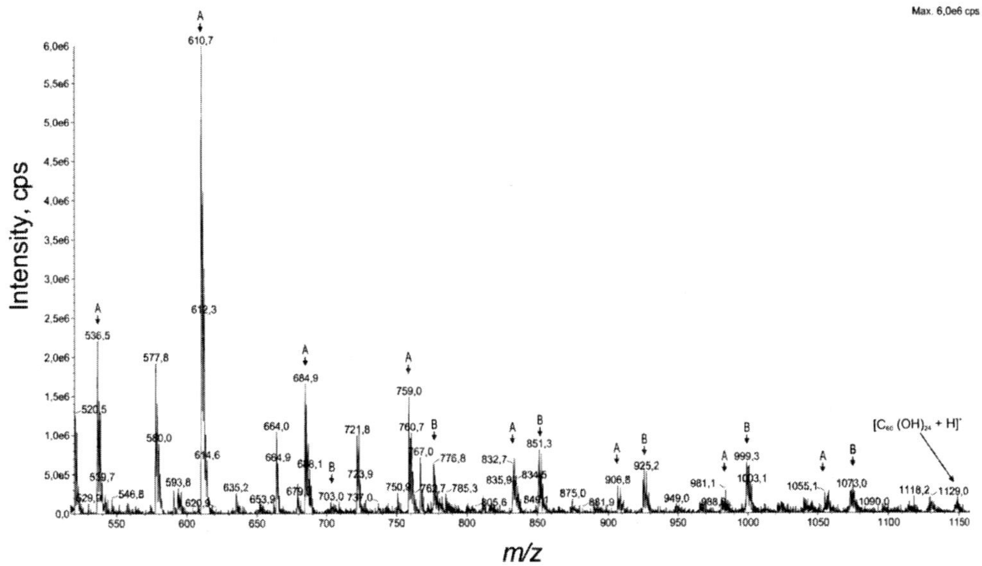

Figure 9. Overview (Q1) scan spectrum of a fullerol sample. Peaks of the two 74 *m/z* series (A and B) are indicated by the arrows. Reproduced with permission from reference [65]. Copyright (2011) American Chemical Society.

As previously commented, LC-MS appears as the technique of choice for the analysis of fullerene compounds in order to satisfy legal requirements for unequivocal identification and confirmation of positive and negative samples and to comply with the identification points system of European Union directive 2002/657/EC [75]. In general, the use of LC-MS/MS with triple quadrupole instruments by recording two selected reaction monitoring (SRM) transitions is sufficient to guarantee confirmation, but pristine fullerenes cannot be fragmented because of their stable bulky structure, so their analysis is difficult as regards to confirmation purposes. Several authors are proposing to monitor two "SRM" transitions by selecting the same ion as both precursor and product ion, although strictly this cannot be considered a transition as no fragmentation occurs. For instance, the monitoring of $[M]^{-\bullet} \rightarrow [M]^{-\bullet}$ for quantification, and the oxidized ions $[MO]^{-\bullet} \rightarrow [MO]^{-\bullet}$ for confirmation has been reported for the analysis of pristine fullerenes in wastewater [45], water [58] and aerosol [52] samples. In other works, ions corresponding to the fullerene isotopic clusters were also used. For instance, Sanchis et al. [44] monitored $[M]^{-\bullet} \rightarrow [M]^{-\bullet}$ and $[M+1]^{-\bullet} \rightarrow [M+1]^{-\bullet}$ for the analysis of pristine fullerenes in river sediments and soils.

Other strategies to guarantee confirmation for this family of compounds have also been proposed. For instance, Núñez et al. [54] used highly selective-selected ion monitoring (H-SIM) by directly monitoring two ions of the pristine fullerene isotopic cluster $[M]^{-\bullet}$ and $[M+1]^{-\bullet}$. However, this strategy can only be applied with instruments allowing enough mass resolving power such as the triple quadrupole used in that work (TSQ Quantum ultra AM,

Thermo Fisher Scientific) that allowed working at enhanced mass resolution (Q at 0.06 *m/z* with a mass resolving power higher than 12,500 FWHM).

In contrast to pristine fullerenes, functionalized fullerenes showed MS/MS fragmentation at relatively high collision energies and, in general, all of them are showing as the most abundant peak of the MS/MS spectra the fullerene bulky structure corresponding to the loss of the whole functional group [44,45,52,53,55,58]. In some cases, additional fragments have also been reported. For instance, the loss of both the propyl acid alkyl ester group $(CH_2)_2COO(CH_x)H$ and the butyl acid alkyl ester group $(CH_2)_3COO(CH_x)H$ for PCBM and PCBB, as well as the loss of the butyl group of the carboxylate ester for PCBB has been described [55]. The cleavage of the carboxylate ester for fullerenes with large alkyl moieties and the loss of the propyl acid alkyl ester group in the MS/MS spectra of PCBM and PCBB have also been described by other authors [53,58]. As the fragmentation for this group of compounds is possible, acquisition is carried out, in general, by monitoring the transition $[M]^{-\bullet} \rightarrow [C_{60}]^{-\bullet}$ for quantitation, and $[M]^{-\bullet} \rightarrow [M]^{-\bullet}$ for confirmation [58]. Other authors proposed to use some isotopic cluster ions for the confirmation ($[M+1]^{-\bullet} \rightarrow [C_{60}+1]^{-\bullet}$) [44,55].

LC AND LC-MS APPLICATIONS

The rapid growth in the development, production, and use of fullerene and functionalized fullerene compounds is driving the development of new analytical methodologies for their analysis. These methods are not only needed from the point of view of product quality controls (such as in the case of cosmetics containing fullerenes), but also for biomedical applications, toxicity tests, and to assess the occurrence, fate and behavior of these compounds in the environment, as their concentration is expected to increase in the future. However, although liquid chromatography (mostly coupled to mass spectrometry) is the technique of choice for that purpose, there are only few works dealing with the analysis of these compounds in real matrices, since most of the studies are working only with standards (see Table 1).

Most of the LC and LC-MS methods proposed in the literature for the analysis of fullerenes are dealing with environmental samples, and only a limited number of works are addressed to other sample matrices such as cosmetics [13,14] or biological fluids and tissues [17,51]. As an example, Benn et al. [14] analyzed C_{60} and C_{70} in commercial products by LC-APCI-MS as a way to provide potential exposure levels for future risk assessment of fullerene technologies. C_{60} was quantified in commercial cosmetics samples used as anti-aging formulations and concentrations ranging from 0.04 to 1.1 µg/g were found while C_{70} was only qualitatively detected in some samples. The authors indicated that a single-use quantity of cosmetic (around 0.5 g) may contain up to 0.6 µg of C_{60}, which can be a probable pathway for human exposure. Regarding bio-analytical applications, it should be considered that water-soluble derivatives of C_{60} fullerene are showing interesting properties in various biological systems but the biological properties of C_{60} by itself are still being evaluated. For this purpose Moussa et al. [51] developed an LC-APCI-MS method using an ion-trap mass analyzer for the quantification of C_{60} in blood, liver and spleen samples to check its possible acute toxicity *in vivo*. Limits of detection down to 0.1 ng injected were obtained.

Regarding environmental applications, Farré et al. [3] recently reviewed the analysis and assessment of the occurrence, the fate and the behavior of nanomaterials (including fullerene compounds) in the environment. Assessing the risks imposed by their extensive use and environmental applications requires a good understanding of their mobility, bioavailability and toxicity. The main processes involving the presence of nanoparticles in the environment are summarized in Figure 10. Despite the amount of research conducted to date, there are still questions unanswered regarding their environmental transformation, bio-uptake and their effects on humans. Functionalized nanoparticles can have very different properties depending on the functionalization and this will have important implications for assessing their environmental behavior. Independently of their source, they all end up in soil or water, either directly or indirectly. In the aquatic environment, they are prone to form aggregates and this plays a key role in determining their stability in the environment and hence the potential exposure and risk presented by these colloids. Aggregated or adsorbed fullerenes will be less mobile, altering their transport properties and the way their surface can participate in sorption/desorption processes.

Regarding fullerene analysis, LC and LC-MS methods have been proposed for pristine fullerenes and some functionalized fullerenes in geological materials [39], soot extracts [40], soils and sediments [43,44,55,57], aerosols [52], and water samples [45,46,53-55,58,59,61], although in some of these works only blank matrices spiked with fullerene compounds are used. To the best of our knowledge, there are still no LC or LC-MS applications reported in the literature dealing with the analysis of water soluble fullerenes such as fullerenols in real environmental samples.

As regards to the presence of fullerenes and fullerene derivatives in aerosols, soils and sediments, Sanchis et al. [52] studied for the first time the occurrence of some selected fullerenes (C_{60}, C_{70}, C_{60}-pyrr, C_{60} pyrrolidine tris-acid ethyl ester (CPTAE), PCBM, and ThCBM) in airborne particulate. The samples were collected in the Mediterranean Sea during two sampling campaigns from Barcelona (Spain) to Istanbul (Turkey) and Alexandria (Egypt), respectively. The analysis of the samples were carried out by LC-ESI-MS(/MS) using a triple quadrupole analyzer achieving sensitivities between 5.4 and 20.9 pg/m^3. The authors analyzed a total of 43 samples covering the different basins of the Mediterranean Sea. Fullerenes were detected in all analyzed samples and quantified in 28 of them. The median of C_{60} and C_{70} fullerenes aerosol phase concentrations were 0.06 ng/m^3 and 0.48 ng/m^3, respectively, for the Mediterranean Sea atmosphere. The same research group recently proposed a similar LC-ESI-MS(/MS) method for the analysis of C_{60}, C_{70} and three C_{60}-fullerene derivatives in river sediments and soils [44] after their extraction by ultrasound-assisted toluene extraction. The method showed limits of detection ranging from 14 to 290 pg/g which makes it suitable for the analysis of environmental samples. 58 soil samples collected from different urban and industrial areas in Saudi Arabia and in river sediments from six different sites in the Llobregat River Basin (Spain) were analyzed. C_{60} was the only fullerene compound detected and quantified in 19% of soils from Saudi Arabia (0.15 to 6.83 ng/g concentration levels) and 33% of sediments from Llobregat River (0.13-0.70 ng/g concentration levels).

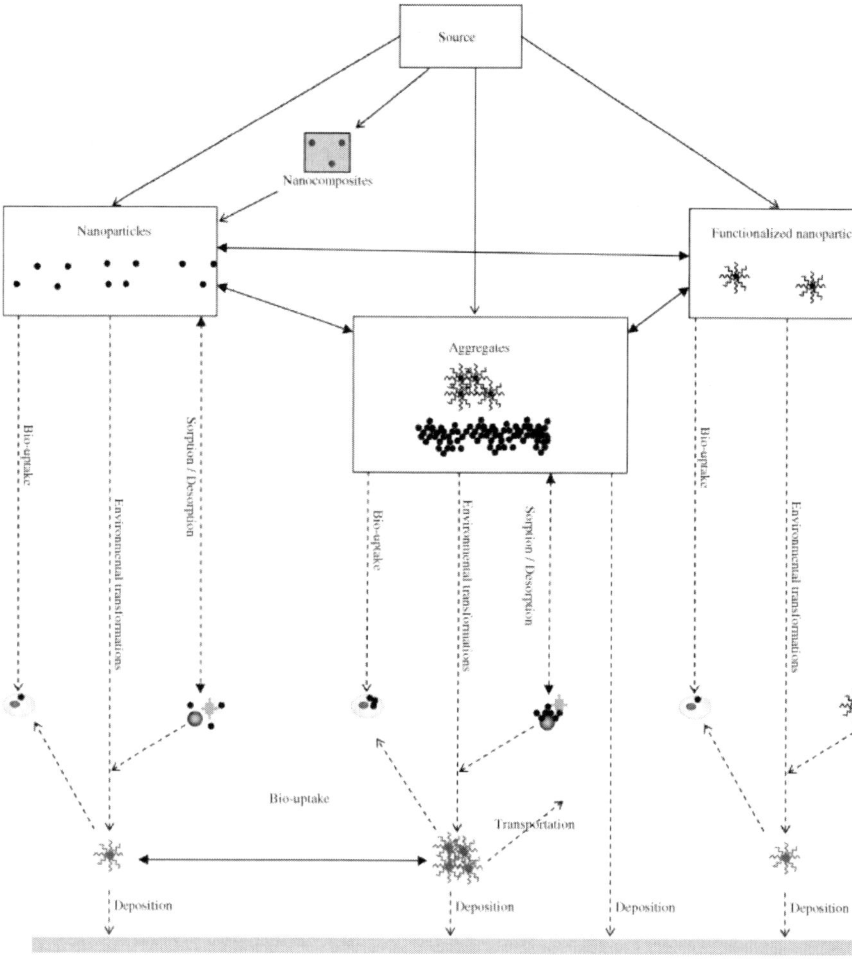

Figure 10. Main processes involving nanoparticles in the environment. Reproduced with permission from reference [3]. Copyright (2011) Elsevier.

Recently, Astefanei et al. [55] reported the presence of C_{60} together with other fullerene compounds in sediment samples collected from several ponds located around Airport of Barcelona (Spain). The authors proposed the use of pressurized liquid extraction using toluene for the extraction of fullerenes, and the extracts were analyzed by LC-APPI-MS(/MS). Seven sediment samples were analyzed, and C_{60} and C_{70} fullerenes were detected in six and five of them, respectively, although only at quantifiable levels in 2 and 3 of them, respectively. One noticeable result of this work is that the authors reported for the first time the presence of C_{60}-fullerene derivatives in sediments. As an example, Figure 11 shows the LC-APPI-MS/MS signals (SRM transitions for quantification and confirmation) of three C_{60}-fullerene derivatives in one of the sediment samples.

The three C_{60}-fullerene derivatives were detected in all sediment samples and at quantifiable levels in most of them. For instance, C_{60}-pyrr was reported in six samples at concentrations between 0.14 and 2.7 ng/Kg. The highest concentration found was for PCBB at 5.1 ng/Kg.

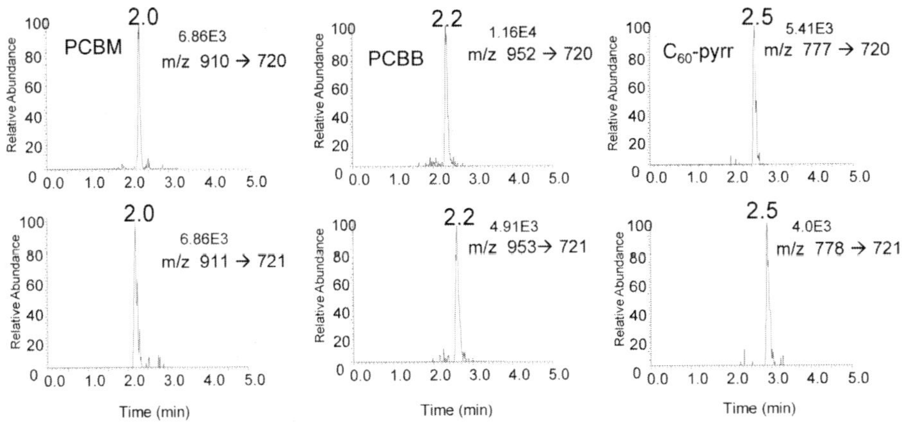

Figure 11. LC-APPI-MS/MS quantification and confirmation signals of three C_{60}-fullerene derivatives reported for the first time in a sediment sample.

The number of publications dealing with the LC or LC-MS analysis of fullerene compounds in environmental waters is higher than is soils and sediments [45,46,53-55,58,59,61]. Only two of them are dealing with the analysis of C_{60} [46,59] while most of works are analyzing several pristine fullerenes and functionalized fullerenes simultaneously [45,53-55,58,61]. For the analysis of fullerene aggregate nC_{60} in wastewater samples, Wang et al. [59] proposed the use of LC with UV detection at 332 nm and LC-APCI-MS using a single quadrupole mass analyzer. Liquid-liquid extraction (LLE) with toluene and solid phase extraction (SPE) using Strata SDB-L cartridges (500 mg, Phenomenex) were proposed for sample treatment and preconcentration. Method detection limits of LLE with LC-UV were 3-4 μg/L for water matrices and 0.64 μg/L in a secondary effluent sample. Both UV-vis spectroscopy and mass spectrometry gave similar sensitivity. Van Wezel et al. [46] developed a sensitive analytical method to quantify nC_{60} and some of its transformation products (by oxidation, reduction and photochemical reactions) in water using accurate mass screening liquid chromatography-hybrid liner ion trap Orbitrap mass spectrometry. The method was able to detect and quantify aqueous concentrations of nC_{60} as low as 5 ng/L. However, despite the high sensitivity of the developed method, no nC_{60} or transformation products were detected in any of the Dutch surface waters they analyzed.

Regarding the simultaneous analysis of pristine and functionalized fullerene compounds in environmental water samples, two recent works are dealing with the determination of up to 8 fullerene compounds. Kolkman et al. [58] proposed the SPE (with C18 cartridges) extraction of fullerenes from water samples with subsequent analysis on a LC-Orbitrap mass spectrometry system. Very low limits of detection ranging from 0.17 ng/L for [70] PCBM to 0.28 ng/L for C_{60} were obtained, with recoveries higher than 78% in surface waters. However, the authors did not apply the method to the analysis of real water samples and one of the disadvantages of this method is the long analysis time (chromatographic run time of 27 min) reported. In contrast, Astefanei et al. [55] proposed a LLE method for the analysis of this family of compounds in river water samples followed by LC-APPI-MS(/MS), achieving method limits of detection ranging from 1.4 pg/L for PCBM to 1.6 ng/L to C_{84}, and with a run chromatographic time lower than 4.3 min. The method was applied to the analysis of 6 pond water samples collected around the airport of Barcelona (Spain). The three C_{60}-fullerene derivatives (C_{60}-pyrr, PCBM, and PCBB), C_{60} and C_{70} were detected in most of them and at quantifiable levels in 3-4 samples. Only C_{60} was detected in all analyzed samples. As regards larger pristine fullerenes (C_{76}, C_{78}, and C_{84}), they were not found in any of the pond water samples analyzed. To the best of our knowledge, the presence of C_{84} in water samples has only been reported once when analyzing water particulate suspended material using a similar LC-APPI-MS method [54], and C_{76} and C_{78} has never been found in environmental samples.

SUMMARY AND FUTURE TRENDS

The use of fullerene nanoparticles is becoming a major, innovative, scientific and economic growth area because of their remarkable physic and chemical properties, but the rapid increase in development, production, and use of this family of compounds have led to call for more information regarding the potential impacts that their continuous release may have on human health and the environment. Liquid chromatography and liquid chromatography-mass spectrometry are without any doubt the best options when dealing with the analysis of this special family of compounds. However, in order to cope with the necessity of performing high number of analyses and in very difficult sample matrixes such as environmental ones, future efforts in reducing analysis time are still required. Sample treatments for these compounds are mostly based on LLE, PLE, and SPE which are time-consuming but still necessary because of their effectiveness. Thus, the best way to reduce analysis time is through the chromatographic separation, and here the use of UHPLC column technology either using sub-2μm particle size totally porous columns or core-shell partially porous columns will play an important role in the future.

Applications of water soluble fullerenes (fullerenols) are expected to increase in medicine, cosmetics and industry due to their appealing properties, but their chromatographic separation is still challenging because of their stronger hydroxylation. Efforts in developing chromatographic methodologies for their separation and their analysis in real samples are still necessary, and it is sure that this will become one of the most promising, although difficult, research topic in the field of fullerene nanoparticles.

Currently, not enough data is available to correctly asses the occurrence, fate and behavior of fullerene nanoparticles in the environment, because very few works are reporting levels for these compounds in environmental samples, so both the scientific community and administrations must dedicate efforts to solve this lack of data if we want to have complete information regarding the potential impacts of fullerenes in human health and the environment. Additionally, more studies regarding fullerene toxicity are still necessary and in this field liquid chromatography techniques will play an important role.

There is still much to do in the field of liquid chromatography applied to the analysis of fullerene compounds and surely the number of scientific publications will increase in the near future.

ACKNOWLEDGMENTS

This work has been funded by the Spanish Ministry of Economy and Competitiveness under the project CTQ2012-30836. Alina Astefanei thanks the Spanish Ministry of Economy and Competitiveness for a Ph.D. grant (FPI).

REFERENCES

[1] Kroto, H. W., Heath, J. R., O'Brien, S. C., Curl, R. F. & Smalley, R. E. (1985). C60: buckminsterfullerene. *Nature*, 318, 162-163.

[2] Hendren, C. O., Mesnard, X., Droge, J. & Wiesner, M. R. (2011). Estimating Production Data for Five Engineered Nanomaterials As a Basis for Exposure Assessment. *Environ. Sci. Technol.*, 45, 4190-

[3] Farre, M., Sanchis, J. & Barcelo, D. (2011). Analysis and assessment of the occurrence, the fate and the behavior of nanomaterials in the environment. *TrAC, Trends Anal. Chem.*, 30, 517-527.

[4] Becker, L., Poreda, R. J. & Bada, J. L. (1996). Extraterrestrial helium trapped in fullerenes in the Sudbury impact structure. *Science*, 272, 249-252.

[5] Becker, L., Poreda, R. J., Hunt, A. G., Bunch, T. E. & Rampino, M. (2001). Impact event at the Permian-Triassic boundary: evidence from extraterrestrial noble gases in fullerenes. *Science (Washington, DC, U. S.)*, 291, 1530-1533.

[6] Heymann, D., Jenneskens, L. W., Jehlicka, J., Koper, C. & Vlietstra, E. (2003). Terrestrial and extraterrestrial fullerenes. *Fullerenes, Nanotubes, Carbon Nanostruct.*, 11, 333-370.

[7] Kronholm, D. & Hummelen, J. C. (2007). Fullerene-based n-type semiconductors in organic electronics. *Mater. Matters (Milwaukee, WI, U. S.)*, 2, 16-19.

[8] Tiwari, S. P., Namdas, E. B., Rao, V. R., Fichou, D. & Mhaisalkar, S. G. (2007). Solution-processed n-type organic field-effect transistors with high ON/OFF current ratios based on fullerene derivatives. *IEEE Electron Device Lett.*, 28, 880-883.

[9] Morana, M., Koers, P., Waldauf, C., Koppe, M., Muehlbacher, D., Denk, P., Scharber, M., Waller, D. & Brabec, C. (2007). Organic field-effect devices as tool to characterize the bipolar transport in polymer-fullerene blends: the case of P3HT-PCBM. *Adv. Funct. Mater.*, 17, 3274-3283.

[10] Guldi, D. M. & Martin, N. (2002). Fullerenes: From Synthesis to Optoelectronic Properties. *Dev. Fullerene Sci.*, 4, 440-

[11] Erkoc, S. (2002). Perspectives of Fullerene Nanotechnology Edited by Eiji Osawa. *Mater. Manuf. Processes*, 17, 881-883.

[12] Kim, Y., Cook, S., Tuladhar, S. M., Choulis, S. A., Nelson, J., Durrant, J. R., Bradley, D. D. C., Giles, M., McCulloch, I., Ha, C. S. & Ree, M. (2006). A strong regioregularity effect in self-organizing conjugated polymer films and high-efficiency polythiophene:fullerene solar cells. *Nat. Mater.*, 5, 197-203.

[13] Chae, S. R., Hotze, E. M., Xiao, Y., Rose, J. & Wiesner, M. R. (2010). Comparison of methods for fullerene detection and measurements of reactive oxygen production in cosmetic products. *Environ. Eng. Sci.*, 27, 797-804.

[14] Benn, T. M., Westerhoff, P. & Herckes, P. (2011). Detection of fullerenes (C60 and C70) in commercial cosmetics. *Environ. Pollut.*, 159, 1334-1342.

[15] Xiao, L., Takada, H., Maeda, K., Haramoto, M. & Miwa, N. (2005). Antioxidant effects of water-soluble fullerene derivatives against ultraviolet ray or peroxylipid through their action of scavenging the reactive oxygen species in human skin keratinocytes. *Biomed. Pharmacother.*, 59, 351-358.

[16] Aschberger, K., Johnston, H. J., Stone, V., Aitken, R. J., Tran, C. L., Hankin, S. M., Peters, S. A. K. & Christensen, F. M. (2010). Review of fullerene toxicity and exposure - Appraisal of a human health risk assessment, based on open literature. *Regul. Toxicol. Pharmacol.*, 58, 455-473.

[17] Xia, X. R., Monteiro-Riviere, N. A. & Riviere, J. E. (2006). Trace analysis of fullerenes in biological samples by simplified liquid-liquid extraction and high-performance liquid chromatography. *J. Chromatogr. , A*, 1129, 216-222.

[18] Tagmatarchis, N. & Shinohara, H. (2001). Fullerenes in medicinal chemistry and their biological applications. *Mini-Rev. Med. Chem.*, 1, 339-348.

[19] Sagman, U. (2002). Application and commercial prospects of fullerenes in medicine and biology. *Perspect. Fullerene Nanotechnol.*, 145-153.

[20] Mashino, T. (2004). Application of fullerenes to medicine. *Farumashia*, 40, 1023-1027.

[21] Bakry, R., Vallant, R. M., Najam-ul-Haq, M., Rainer, M., Szabo, Z., Huck, C. W. & Bonn, G. K. (2007). Medicinal applications of fullerenes. *Int. J. Nanomed.*, 2, 639-649.

[22] Moghimi, S. M., Hunter, A. C. & Murray, J. C. (2005). Nanomedicine: current status and future prospects. *FASEB J.*, 19, 311-330.

[23] Husebo, L. O., Sitharaman, B., Furukawa, K., Kato, T. & Wilson, L. J. (2004). Fullerenols Revisited as Stable Radical Anions. *J. Am. Chem. Soc.*, 126, 12055-12064.

[24] Da Ros, T. & Prato, M. (1999). Medicinal chemistry with fullerenes and fullerene derivatives. *Chem. Commun. (Cambridge)*, 663-669.

[25] Chen, B. H., Huang, J. P., Wang, L. Y., Shiea, J., Chen, T. L. & Chiang, L. Y. (1998). Synthesis of octadecahydroxylated C70. *Synth. Commun.*, 28, 3515-3525.

[26] Andrievsky, G. V., Klochkov, V. K., Karyakina, E. L. & Mchedlov-Petrossyan, N. O. (1999). Studies of aqueous colloidal solutions of fullerene C60 by electron microscopy. *Chem. Phys. Lett.*, 300, 392-396.

[27] Jafvert, C. T. & Kulkarni, P. P. (2008). Buckminsterfullerene's (C60) Octanol-Water Partition Coefficient (Kow) and Aqueous Solubility. *Environ. Sci. Technol.*, 42, 5945-5950.

[28] Prato, M. (1997). [60]Fullerene chemistry for materials science applications. *J. Mater. Chem.*, 7, 1097-1109.

[29] Prato, M. (1999). Fullerene materials. *Top. Curr. Chem.*, 199, 173-187.

[30] Wudl, F. (2002). Fullerene materials. *J. Mater. Chem.*, 12, 1959-1963.

[31] Andrievsky, G., Klochkov, V. & Derevyanchenko, L. (2005). Is the C60 Fullerene Molecule Toxic?! *Fullerenes, Nanotubes, Carbon Nanostruct.*, 13, 363-376.

[32] Gharbi, N., Pressac, M., Hadchouel, M., Szwarc, H., Wilson, S. R. & Moussa, F. (2005). [60]Fullerene is a Powerful Antioxidant in Vivo with No Acute or Subacute Toxicity. *Nano Lett.*, 5, 2578-2585.

[33] Aoshima, H., Yamana, S., Nakamura, S. & Mashino, T. (2010). Biological safety of water-soluble fullernes evaluated using tests for genotoxicity, phototoxicity, and pro-oxidant activity. *J. Toxicol. Sci.*, 35, 401-409.

[34] Seda, B. C., Ke, P. C., Mount, A. S. & Klaine, S. J. (2012). Toxicity of aqueous C70-gallic acid suspension in Daphnia magna. *Environ. Toxicol. Chem.*, 31, 215-220.

[35] Lovern, S. B. & Klaper, R. (2006). Daphnia magna mortality when exposed to titanium dioxide and fullerene (C60) nanoparticles. *Environ. Toxicol. Chem.*, 25, 1132-1137.

[36] Lyon, D. Y., Adams, L. K., Falkner, J. C. & Alvarez, P. J. J. (2006). Antibacterial activity of fullerene water suspensions: Effects of preparation method and particle size. *Environ. Sci. Technol.*, 40, 4360-4366.

[37] Song, M., Liu, S., Yin, J. & Wang, H. (2011). Interaction of human serum album and C60 aggregates in solution. *Int. J. Mol. Sci.*, 12, 4964-4974.

[38] Jinno, K., Saito, Y., Chen, Y. L., Luehr, G., Archer, J., Fetzer, J. C. & Biggs, W. R. (1993). Separation of C60 and C70 fullerenes on methoxyphenylpropyl bonded stationary phases in microcolumn liquid chromatography. *J. Microcolumn Sep.*, 5, 135-140.

[39] Heymann, D., Chibante, L. P. F. & Smalley, R. E. (1995). Determination of C60 and C70 fullerenes in geologic materials by high-performance liquid chromatography. *J. Chromatogr. A*, 689, 157-163.

[40] Anacleto, J. F. & Quilliam, M. A. (1993). Liquid chromatography/mass spectrometry investigation of the reversed-phase separation of fullerenes and their derivatives. *Anal. Chem.*, 65, 2236-2242.

[41] Li, Z., Wu, X., Zeng, G., Sakai, M. & Muramatsu, Y. (1996). Separation and analysis of fullerenes by high-performance liquid chromatography. *Fenxi Kexue Xuebao*, 12, 220-222.

[42] Xiao, J. & Meyerhoff, M. E. (1995). High-performance liquid chromatography of C60, C70, and higher fullerenes on tetraphenylporphyrin-silica stationary phases using strong mobile phase solvents. *J. Chromatogr. , A*, 715, 19-29.

[43] Perez, R. A., Albero, B., Miguel, E., Tadeo, J. L. & Sanchez-Brunete, C. (2013). A rapid procedure for the determination of C60 and C70 fullerenes in soil and sediments by ultrasound-assisted extraction and HPLC-UV. *Anal. Sci.*, 29, 533-538.

[44] Sanchis, J., Bozovic, D., Al-Harbi, N. A., Silva, L. F., Farre, M. & Barcelo, D. (2013). Quantitative trace analysis of fullerenes in river sediment from Spain and soils from Saudi Arabia. *Anal. Bioanal. Chem.*, 405, 5915-5923.

[45] Farre, M., Perez, S., Gajda-Schrantz, K., Osorio, V., Kantiani, L., Ginebreda, A. & Barcelo, D. (2010). First determination of C60 and C70 fullerenes and N-methylfulleropyrrolidine C60 on the suspended material of wastewater effluents by liquid chromatography hybrid quadrupole linear ion trap tandem mass spectrometry. *J. Hydrol.*, 383, 44-51.

[46] van Wezel, A. P., Moriniere, V., Emke, E., ter Laak, T. & Hogenboom, A. C. (2011). Quantifying summed fullerene nC60 and related transformation products in water using LC LTQ Orbitrap MS and application to environmental samples. *Environ. Int.*, 37, 1063-1067.

[47] Chen, Z., Westerhoff, P. & Herckes, P. (2008). Quantification of C60 fullerene concentrations in water. *Environ. Toxicol. Chem.*, 27, 1852-1859.

[48] Isaacson, C. W., Usenko, C. Y., Tanguay, R. L. & Field, J. A. (2007). Quantification of Fullerenes by LC/ESI-MS and Its Application to in Vivo Toxicity Assays. *Anal. Chem.*, 79, 9091-9097.

[49] Kawano, S. i., Murata, H., Mikami, H., Mukaibatake, K. & Waki, H. (2006). Method optimization for analysis of fullerenes by liquid chromatography/atmospheric pressure photoionization mass spectrometry. *Rapid Commun. Mass Spectrom.*, 20, 2783-2785.

[50] Zarzycki, P. K., Ohta, H., Saito, Y. & Jinno, K. (2006). Chromatographic behavior of C60 and C70 fullerenes at subambient temperature with n-alkanes mobile phases. *Chromatographia*, 64, 79-82.

[51] Moussa, F., Pressac, M., Genin, E., Roux, S., Trivin, F., Rassat, A., Ceolin, R. & Szwarc, H. (1997). Quantitative analysis of C60 fullerene in blood and tissues by high-performance liquid chromatography with photodiode-array and mass spectrometric detection. *J. Chromatogr. , B: Biomed. Sci. Appl.*, 696, 153-159.

[52] Sanchis, J., Berrojalbiz, N., Caballero, G., Dachs, J., Farre, M. & Barcelo, D. (2012). Occurrence of Aerosol-Bound Fullerenes in the Mediterranean Sea Atmosphere. *Environ. Sci. Technol.*, 46, 1335-1343.

[53] aggregates in water by ultrasound-assisted dispersive liquid-liquid microextraction with liquid chromatography-atmospheric pressure photoionization-tandem mass spectrometry. *J. Chromatogr. , A*, 1223, 15-23.

[54] Nunez, O., Gallart-Ayala, H., Martins, C. P. B., Moyano, E. & Galceran, M. T. (2012). Atmospheric Pressure Photoionization Mass Spectrometry of Fullerenes. *Anal. Chem. (Washington, DC, U. S.)*, 84, 5316-5326.

[55] Astefanei, A., Núñez, O. & Galceran, M. T. (2014). Analysis of Fullerenes and C60-Fullerene Derivatives in Environmental Samples by Liquid Chromatography-Atmospheric Pressure Photoionization-Mass Spectrometry. *J. Chromatogr. A (submitted)*

[56] Ohta, H., Saito, Y., Nagae, N., Pesek, J. J., Matyska, M. T. & Jinno, K. (2000). Fullerenes separation with monomeric type C30 stationary phase in high-performance liquid chromatography. *J. Chromatogr. A*, 883, 55-66.

[57] Carboni, A., Emke, E., Parsons, J. R., Kalbitz, K. & de Voogt, P. (2014). An analytical method for determination of fullerenes and functionalized fullerenes in soils with high performance liquid chromatography and UV detection. *Anal. Chim. Acta*, 807, 159-165.

[58] Kolkman, A., Emke, E., Baeuerlein, P. S., Carboni, A., Tran, D. T., ter Laak, T. L., van Wezel, A. P. & de Voogt, P. (2013). Analysis of (Functionalized) Fullerenes in Water Samples by Liquid Chromatography Coupled to High-Resolution Mass Spectrometry. *Anal. Chem.*, 85, 5867-5874.

[59] Wang, C., Shang, C. & Westerhoff, P. (2010). Quantification of fullerene aggregate nC60 in wastewater by high-performance liquid chromatography with UV-vis spectroscopic and mass spectrometric detection. *Chemosphere*, 80, 334-339.

[60] Isaacson, C. W. & Bouchard, D. (2010). Asymmetric flow field flow fractionation of aqueous C60 nanoparticles with size determination by dynamic light scattering and quantification by liquid chromatography atmospheric pressure photo-ionization mass spectrometry. *J. Chromatogr. , A*, 1217, 1506-1512.

[61] Bouchard, D. & Ma, X. (2008). Extraction and high-performance liquid chromatographic analysis of C60, C70, and [6,6]-phenyl C61-butyric acid methyl ester in synthetic and natural waters. *J. Chromatogr. , A*, 1203, 153-159.

[62] Yu, Q. W., Feng, Y. Q., Shi, Z. G. & Yang, J. (2005). HPLC separation of fullerenes on the stationary phases of two Lewis bases modified magnesia-zirconia. *Anal. Chim. Acta*, 536, 39-48.

[63] Yu, Q. W., Shi, Z. G., Lin, B., Wu, Y. & Feng, Y. Q. (2006). HPLC separation of fullerenes on two charge-transfer stationary phases. *J. Sep. Sci.*, 29, 837-843.

[64] Coutant, D. E., Clarke, S. A., Francis, A. H. & Meyerhoff, M. E. (1998). Selective separation of fullerenes on hydroxyphenyl-triphenylporphyrin-silica stationary phases. *J. Chromatogr. A*, 824, 147-157.

[65] Chao, T. C., Song, G., Hansmeier, N., Westerhoff, P., Herckes, P. & Halden, R. U. (2011). Characterization and Liquid Chromatography-MS/MS Based Quantification of Hydroxylated Fullerenes. *Anal. Chem.*, 83, 1777-1783.

[66] Silion, M., Dascalu, A., Pinteala, M., Simionescu, B. C. & Ungurenasu, C. (2013). A study on electrospray mass spectrometry of fullerenol C60(OH)24. *Beilstein J. Org. Chem.*, 9, 1285-1295, No.

[67] Gross, B., Schurig, V., Lamparth, I. & Hirsch, A. (1997). Enantiomer separation of [60]fullerene derivatives by micro-column high-performance liquid chromatography using (R)-(-)-2-(2,4,5,7-tetranitro-9-fluorenylideneaminooxy) propionic acid as chiral stationary phase. *J. Chromatogr. A*, 791, 65-69.

[68] Cunliffe, J. M. & Maloney, T. D. (2007). Fused-core particle technology as an alternative to sub-2-μm particles to achieve high separation efficiency and low backpressure. *J. Sep. Sci.*, 30, 3104-3109.

[69] Staub, A., Zurlino, D., Rudaz, S., Veuthey, J. L. & Guillarme, D. (2011). Analysis of peptides and proteins using sub-2μm fully porous and sub 3-μm shell particles. *J. Chromatogr. A*, 1218, 8903-8914.

[70] Schuster, S. A., Boyes, B. E., Wagner, B. M. & Kirkland, J. J. (2012). Fast high performance liquid chromatography separations for proteomic applications using Fused-Core silica particles. *J. Chromatogr. A*, 1228, 232-241.

[71] Jinno, K., Uemura, T., Ohta, H., Nagashima, H. & Itoh, K. (1993). Separation and identification of higher molecular weight fullerenes by high-performance liquid chromatography with monomeric and polymeric octadecylsilica bonded phases. *Anal. Chem.*, 65, 2650-2654.

[72] Mrzel, A., Mertelj, A., Omerzu, A., Copic, M. & Mihailovic, D. (1999). Investigation of Encapsulation and Solvatochromism of Fullerenes in Binary Solvent Mixtures. *J. Phys. Chem. B*, 103, 11256-11260.

[73] Alargova, R. G., Deguchi, S. & Tsujii, K. (2001). Stable colloidal dispersions of fullerenes in polar organic solvents. *J. Am. Chem. Soc.*, 123(43), 10460-10467.

[74] Cosmosil Buckyprep column, Nacalai USA, Inc. Available at: http://www.nacalaiusa.com/product.php?id=22

[75] European Commission (2002) Commission Decision of 12 August 2002 implementing Council Directive 96/23/EC concerning the performance of analytical methods and the interpretation of results. European Commission, Brussels.

[76] Robb, D. B., Covey, T. R. & Bruins, A. P. (2000). Atmospheric pressure photoionization: an ionization method for liquid chromatography-mass spectrometry. *Anal. Chem.*, 72, 3653-3659.

[77] Ross, M. S. & Wong, C. S. (2010). Comparison of electrospray ionization, atmospheric pressure photoionization, and anion attachment atmospheric pressure photoionization for the analysis of hexabromocyclododecane enantiomers in environmental samples. *J. Chromatogr. A*, 1217, 7855-7863.

[78] Raffaelli, A. & Saba, A. (2003). Atmospheric pressure photoionization mass spectrometry. *Mass Spectrom. Rev.*, 22, 318-331.
[79] Song, L., Wellman, A. D., Yao, H. & Bartmess, J. E. (2007). Negative Ion-Atmospheric Pressure Photoionization: Electron Capture, Dissociative Electron Capture, Proton Transfer, and Anion Attachment. *J. Am. Soc. Mass Spectrom.*, 18, 1789-1798.
[80] Song, L., Dykstra, A. B., Yao, H. & Bartmess, J. E. (2009). Ionization mechanism of negative ion-direct analysis in real time: A comparative study with negative ion-atmospheric pressure photoionization. *J. Am. Soc. Mass Spectrom.*, 20, 42-50.
[81] McEwen, C. N. & Larsen, B. S. (2009). Ionization Mechanisms Related to Negative Ion APPI, APCI, and DART. *J. Am. Soc. Mass Spectrom.*, 20, 1518-1521.
[82] Kauppila, T. J., Kotiaho, T., Kostiainen, R. & Bruins, A. P. (2004). Negative ion-atmospheric pressure photoionization-mass spectrometry. *J. Am. Soc. Mass Spectrom.*, 15, 203-211.
[83] Kozlovski, V., Brusov, V., Sulimenkov, I., Pikhtelev, A. & Dodonov, A. (2004). Novel experimental arrangement developed for direct fullerene analysis by electrospray time-of-flight mass spectrometry. *Rapid Commun. Mass Spectrom.*, 18, 780-786.
[84] Shilova, E. A., Lyakhovetsky, Y. I., Belokon, A. I., Ponomareva, T. V., Gasanov, R. G., Tumanskii, B. L. & Nekrasov, Y. S. (2011). Homolytic reactive mass spectrometry of fullerenes: interaction of C60 and C70 with ketones in the electron impact ion source of a mass spectrometer and the comparison of results with those of photochemical reactions of C60 with several ketones in solution. *J. Am. Soc. Mass Spectrom.*, 22, 2021-2032.
[85] Chiang, L. Y., Bhonsle, J. B., Wang, L., Shu, S. F., Chang, T. M. & Hwu, J. R. (1996). Efficient one-flask synthesis of water-soluble [60]fullerenols. *Tetrahedron*, 52, 4963-4972.

In: Fullerenes
Editor: Shannon B. Ellis
ISBN: 978-1-63321-385-2

Chapter 3

Physicochemical Properties and Biological Activity of Polystyrene/Fullerene Composites

Olga V. Alekseeva, Nadezhda A. Bagrovskaya and Andrew V. Noskov
G. A. Krestov Institute of Solution Chemistry,
Russian Academy of Sciences, Ivanovo, Russia

Abstract

The present chapter includes the detailed study of the physicochemical properties, structure, and antioxidant effect of both polystyrene films and polystyrene/fullerene composite films.

The glass transition temperatures, T_g, and specific heat capacities, C_P, in range of 293-423 K have been determined for the films by DSC technique. The plasticization of the polymer is observed in thermal properties of the films under influence of small fullerene additions. The values of T_g and C_P decrease and thermal coefficient of heat capacity, b, increase as fullerene content increases up to 0.02 wt. %. The effect of interaction between polymer and fullerene molecules on thermal properties becomes evident at higher fullerene content in range from 0.02 to 0.1 wt. %.

Electrical properties of the films were investigated. Based on the results of measurements of capacitance and dielectric loss tangent can be conclude that the fullerene-polystyrene composites studied are nonpolar. We revealed dependence of the dielectric constant on the film composition is nonmonotonic with a minimum if there is 0.035 wt. % of C_{60}. It is found that the doping polystyrene with small amounts of fullerenes results neither an appreciable increase in the static conductivity nor occurrence of the dielectric relaxation processes.

The structural characteristics of polystyrene films filled with fullerenes were researched by X-ray diffraction technique. It was revealed there is the intermediate-range order in these materials. Insertion of fullerenes into polymer matrix increases a correlation length. It is established by small-angle X-ray scattering that composites containing fullerenes in the amount 0.01 wt. % are characterized by the greatest mobility of the segments in the modified polymer chains.

Solvatochromic effect in the absorption spectra of fullerene in weakly polar solvents and in polymer-containing solution was revealed by electron spectroscopy. It is suggested that there is intermolecular interactions of macromolecules of polystyrene and fullerene in studied solutions and polystyrene/fullerene composite films.

Kinetics of adsorption of Cu(II), Zn(II), and Cd(II) ions by both original polystyrene and polystyrene films filled with fullerenes were studied. It was found that modification of polystyrene increases the adsorption capacity of polymer material.

Effect of fullerene-containing polystyrene nanocomposites on free-radical processes in blood serum has been investigated in vitro. The parameters of lipid peroxidation in native serum after adding nanomaterials were determined by chemiluminescent analysis and spectrophotometry. It was revealed that fullerene-containing nanocomposites can manifest antioxidant activity in blood serum.

INTRODUCTION

The development of polymeric composites with controllable structure and properties is one of priority lines of the modern chemistry and materials science. Harsh conditions of polymer operation in power engineering and in chemical, petroleum, and pulp-and-paper industries impose stringent requirements upon the properties of the polymers (hardness, strength, and electrophysical parameters). Effective way to meet these requirements is insertion of fillers into polymer matrix. Insertion of carbon fillers such as fullerenes, carbon nanotubes etc can lead to creation of materials with improved physical and chemical properties and the main service characteristics (mechanical, electrical) [1-3].

The discovery of soccer-ball shaped Buckminster fullerene in 1985 [4] was an exciting and unexpected discovery that established an entirely new branch of chemistry. Since a new direction of fullerene science has been developed and the related studies in this direction have concerned the preparation of fullerene-containing polymers, which combine the unique characteristics of fullerene with the useful properties of matrix polymers [5].

Scientific interest in the doping of polymers with fullerene is likely related to the simplicity of fullerene incorporation, either in its native form or as solutions in organic solvents, and to the use of minor amounts of modifying agents. Furthermore, less significant changes in the electron structure and, hence, less abrupt changes in the characteristics of fullerene molecules in the absence of any covalent interaction between fullerene molecules and polymer chain fragments were expected. These modifiers, when introduced in small amounts (up to several percent), can be nucleating agents and can affect the structure of the polymer.

Polysterene (PS) is well-known film-forming polymer often used for different modifications with low molecular compounds of special properties, including fullerenes. PS is well dissolved in benzene, toluene, o-xylene etc. which are also solvents for fullerenes. This dissolution peculiarity allows to employ widespread procedure of polymer/fullerene composite formation that consists in casting of the solution containing polystyrene and fullerene followed by the evaporation of solvent. That such a procedure we used.

The present chapter includes the detailed study of the physicochemical properties, structure, and antioxidant effect of both polystyrene films and polystyrene/fullerene composite films. We also report results on intermolecular interactions of macromolecules of polystyrene and fullerene in solutions and polystyrene/fullerene composite films. Findings have very important practical significance for materials science, because polystyrene is used often for various modifications with low molecular compounds, including fullerenes.

Polystyrene (Aldrich, US; $M_n=1.4\cdot10^5$, $M_w/M_n=1.64$), fullerenes C_{60} (NeoTechProduct Ltd, Russia) and fullerenes mix $C_{60}+C_{70}$ (Fullerene Technologies Ltd., Russia) were used. For fabrication of films a solvent casting of perspective components from solutions was employed. Preliminary purification of organic solvent (o-xylene, toluene and chloroform[1]) was made by standard techniques [6].

We produced the polystyrene films as follows. Polystyrene batch was dissolved in organic solvent (17 wt. % of PS) and the solution was stirred for about 1 day before being cast into thin films. After casting the solvent was slowly evaporated at room temperature over several days to produce the polystyrene films.

Fullerene-polystyrene composition films have been fabricated as follows. Fullerene batches were dissolved in solvent at various concentrations. Then polystyrene batches were dissolved in all obtained solutions (17 wt. % of PS) and the mixed solutions were stirred for about 1 day before being cast into thin films. After casting the solvent was slowly evaporated at room temperature over several days to produce the composite films. By this technique, we prepared some of composites samples with various fullerene percentages, x, in the form of film.

The samples obtained have been examined by optical microscope «Boetius» (Germany). We found both polystyrene films and fullerene-polystyrene composite films transparent, that is, the films are homogeneous on the optical level. Unmodified polystyrene samples were colorless, whereas the fullerene-polystyrene composite films were light purple. The intensity of color depended on the content of fullerene in the composite.

A thickness of the films was measured by micrometer with uncertainty of 5 μm. As a rule the thickness of the films amounts from 60 to 80 μm. But for dielectric measurements we prepared films having thickness of near 1 mm.

1. DSC Investigation of the Polystyrene Films Filled with Fullerene

To study effect of small additives of fullerenes ($C_{60}+C_{70}$ mix) on thermal and thermodynamic properties of polymer, we used the differential scanning calorimetry technique. DSC measurements were performed using DSC 204 F 1 apparatus (Netzsch, Germany) in argon atmosphere (15 $cm^3\cdot min^{-1}$). A stack of films with a diameter of 5 mm was placed in a press-fitted aluminum crucible covered the pierced lid. Mass of samples was from 4 to 10 mg. The samples were undergone first heating up to 423 K with a scan rate of 10 $K\cdot min^{-1}$ to remove volatile substances from the polymer and cooled down to 283 K by means of liquid nitrogen. Second heating of the simples was carried out according to the following sequence: heating up to 293 K; isothermal mode at 293 K for 10 minutes; and heating up to 423 K with a scan rate of 10 $K\cdot min^{-1}$. The glass transition temperatures and specific heat capacities were determined from data of the second heating. The reference aluminium crucible was empty. Sapphire disk with a mass of 12.5 mg was used as standard of specific heat capacity. All measurements were performed relative to the base line obtained with two

[1] Casting of toluene and chloroform was used only at research of antioxidant activity of films.

empty crucibles. Three measurements required for each composite (base line, sample and standard) were carried out on the same day. Calibration of the calorimeter was performed by measuring the temperatures and heat effects of phase transitions for the 11 standard substances in the temperature range from 187 to 749 K [7].

The content of volatile substances in the samples studied was determined by thermogravimetry using Netzsch microthermobalance TG 209 F 1 under similar conditions of experiment, i.e., heating up to 423 K with a scan rate of 10 K·min^{-1}. A mass loss about 2 % is observed due to evaporation of volatile substances (styrene oligomers and solvent traces) from the films. The error in determination of the mass loss was $1 \cdot 10^{-3}$ mg. Thermogravimetric data were taken into account to correct the mass of samples at second heating in DSC experiments. Each DSC measurement was repeated three times and the results were given as mean values ± standard deviations. Reproducibility for value of the glass transition temperature was 0.5 K, and for value of the change of the specific heat capacity during phase transition – 0.02 J g^{-1} K^{-1}.

Typical DSC curves for the films of original polystyrene and filled composite with fullerene are shown in Figure 1. It is seen that for all examined polymeric materials there is a reversible phase transition from the glassy state to elastic one, which manifests itself as a step of heat flow in endothermic direction. We have been characterized this phase transition by the following parameters:

- T_1 is the extrapolated temperature of the phase transition onset;
- T_2 is the extrapolated temperature of the phase transition end;
- $\overline{T}$ is the average temperature of the phase transition;
- T_g is the temperature of DSC curve inflection taken as the glass transition point;
- $\Delta T = T_2 - T_1$ is the temperature interval in which the phase transition occurs;
- ΔC_P is the change of the specific heat capacity during phase transition.

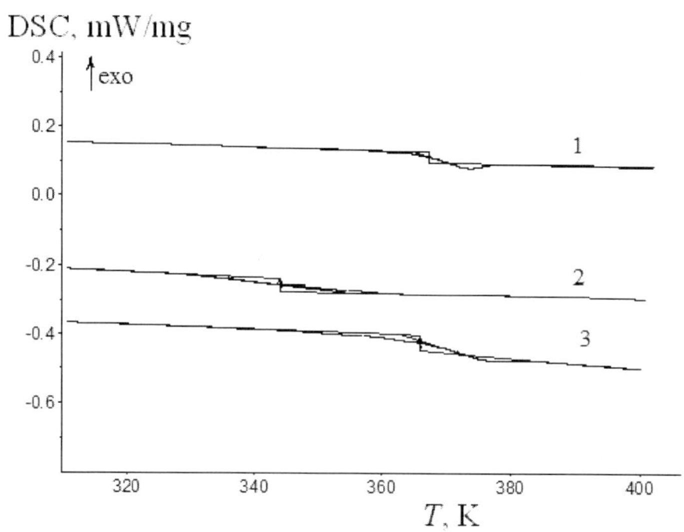

Figure 1. DSC curves of films: *1* – PS; *2* – PS/C_{60}+C_{70} composite with 0.01 wt. % of fullerene; *3* – PS/C_{60}+C_{70} composite with 0.1 wt. % of fullerene.

The average characteristics of the glass transition obtained from tree experiments for each polymer films are presented in Table 1. It was found that the C_{60}+C_{70} addition affects all characteristic temperatures and temperature range of phase transition, ΔT, for the composite materials.

The concentration dependence of ratio of T_g values for original and filled polystyrene is shown in Figure 2. Small additions of fullerene sharply reduce the glass transition temperature of the composite in comparison with original polymer film. The minimum value of T_g is observed when the concentration of fullerenes is equal to 0.01 wt. %. Apparently, in this case, there is the plasticization of polystyrene with the fullerene [8, 9]. Molecules of C_{60}+C_{70} embed between the chains and weaken interchain interactions in the polystyrene. It results in increasing mobility of polymer chain segments and reducing value of T_g. The noticeable extension of the temperature range for phase transition, ΔT=T_2-T_1, at low concentration of fullerene (Table 1) confirms an evident plasticization effect of these additives on thermal properties of the composite.

The glass transition temperature increases with further increase in the C_{60}+C_{70} content. When the fraction of fullerene is equal to 0.1 wt. %, the T_g value for the composite exceeds that for the original polystyrene.

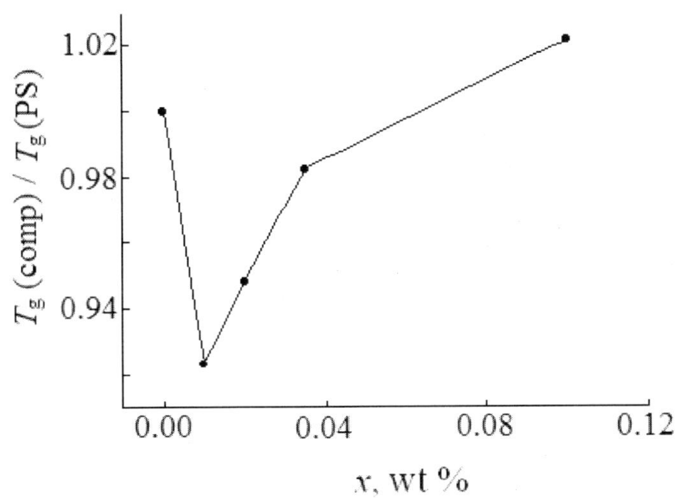

Figure 2. Ratio of glass transition temperatures for composite films, T_g(comp), to one of original polystyrene film, T_g(PS), vs. fullerene content.

Table 1. Parameters of phase transition from glassy state to elastic one for the original and filled polystyrene films with various contents of C_{60}+C_{70}.

x, wt. %	T_1, K	$\overline{T}$, K	T_g, K	T_2, K	ΔT, K	ΔC_P, J g^{-1} K^{-1}
0	357.0	361.5	363.8	365.8	8.8	0.30
0.01	328.1	343.8	335.8	355.1	27.0	0.23
0.02	341.6	348.6	345.0	359.3	17.7	0.25
0.035	347.0	358.4	357.4	366.5	19.5	0.26
0.1	365.3	369.1	371.6	373.5	8.2	0.29

It is known that fullerene molecules are capable of strong intermolecular interaction due to the large number of conjugated double bonds [9]. Intermolecular interactions of polystyrene and fullerene are emerged on the values of T_g in the films with additions of more than 0.01 wt. %. When the concentration of fullerenes is equal to 0.1 wt. % and above, the effect of intermolecular interactions is dominant, the probability of interaction between fullerene molecules in the composite increases, that is manifested in increasing glass transition temperature as compared with the original polystyrene and, apparently, can lead to the physical cross-linking of polystyrene chains. Vice versa, at extremally low concentration (less then 0.01 wt. %) the probability of interaction between fullerene and polystyrene molecules is very small because of there is one fullerene molecule per 55-60 polystyrene macromolecules (or one fullerene molecule per 70000-80000 monomeric unit) in the composite material. In this regime, weakening of interchain interactions in the polystyrene is dominant.

In refs [10, 11], we used electron microscopy and X-ray diffraction to study the influence of the fullerene additives on structure of the polystyrene films formed by evaporation of the solvent. We concluded in films without fullerenes, packing of straightened chains parallel to each other predominated. This shows that a polystyrene molecule, which had the shape of a coil in solution, straightened and stretched itself along the surface of an aggregate when attached to it. When the concentration of fullerenes is equal to 0.035 wt. % (one fullerene per 7–10 polystyrene macromolecules), attachment to aggregates occurred similarly. However, at such concentrations the intermolecular interactions between polystyrene and fullerene are appreciable and under the influence of fullerene molecules, polystyrene molecules straightened with the formation of ordering elements in the arrangement of chains [10].

Based on the results of DSC studies we identified the specific heat capacities, C_P, for the films of original polystyrene and polystyrene doped additive mixture $C_{60}+C_{70}$ (0.01÷0.1 wt. %), in the range of 290-420 K. The specific heat capacity values for original polystyrene are equal to 1.23±0.03 and 1.95±0.03 $J{\cdot}g^{-1}{\cdot}K^{-1}$ at 298 and 398 K, respectively. These results agree well with reference data which amount to 1.224 and 1.924 $J{\cdot}g^{-1}{\cdot}K^{-1}$, respectively [7].

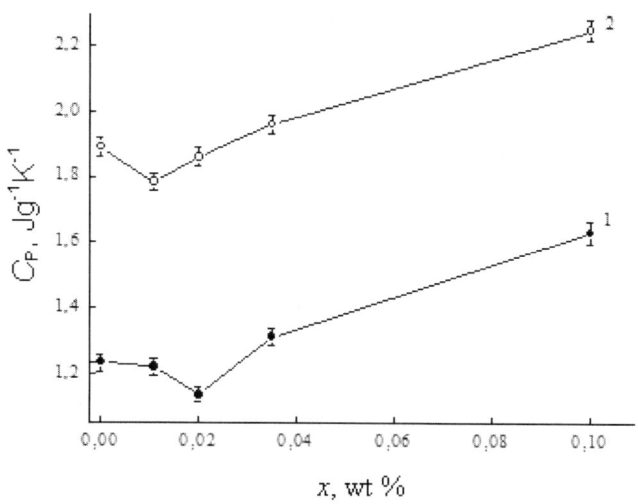

Figure 3. Dependence of specific heat capacity of composite on fullerene content: *1* – glassy state at 298 K; *2* – elastic state at 398 K.

The repeatability of C_P values for the composite films amounted to 0.03 $J{\cdot}g^{-1}{\cdot}K^{-1}$ also as determined from tree measurements for each sample. It can be seen in Figure 3 the dependence of the specific heat capacity of the composite on the fullerene content is non-monotone. The specific heat capacity at 298 K slightly decreases from 1.23 to 1.12 J g^{-1} K^{-1} when the content of C_{60}+C_{70} increases up to 0.02 wt. %. Further enhancement of fullerene content leads to the considerable increase in heat capacity to 1.60 J g^{-1} K^{-1} (curve 1).

The dependence of the heat capacity on the content of C_{60}+C_{70} for filled polymer materials in the elastic state at 398 K passes through a minimum also at a concentration of fullerenes equal to 0.01 wt. % (Figure 3, curve 2). The observed changes in the specific heat capacity confirm the conclusion indicated above about the plasticization of polystyrene by small additions of fullerene. The increase in heat capacity at the content of fullerene more than 0.02 wt. % can be associated with the manifestation of the polystyrene-fullerene interactions that is consistent with the regularities in change of the glass transition temperatures [8].

It can be seen in Figure 4 the specific heat capacity of the composites monotonically increases with temperature and increases abruptly at the phase transition. The dependence "measured values of C_P vs temperature" was described by a linear equation:

$$C_P = a + bT \tag{1}$$

The temperature dependence coefficients of the specific heat capacity, b, for the composites at various contents of fullerenes are presented in Figure 5. It can be seen that value of b for samples in the glassy state in the temperature range 298÷328 K increases sharply with increasing fullerene content up to 0.02 wt. % and decreases with further increase in the concentration of additives (curve 1). The observed concentration dependences of C_p and b values reflect the competing effects of two factors. Namely, the plasticizing effect of fullerenes manifests itself with small additions. At this the values of C_p decrease (Figure 3) and the values of b increase (Figure 5).

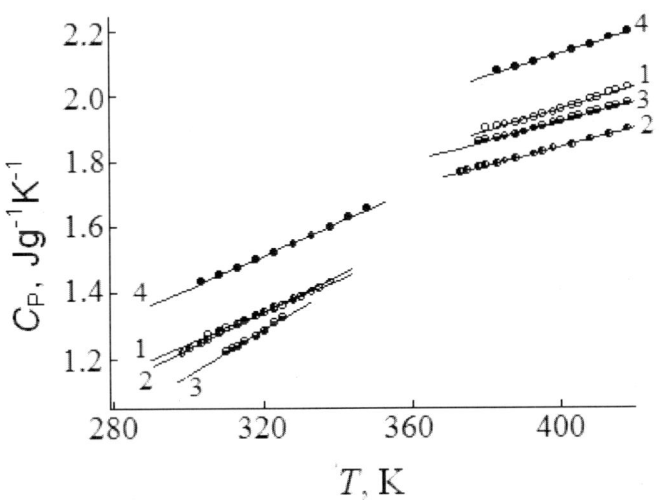

Figure 4. Temperature dependence of specific heat capacity for composite films with various content of fullerene, wt. %: 0 (*1*); 0.01 (2); 0.02 (*3*); 0.035 (*4*).

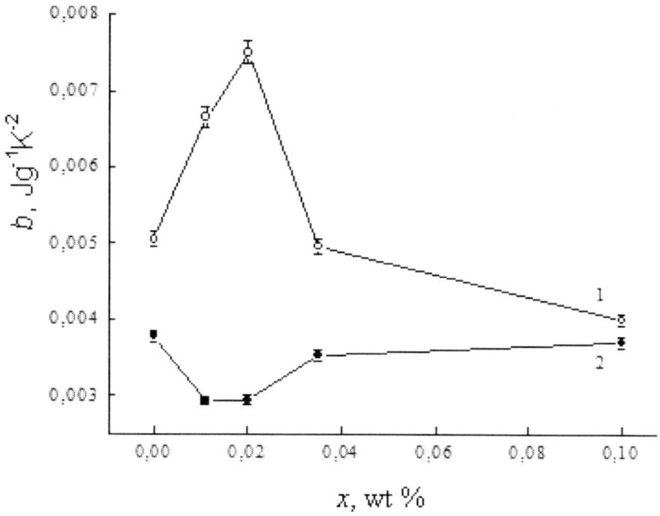

Figure 5. Coefficient of temperature dependence of specific heat capacity for composite films with various content of fullerene: *1* - glassy state in temperature range of 298÷328 K; *2* - elastic state in range of 368÷423 K.

The effect of interactions between fullerene and polymer molecules dominates when the fullerene concentrations are more than 0.02 wt. %. At this the values of C_p increase (Figure3) and *b* decrease (Figure5). The concentration dependence of *b* values for the samples in the elastic state for the range 368÷423 K is weak (Figure 5, curve 2).

Note an analogous non-monotone dependence of physical and chemical parameters of composites on the fullerene concentration was found for crystallinity degree of some polymer materials [3].

2. Dielectric Parameters of Polystyrene/Fullerene Composite Films

It is known because of the absence of polar groups, the electrical conductivity of polystyrene is extremely low (no more than 10-14 Ohm-1·m-1) [12], which makes extensive use of it, e.g., in production of capacitors, of casings for household devices, and of insulation for current leads.

However, doping of polystyrene with fullerenes can deteriorate its dielectric characteristics. In particular, Adamopoulos et al. [13] studied the electrical properties of thin polystyrene films doped with 23-60 wt. % of C_{60}. They found that an increase in the fullerene concentration leads to a monotonic increase in the high-frequency dielectric constant of the material from 2.4 to 2.8 and to a considerable increase in the electron mobility (by 2 orders of magnitude). Nourdine et al. [14] synthesized fullerene-containing polystyrene-based polymers and revealed a stepwise increase in the electrical conductivity (percolation threshold) at 4 vol. % of C_{60}.

A still greater effect is attained on doping of PS with another allotropic modification of carbon, graphene. A stepwise increase in the electrical conductivity occurs on introducing into the polymer graphene in an amount as low as 0.1 vol % [15].

Thus, when developing a composite material, it should be taken into account that modification may improve some properties of the polymer but simultaneously deteriorate other properties. Therefore, studies aimed to elucidate the composition-property relationship and to reveal the optimal dopant concentration become very topical.

In this section we determined the dielectric parameters of polystyrene films filled with small (up to 1 wt. %) additions of fullerenes C_{60}. To determine these characteristics of composites (capacitance, *C*, and dielectric loss tangent, $tan\delta$) we used Solartron 1255 frequency response analyzer (UK). The numerical values of the parameters were obtained at room temperature using a two-electrode cell with round clamp electrodes 19.8 mm in diameter.

The dielectric constant (ε') of the film substance was calculated using the flat-plate capacitor formula:

$$C = \frac{\varepsilon'\varepsilon_0 S}{d}, \tag{2}$$

where *S* is the electrode area, *d* is the film thickness, and $\varepsilon_o = 8.854\cdot 10^{-12}$ $F\cdot m^{-1}$ is the electric constant. The ac resistivity was determined by the relationship

$$\rho = \left(2\pi f \varepsilon'\varepsilon_0 \tan\delta\right)^{-1}, \tag{3}$$

where *f* is the frequency of alternating current.

The frequency dependences of the capacitance of the formed polystyrene and composite films with various fullerene contents are shown in Figure 6. It can be seen for all the samples studied, the capacitance does not noticeably change in the frequency range $10^4 \div 3.2\cdot 10^6$ Hz, indicating that the composites are nonpolar or weakly polar [12, 16].

Figure 7 shows the frequency dependences of the dielectric constant, ε', of the composites with various fullerene contents, calculated by formula (2) taking into account the measured film thickness Table 2. It can be expected on base of proportionality between *C* and ε', the shape of the $\varepsilon'(\log f)$ curves (Figure 7) correlates with the shape of the $C(\log f)$ curves (Figure 6). In the frequency interval $10^4 \div 3.2\cdot 10^6$ Hz, the dielectric constant does not change noticeably. The ε' values obtained are well consistent with the published data. In particular, Skanavi [12] reported for unmodified polystyrene the dielectric constants in the interval 2.5÷2.9.

It can be seen in Figure 7, for the composites containing 0.035 wt. % of fullerenes, the dependence $\varepsilon'(\log f)$ lies below the other dependences, i.e., the dielectric constant passes through a minimum at 0.035 wt. % of C_{60}.

We should emphasize that the results are not consistent with the so-called effective medium models, which have been successfully used to describe the dielectric parameters of nanocomposite materials. In particular, in accordance with the Maxwell-Garnett approximation, the dielectric constant should increase monotonically with filler volume fraction [17]. There is a minimum (Figure 7 and Table 3), which indicates misuse of this approximation to fullerene-containing films investigated in the present. Apparently, in this case are not met the basic assumptions of the Maxwell-Garnett model. Namely, the fullerene molecules cannot be considered as rare inclusions that do not interact together and with fragments of polystyrene chains. Because of this interaction, the nanocrystalline regions can be formed in the original amorphous polymer resulting in a distortion of classical monotonic dependence "dielectric constant vs film composition".

It should be noted that the extrema at small additions of carbon nanoparticles were also found for other physicochemical properties of modified polymers [3, 18-21]. Badamshina and Gafurova [3] suggest for these facts the following explanation: whereas at low concentrations fullerene forms small clusters with fragments of molecular chains of polystyrene, with an increase in the concentration these clusters coalesce into coarse aggregates which are poorly soluble in the polymer and are inefficient as modifiers.

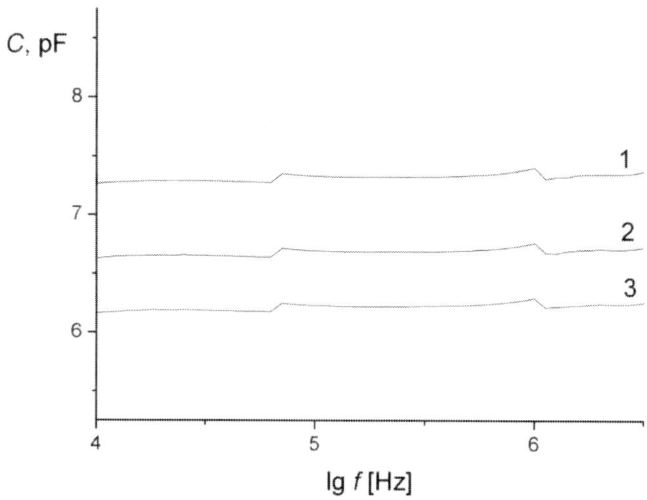

Figure 6. Frequency dependences of the capacitance of composite films with various contents of C_{60}, wt. %: 0 (*1*), 0.1 (*2*), 0.035 (*3*).

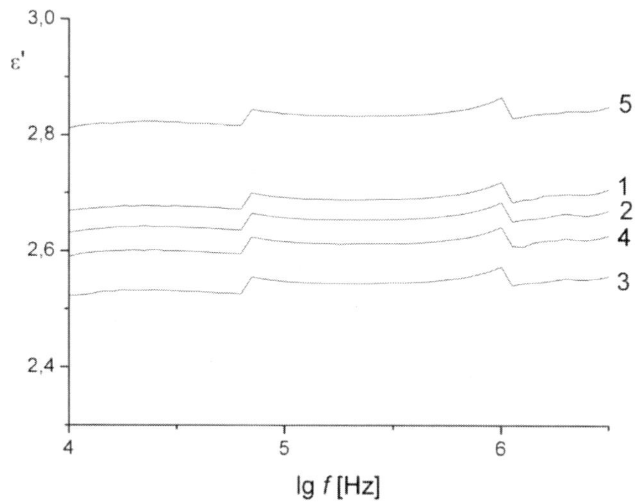

Figure 7. Frequency dependences of the dielectric constant, ε', of composite films with various contents of C_{60}, wt. %: 0 (*1*), 0.01 (*2*), 0.035 (*3*), 0.1 (*4*), 0.5 (*5*).

Table 2. Thickness and specific resistivity of PS/C_{60} composite films as functions of C_{60} content

C_{60} content, wt. %	d, mm	ρ, Ohm·m
0.0	1.01	$2.40 \cdot 10^{12}$
0.001	1.00	$2.25 \cdot 10^{12}$
0.01	1.08	$3.11 \cdot 10^{12}$
0.035	1.13	$9.65 \cdot 10^{11}$
0.1	1.08	$4.90 \cdot 10^{11}$
0.5	1.09	$3.14 \cdot 10^{12}$
1.0	1.05	$1.68 \cdot 10^{12}$

Table 3. Dielectric parameters of PS/C_{60} composites as functions of C_{60} content

C_{60} content, wt. %	ε'	$N\beta$	$\tan\delta$
0.0	2.685	1.079	$(1.75 \div 15.01) \cdot 10^{-4}$
0.001	2.730	1.097	$(1.84 \div 40.40) \cdot 10^{-4}$
0.01	2.655	1.067	$(1.37 \div 11.73) \cdot 10^{-4}$
0.035	2.540	1.018	$(4.61 \div 24.00) \cdot 10^{-4}$
0.1	2.610	1.048	$(8.83 \div 23.87) \cdot 10^{-4}$
0.5	2.835	1.139	$(1.27 \div 59.08) \cdot 10^{-4}$
1.0	2.730	1.098	$(8.50 \div 24.60) \cdot 10^{-4}$

Nonmonotonic composition-property dependences were also observed for liquid-phase fullerene-containing systems. In particular, Ginzburg et al. [22] observed inversion of the concentration dependence of ε' for solutions of C_{60} in p-xylene, with the dielectric constant varying within the range covering several percents. This anomaly is attributed to structural changes induced by introduction of fullerenes into the aromatic solvent.

It is known that electric field results polarization of a dielectric. Polystyrene is a solid amorphous dielectric containing no polar groups. In such polymers, the main kind of polarization is deformation (electronic) polarization caused by deformation of the electronic shells and appearance of induced dipoles. The contribution of the other kinds of polarization (ionic and orientation) is negligibly low because of the absence of free ions and polar molecules. However, the typical frequencies of the relaxation peaks corresponding to the electron polarization are in the visible range of the electromagnetic spectrum ($10^{14} \div 10^{15}$ Hz). Therefore, the experimental data obtained by Solartron 1255 frequency response analyzer, do not allow to confirm (or disprove) the corresponding peaks for the fullerene-polystyrene composites.

Quantitative measure of polarization that characterizes the response of matter to an external electric field is the molecular polarizability, β. For non-polar dielectric the polarizability associates with permittivity by means of the Clausius-Mossotti equation:

$$\frac{\varepsilon' - 1}{\varepsilon' + 2} = \frac{N\beta}{3}, \qquad (4)$$

where N is number of induced dipoles.

Table 3 shows the values of the product $N\beta$ calculated using the formula (4) for the researched composite films with different content of fullerenes. Apparently the revealed complex nature of the dependence “the value of $N\beta$ vs the film composition” is due to many factors, one of which is an unusually high polarizability of fullerene molecules [23].

A decrease in the polarization with time (dielectric relaxation) can be quantitatively characterized by the position of the peak in the frequency dependence of the dielectric loss tangent. For the fullerene-containing polystyrene films examined in this study, these peaks were not observed at room temperature. Apparently, the induced dipoles arise very rapidly under these conditions, and the dielectric loss is low. This is confirmed, in particular, by the results of measuring the dielectric loss tangent: In the frequency range $10^2 \div 10^6$ Hz, tanδ value is $10^{-3} \div 10^{-4}$. At higher temperatures, the frequency dependence of tanδ value can be expected to pass through a maximum. Such a maximum was found for fullerene-containing composites in the regions of transition to the hyperelastic state [9].

According to the results of thermochemical studies (section 1 of present chapter), small (up to 0.01 wt. %) additions of fullerenes exert plasticizing effect on polystyrene. It can lead to changes in the dielectric constant and in the dielectric loss in the polymer [16]. The lack of significant changes in these characteristics in the materials that we studied can be attributed to high symmetry and nonpolarity of dopant molecules (C_{60}). It should be noted that the estimated values of the dielectric loss tangent coincide in the order of magnitude with those for unmodified polystyrene, given in ref. [12]. In accordance with ref. [12], it can be stated that the dielectric loss under these temperature conditions in the low-frequency range is of ohmic nature. The polymeric materials resistivity calculated by the tanδ values (Eq. (3)) at a frequency of 100 Hz (Table 2), indicates that doping of polystyrene with up to 1 wt. % of fullerenes does not give rise to noticeable conductivity. It is noted in ref. [12], for polystyrene polarization is purely electronic, and dielectric losses consist of conduction losses and additional high-frequency losses. In this regard, we researched frequency dependencies of the conductivity for polystyrene/fullerene composite films. It can be seen, for all the samples the specific conductivity, σ, increases with value of f (Figure 8 (a-d), which is typical for dielectric materials. In the frequency range from 25 to 10^6 Hz, the specific conductivity increases by 4-5 orders of magnitude.

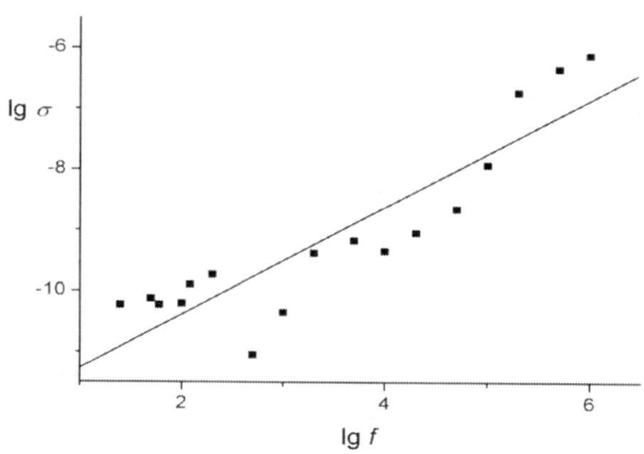

a

Figure 8. (Continued).

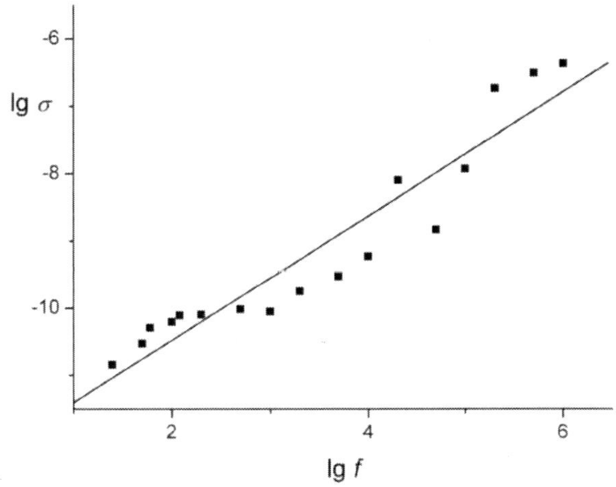

b

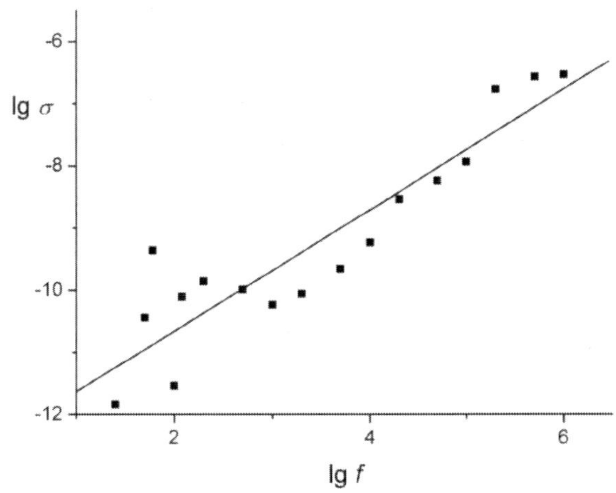

c

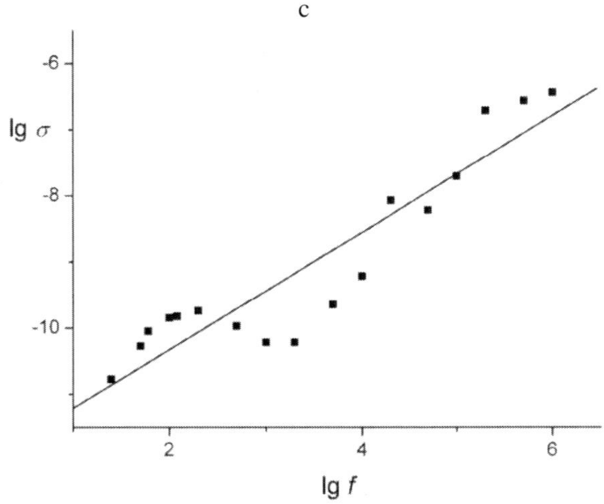

d

Figure 8. Frequency dependence of the conductivity of the polystyrene/fullerene films with various concentrations of C_{60}, wt. %: 0 (a); 0.01 (b); 0.1 (c); 1 (d). σ is the specific conductivity, $Ohm^{-1}\cdot m^{-1}$; f is the ac frequency, Hz.

Table 4. Fitting of linear regression equations for the experimental frequency dependencies of the conductivity for polystyrene/fullerene composite films

C_{60} content, wt. %	Regression equation	R^2
0.0	y=0.8822x-12.150	0.805
0.01	y=0.9236x-12.329	0.900
0.10	y=0.9732x-12.606	0.842
1.00	y=0.8854x-12.099	0.858

The obtained dependences are linearized by logarithmic coordinates with satisfactory correlation coefficients (Table 4). This means that $\sigma \sim f^n$ (where $n<1$), and the system can be studied using the model described by Maxwell-Wagner with a very wide distribution of conductive and capacitive domains (conductive particles are dispersed in a nonconducting matrix) [24].

3. X-Ray Diffraction Analysis

Structure of both PS films and PS/C_{60} composite films was evaluated by X-ray diffraction (XRD) measurements on the base of Debay-Sherrer method. XRD patterns of film samples were obtained by X-ray diffractometer DRON-UM1 (Russia) equipped with MoK_α radiation that monochromate by the Zr-filter, λ=0.071 nm. X-ray diffractometer was modernized for substances in condensed and polycrystalline state. The voltage and the current of the X-ray tubes were 40 kV and 40 mA, respectively. A scan rate of 0.04 degree/s was used.

We investigated structure of the original polystyrene and fullerene-containing composites by the X-ray diffraction technique in both wide angles (WAXS) and small angles (SAXS).

WAXS

XRD patterns of both PS films and PS/C_{60} composite films with 0.03, 0.1, 0.5, and 1 wt. % of C_{60} are given in Figure 9. XRD of solid C_{60} reported in ref. [25]. Comparing these data it can be seen the reflexes associated with fullerene are absent in XRD patterns of composite films Figure 9. Apparently, this is due to the concentration of C_{60} in the studied films is inadequate for the occurrence of reflex. This finding correlates to the data of ref. [26], in which there is a comparison of wide-angle diffraction patterns of origin polymethylmethacrylate (PMMA), composites PMMA/C_{60} (1 wt. % of C_{60}), and PMMA/C_{60} (10 wt. % of C_{60}). It is found in ref. [26] that diffraction pattern does not change under insertion of fullerene (up to 1 wt. %) into the polymer matrix. For opinion of the authors, the absence of these changes is due to the weak aggregation of C_{60}. But in diffraction pattern of PMMA/C_{60} composite (10 wt. % of C_{60}) there are additional peaks associated with fullerene aggregates [26].

It is shown in Figure 9, both PS films and PS/C_{60} composite films exhibit broad diffraction peak (halo). Its location does not depend on the film composition. Abscissa of maximum, $2\theta_m$, is equal to 5.2 degree.

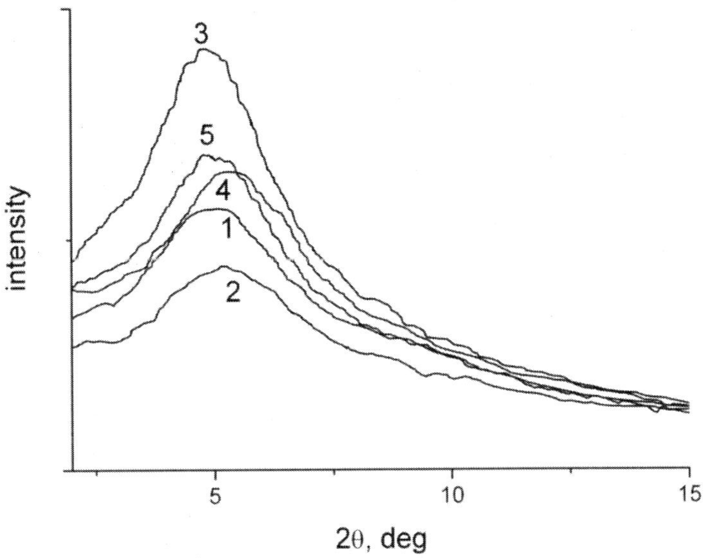

Figure 9. X-ray diffraction patterns for the films with various concentrations of C_{60}, wt. %: 0 (*1*); 0.03 (*2*); 0.1 (*3*); 0.5 (*4*); 1 (*5*).

In the reciprocal space, the position of peak appears at 8.029 nm^{-1}. We calculated the repetitive characteristic distance in structural arrangement, *h*, using the relation [27]:

$$2h\sin\theta_m = K\lambda\,, \tag{5}$$

where *K*=1.2÷1.3. We obtained the value of *h* is equal to 0.94÷1.02 nm for all studied composites.

The broad peak in the X-ray diffraction patterns of films is an indication of the existence of the intermediate-range order [28] in these materials. The full width at half maximum, Δ*s*, makes it possible to specify a correlation length (a scale of the intermediate-range ordering) in the disordered phase, *L*, using the relation:

$$L = \frac{2\pi}{\Delta s}\,, \tag{6}$$

where $s = 4\pi\sin\theta/\lambda$ is the modulus of dispersion vector.

Table 5. X-ray diffraction results of the PS and PS/C_{60} films

C_{60} content, wt. %	Δ*s*, nm^{-1}	*L*, nm
0	15.53	0.40
0.03	16.12	0.39
0.1	8.80	0.71
0.5	9.29	0.68
1.0	8.98	0.69

The values of L calculated by Eq. (6) are given in Table 5. The obtained data demonstrate the following. When the concentration of fullerenes in the film does not exceed 0.03 wt. %, the value of L is 0.39÷0.42 nm. Greater correlation length (L=0.68÷0.71 nm) occurs if concentration of C_{60} is 0.1÷1 wt. %. Thus insertion of fullerenes into polystyrene promotes "local ordering in the disordered phase". Note such conclusion is consistent with findings of ref. [9]. It found in ref. [9], if the content of fullerenes is more than 0.1 wt. %, the effects of the intermolecular interaction of fullerene with polystyrene dominate, which may result to cross-linking of the polymer chains.

SAXS

The obtained diffraction patterns in small angles were quantitatively analyzed within the Kratky–Porod model that was proposed to describe the conformational movements in high-molecular substances - persistent (wormlike) polymer chains including both very rigid and extremely flexible macromolecules [29]. We represented the experimental data in terms of Kratky coordinates (s, $s^2I(s)$), where $I(s)$ is the scattering intensity.

Figure 10 shows SAXS patterns for polystyrene films with various fullerene contents, and Figure 11 shows the same diffraction patterns in terms of Kratky coordinates. It is noted in ref. [29] for an ideal persistent chain, in the small-angle scattering curve represented in terms of Kratky coordinates we should observe a smooth transition from the Gaussian approximation (at small s values) to a "rigid rod" approximation (at large s values). We can see in Figure 11 like this smooth transition does not observe for the majority of composite films: the dependences mentioned have maxima and minima. The nonmonotonicity of these dependences is apparently caused by the limitation of the interior rotation angle of the polymer units [29, 30].

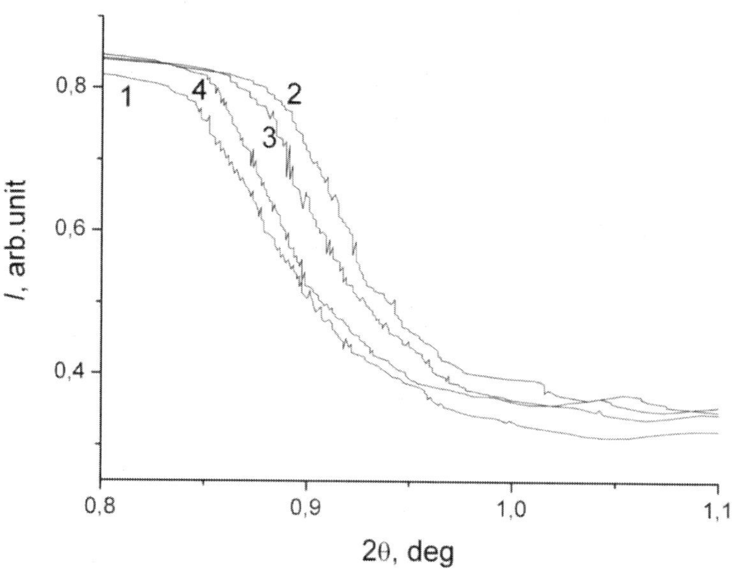

Figure 10. X-ray patterns of small-angle scattering depending on the fullerene concentration, wt. %: 0 (*1*); 0.0004 (*2*); 0.01 (*3*); 0.03 (*4*).

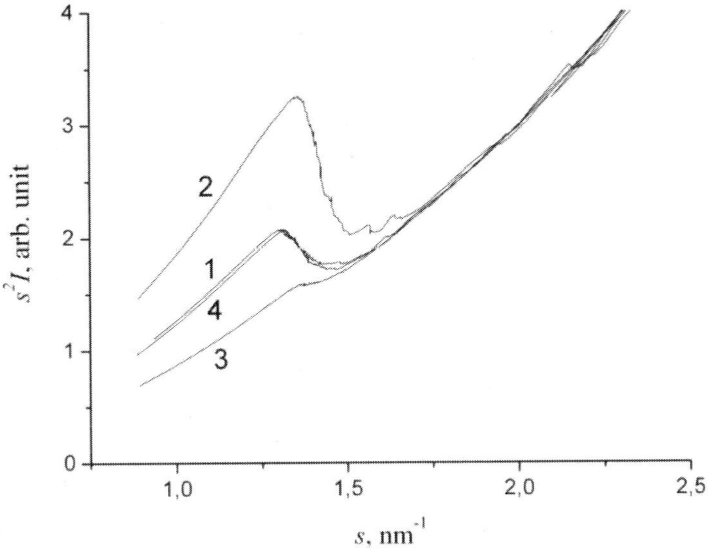

Figure 11. X-ray patterns of small-angle scattering in terms of Kratky coordinates depending on the fullerene concentration, wt. %: 0 (*1*); 0.0004 (*2*); 0.01 (*3*); 0.03 (*4*).

The exception is a composite containing 0.01 wt. % of fullerene (Figure 11, curve 3). The Kratky dependence corresponding to this film does not have the expressed extrema and is characterized by inflection. This composite, according to the model conceptions [29], does not seem to have the restrictions mentioned and all interior rotation angles are statistically equiprobable. The result obtained easily correlates with the conclusions drawn above (section 1) based on the data of thermochemical experiments that, of all the samples under investigation, composites with 0.01 wt. % of fullerene are characterized by greater mobility of units.

In the region of relatively large s values (s>1.75 nm^{-1}), the curves corresponding to various fullerene concentrations in polymer are linear and almost coincide. The dependence s–s^2I corresponds to scattering by rigid rod [29].

It can be seen in Figure 11) the point of transition to linear asymptotics, $s^*\approx 1.5$ nm^{-1}, is independent on film composition. This value allows one to estimate the persistent length of chain, a, and Kuhn segment, l if one uses the following equations [29]:

$$a = \frac{1.91}{s^*}, \quad (7)$$

$$l = 2a \quad (8)$$

For all the polymer compositions studied, the values of the chain persistent length and Kuhn segment were $a\approx 1.27$ nm and $l\approx 2.54$ nm. The data obtained correlate well with the literature. In particular, according to [16, 31], for unmodified polystyrene, the persistent length of the chain is 1.0÷1.4 nm depending on conditions.

4. Effect of Solvent on Optical Properties of Fullerene in Solutions and Polymer Composites Based on Polystyrene

Currently, there are a large number of articles devoted to the study of the electronic absorption spectra of fullerenes in aliphatic and aromatic solvents [32-36]. Authors of listed references identified the maxima of the absorption bands of C_{60} in solutions, measured extinction coefficients, found solvatochromic effect. It was shown C_{60} has four absorption maxima in the wavelength ranges 326÷340, 395÷400, 515÷517, 568÷570 nm differing in intensity. However, opinions of various researchers differ in regarding to the state of fullerene in solutions. In ref. [33] on the basis of the linear concentration dependence of the absorbance the authors suggest fullerene is as monomer form at very low concentrations of C_{60} (less than 10^{-5} mol/l). But for more concentrated solutions deviation from the law of Lambert-Beer-Bouguer found which explained by aggregation of the fullerene molecules [36].

The study of the fullerene electronic spectra in the polymer solutions is important due to the production method of polymer/fullerene composites with noncovalent bond. Such composites can be prepared by evaporation of the organic solvent solution that contains both components (polymer and fullerene). However, the interaction nature between polymers and fullerene in these materials has been insufficiently studied. The authors of ref. [37] are believed in the polystyrene/fullerene composites (1÷20 wt. % of C_{60}) prepared from solutions of benzene and toluene, it is possible the coordination bond of polymer macromolecules with fullerene generates and inhomogeneous C_{60}/polystyrene (or C_{60}/solvent) multi-adducts form. On the base of dielectrometrical study it is shown the fullerenes cluster state remains in the polystyrene/fullerene composite films [9]. Lack of systematic studies of physical and chemical properties of fullerene-containing composites formed from various solvents, does not allow to determine unambiguously the interaction nature between polymers and fullerene in such systems.

In present section we describe the findings on influence of the solvent nature on the optical properties of fullerenes in solutions and films. For clarification nature of the interaction between polymer and fullerene, we used the ultraviolet and infrared spectroscopy techniques.

Optical absorption spectra of fullerene in the organic solvents were recorded in the wavelength range 200÷500 nm in quartz cell (size is equal to 1.0 cm) by the Spectrophotometer U-2001 (Hitachi, Japan) with a working range of 190÷1100 nm. A deuterium lamp was as light source for UV zone; photometric accuracy is ±0,002 Abs. Optical absorption spectra of both polystyrene films and polystyrene/fullerene films were recorded by the Spectrophotometer U-2001 (Hitachi, Japan) with a reflection attachment Smart multi-bounce hatr, crystal is ZnSe 45°.

IR spectra of films were recorded by Avatar 360 FT-IR ESP spectrometer (Termo Nicolet, US).

Identification of interaction nature between fullerene and polymer in solutions and films was carried out by the UV- and IR-spectroscopy. Figure 12 shows the electronic absorption spectra of fullerene in slightly polar aromatic and aliphatic solvents. It can be seen for all media in the ultraviolet region of the fullerene spectrum there is an intense absorption band at λ=330-336 nm corresponding to the symmetry-allowed π-π* electronic transitions. In the visible region the weak bands are in the range of 400-500 nm.

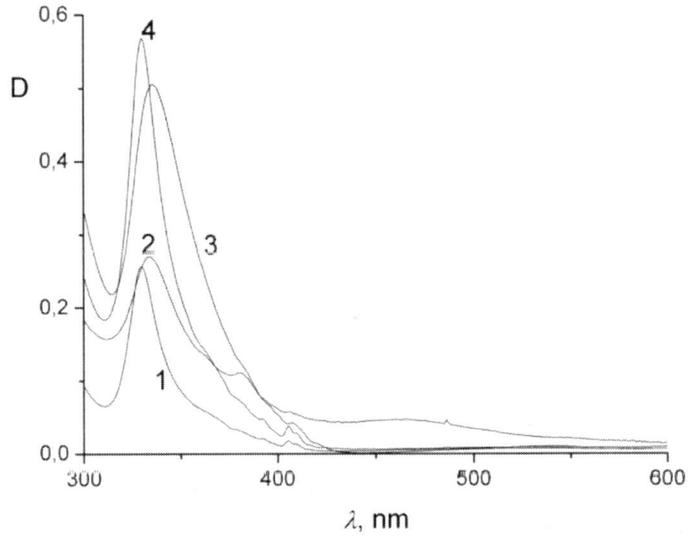

Figure 12. Fullerene electronic absorption spectra in various solvents: *1* - chloroform; *2* – toluene; 3 - o-xylene; 4 - carbon tetrachloride. Concentration of C_{60}+C_{70} is equal to 5.56 μmol/l.

Influence of the solvent nature on the optical spectra of fullerene is shown as solvatochromic effect. When replacing of aromatic solvents (o-xylene, toluene) by aliphatic ones (chloroform, carbon tetrachloride) there are hypsochromic shift by 6 nm for the band at λ=336 nm and the absorption intensity change.

An important factor affecting the reactivity of fullerenes in solutions is their liability to the association. We researched the electronic absorption spectra of fullerene in o-xylene when the C_{60}+C_{70} concentration range is 5÷60 μmol/l.

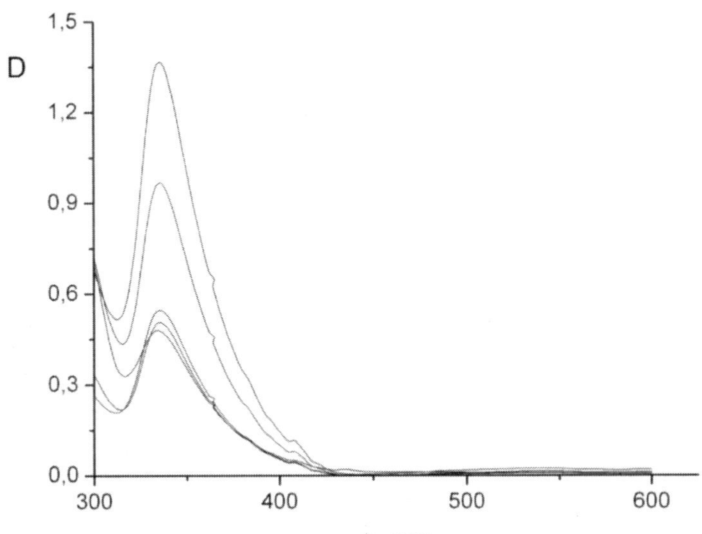

Figure 13. Fullerene electronic absorption spectra in o-xylene for the fullerene concentration in range from 5 to 60·μmol/l.

We founded both position of the bands maxima and their contours are not changed with the increase of the fullerene content in solution (Figure 13).

Concentration dependence of absorbance, *D*, for the fullerene solutions in o-xylene is linear (correlation coefficient is equal to 0.92) at λ=336 nm, which is consistent with the law of Lambert-Beer-Bouguer (Figure 14). That is the molar absorption coefficients are almost identical regardless of the fullerene concentration. These data indicate fullerenes in solutions are not in an aggregated form for this concentration range. It is likely that donor-acceptor complexes between the fullerenes molecules and the o-xylene molecules are generated, which prevents formation of typical crystalline fullerene structures in solution.

Optical properties of fullerene for different liquid systems without and with polystyrene are represented in Table 6. It can be seen there are changes in the electronic-absorption spectrum of fullerene when replacing of the C_{60}+C_{70} solutions in organic solvents by the systems containing polystyrene. Significant difference of the electronic-absorption spectrum of fullerene in the polymer-containing systems from the fullerene spectrum in organic solvents is the reduced intensity of the analytical absorption bands while maintaining the position of the peak. Such tendency is for all concentrations of C_{60}+C_{70}. Polystyrene does not have the strong absorption bands in this wavelength range.

It should be noted that the influence of the polymer on the optical properties of fullerene in chloroform manifests to a lesser extent as compared with other solvents (Table 6).

On the basis of the electronic spectra it can be assumed that in the polymer-containing systems the polystyrene macromolecules affect to the solvation shell of the fullerene to form intermolecular complexes with solvent.

Additional information about interaction nature of fullerene with polystyrene was obtained from the analysis of the vibrational spectra of the original and modified films. As an example, we analyzed the IR spectra of the original PS film and the modified PS films formed by casting from o-xylene (Figure 15). In the polystyrene IR spectrum the bands of the C=C stretching vibrations of the benzene ring appear as doublets in ranges of 1600-1585 cm^{-1} and 1500-1400 cm^{-1}.

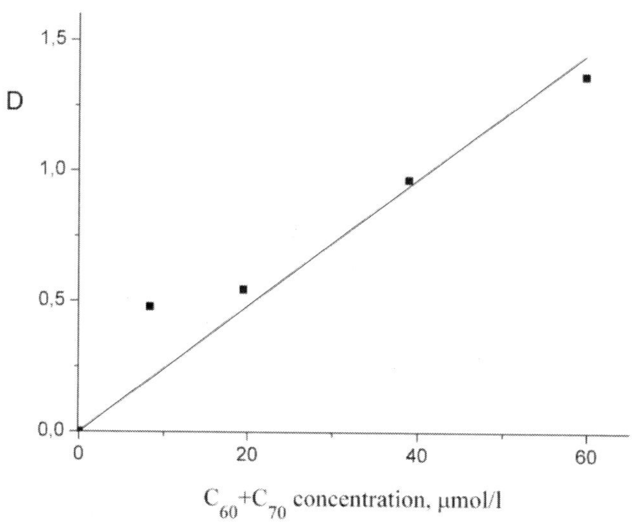

Figure 14. Concentration dependence of absorbance for the fullerene solutions in o-xylene; λ=336 nm.

Table 6. Optical properties for solutions of both C_{60}+C_{70} and PS/C_{60}+C_{70}

solvent	C_{60}+C_{70} content, μmol/l	C_{60}+C_{70}/solvent		PS/C_{60}+C_{70}/solvent	
		λ, nм	*D*	λ, nм	*D*
o-xylene	5.56	336 411	0.505 0.034	336 408	0.457 0.043
	8.33	334 408	0.516 0.048	334 408	0.479 0.041
	59.72	336 408	1.366 0.116	336 408	1.249 0.109
toluene	5.72	335 381 467	0.270 0.111 0.047	335 381 467	0.251 0.096 0.043
	8.72	335 382 470	0.638 0.223 0.132	335 382 470	0.463 0.208 0.095
chloroform	4.72	330 410	0.257 0.013	330 410	0.242 0.011
CCl_4	5.50	330 409	0.568 0.028	330 409	0.442 0.020

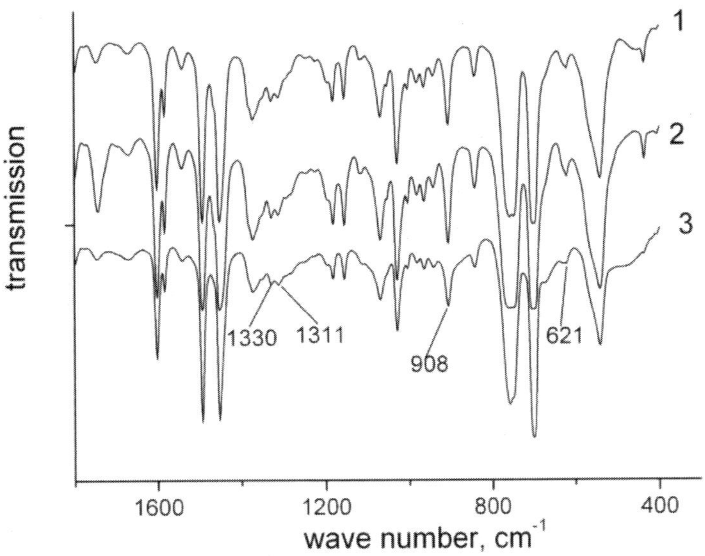

Figure 15. IR spectra for the films with various concentrations of C_{60}+C_{70}, wt. %: 0 (*1*); 0.02 (*2*); 0.035 (*3*).

The bands corresponding to planar deformation vibrations of C-H bonds are in the region of 1300-1000 cm^{-1}. Out-of-plane deformation vibrations of C-H bonds are at 900-675 cm^{-1} [38]. Bands corresponding to valent vibrations in the phenyl rings are usually very sensitive to the possibility of conjugation π-electron density of the ring with reactive groups. In the IR spectrum of the fullerene molecule four vibrations with the absorption bands corresponding to 527, 577, 1183, and 1429 cm^{-1} are active [23]. Absorption bands at 1429 cm^{-1} are the most sensitive to the charge transfer.

Table 7. The relative intensities of the absorption bands in the IR spectra of PS and composite films

C_{60}+C_{70} content, wt. %	D_{1452}/D_{908}	D_{1601}/D_{908}
0	2.44	2.13
0.01	2.19	2.08
0.035	3.16	2.28
0.10	2.81	2.31
1.0	3.45	2.40

When analysis of IR spectroscopy of the modified polystyrene samples we founded in spectra of the composites the new bands do not appear. However in spectrum of film containing 0.035 wt. % of C_{60}+C_{70} there are a change of contours of the absorption bands in the range of 700÷400 cm^{-1} and displacement of the band maximum at 621 cm^{-1} by 7 cm^{-1} toward the higher frequencies. Quantitative interpretation of IR spectra is difficult due to overlapping of the bands corresponding to the vibrations of the phenyl ring and fullerene (1451 and 1429 cm^{-1}, respectively). Therefore semi quantitative analysis of IR spectra of the studied films by the method of a base line and internal standard [39] was carried out. We determined the ratios of absorbance for characteristic bands to absorbance for the C-H bond at 908 cm^{-1} selecting as the internal standard (Table 7).

Analysis showed in the IR spectra of the composite a significant change in modes relative intensities compared to the same parameters in the spectrum of polystyrene film is observed. This indicates the electronic structure of the phenyl rings varies considerably. We can assume non-covalent interaction of donor polystyrene macromolecule with fullerene acceptor molecule occurs.

It is known [38] that the molecular chains of polystyrene possess helical conformation that attributes by doublet at 1330 and 1311 cm^{-1} in the vibrational spectrum of the polymer. In the spectra of fullerene-containing films, change in the ratio of intensities of these bands occurs. The observed effect is probably associates with a change in helical conformation of polystyrene macromolecules under the influence of the fullerene.

5. Adsorption Kinetics of the Heavy Metal Ions by Original Polystyrene and Polystyrene/Fullerene Composite Films

Various forms of fullerene (C_{60}, C_{70}, fullerene nanotubes, fullerene soot) exhibit adsorption activity to organic solvents [40, 41] and metal ions [42]. This property of nanoparticles is used in modification of the surface of polymer materials with the purpose of increasing their surface activity [43, 44]. In this section we report on the effect of modification of polystyrene with fullerenes on its sorption properties with respect to heavy metal ions. Findings have very important practical significance because of styrene/divinylbenzene copolymers are raw materials for commercial manufacturing cationites, and the development of sorbent based on polystyrene filled with fullerenes may allow significant extension of the choice of sorbents.

The modification of polystyrene with fullerene may be suggested to lead to an increase in the specific surface of the produced films and to an enhancement in the sorption activity of the composites. In addition, the adsorption characteristics of composite can be used for quantitative assessment of how the properties of the polymer surface alter after modification.

Kinetic characteristics are one of the main criteria in the selection of sorbents for extracting metal ion out of various aqueous solutions. In this section the sorption equilibrium time was determined by us on base of the kinetic curves which characterize the increase in the metal ions concentration in the original polystyrene film and in the samples filled with fullerene C_{60} occurring during sorption.

The sorption properties of polystyrene and modified polystyrene films with respect to Cu(II), Ni(II), Zn(II), and Cd(II) ions were studied under static conditions at 293 K. The sorption kinetics of heavy metal ions out of aqueous solutions of sulfates by the original and modified polystyrene films was studied by ion exchange between the sorbent and a solution in restricted volume [45].

Polymer batches of 0.05 g each were placed into test tubes; then, 5 ml of an aqueous metal sulfate solution were added to each of them and held under stirring for from 10 min to 1 day. The initial concentration, C_0, of heavy metal ions in aqueous solutions was in range of $(7 \div 9) \cdot 10^{-5}$ mol/l. Then after a certain periods of time, t, the solutions were filtered and the residual concentrations of metal ions, C_t, were determined using Saturn 3P-1 spectrometer (Belarus). The amount of metal ions adsorbed for a time t, A_t, was calculated taking into account the magnitudes of both polymer batches mass and solution volume in accordance with the following formula:

$$A_t = 0.1\left(C_0 - C_t\right) \text{, mol/g} \tag{9}$$

Using experimental results obtained we plotted the kinetic dependences that characterize sorption of copper ions by the original polystyrene and by the composite films with various amounts of C_{60} (Figs 16 and 17). It can be seen the most amount of Cu(II) ions is absorbed by the original polystyrene film for approximately 35 min and then this value almost does not change for 24 h of polymer being in contact with solution (Figure 16, curve 1). Doping of 0.1 wt. % of fullerene into polystyrene matrix facilitates a certain improvement in the kinetic characteristics of the composite in comparison with the characteristics of the original polystyrene film, which is confirmed by the reduction in the equilibrium time down to 20 min (Figure 16, curve 5).

It can be seen in Figure 16 the sorption activity of the composites depends on the fullerene content in the film. The composite containing 0.01 wt. % of C_{60} and the original polystyrene film absorb the nearly equal amount of Cu(II) ions. As the fraction of carbon nanoparticles increases to 0.03 wt. %, the efficiency of sorption grows in 4.5 times in comparison with the original polystyrene film. Further increase in the fullerene concentration up to 0.1 wt. % has almost no influence on the efficiency of the sorption process.

For the original polystyrene the extraction degree of Cu(II) ions is equal to 12 %, and it is equal to 60 % in case of sorption by the fullerene-containing polymer. Most likely, the improvement in the sorption-kinetic characteristics of the modified polymer occurs due to the increase in the specific surface of the composites and the growth in number of active sorption sites.

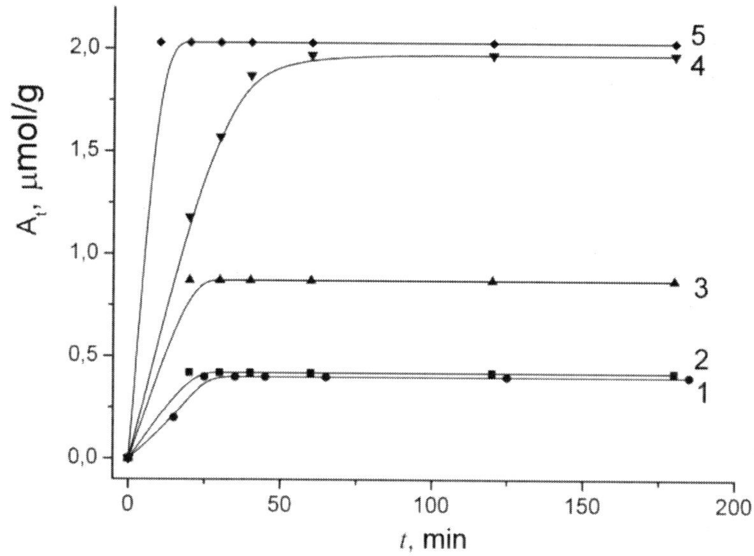

Figure 16. Adsorption kinetics of Cu(II) ions by polystyrene/fullerene composite films containing various amount of C_{60}, wt. %: 0 (*1*); 0.01 (*2*); 0.02 (*3*); 0.03 (*4*); 0.1 (*5*).

The selectivity of the composite with 0.1 wt. % of C_{60} with respect to various metal ions is illustrated by the kinetic curves of sorption (Figure 17). According to these data the composite sorbs the copper and cadmium ions most effectively and exhibits less activity with respect to the zinc ions. It should be note that the rate of sorption of Ni(II), Zn(II), and Cd(II) ions is somewhat lower than of Cu(II) ions. The selectivity of sorption by this composite is represented with the following order: Cu(II)≥Cd(II)>Ni(II)>Zn(II).

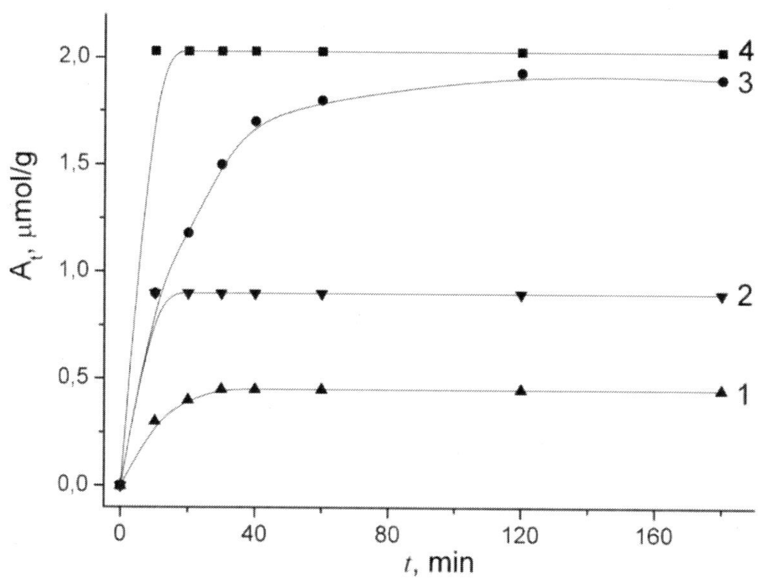

Figure 17. Adsorption kinetics of the various metal ions by polystyrene/fullerene composite films containing 0.1 wt. % of C_{60}: *1* – Zn(II); *2* – Ni(II); *3* – Cd(II); *4* – Cu(II).

Thus it was found that filling of polystyrene with C_{60} particles results the composite films with improved sorption-kinetic characteristics with respect to heavy metal ions, which can be recommended for the extraction of trace heavy metals out of aqueous solutions of electrolytes and wastewater.

6. Antioxidant Activity of Polystyrene and Polystyrene/Fullerene Nanocomposites

Nowadays a wide specter of fullerene derivates is synthesized, having anticancer, antiviral, antibacterial, neuroprotective and antioxidant activities [46]. Biological abilities of fullerenes are due to lipophilic properties which facilitate cell penetration and lack of electrons, helping to react with free radicals and generate active oxygen species [47]. The mechanism of biological role of fullerene is shown to depend of its aggregate form: crystalline, colloid or soluble organic complex [48, 49]. Soluble organic fullerene complex has highest bioactivity [48]. The authors explain this effect by low association of nanocarbonic particles. It is considered that integration of fullerenes into polymer matrix can produce biocomposites which have medical potential as drug transporters, antiseptics and antioxidants. Today regulation of free-radical processes is adjusted by both natural and synthetic pharmaceutical compositions [50]. As any other medicine some antioxidants may produce adverse events. So finding of safe preparations with high antioxidant activity is still actual. We researched influence of the PS/C_{60}+C_{70} nanocomposites on free-radical oxidation of lipids in biologic fluid (blood serum) in vitro.

Subject of research was native blood serum of 10 patients managed in V.N. Gorodkov Research Institute of Maternity and Childhood (Ivanovo, Russia). Specimen of original or composite film (size 1.5 cm^2, weight 5 mg) with certain fullerene concentration (0.01, 0.03, 0.10 or 1.00 wt. %) was put into blood serum (1 ml). System was incubated for 1 hour at 4°C. Then, the film specimen was removed from biologic fluid.

The parameters of lipid peroxidation in serum after exposure of the film nanomaterials were determined by chemiluminescent analysis and spectrophotometry.

The induced chemiluminescence (ChL) tests were performed by BChL-07 luminometer (Medozons, Russia). We used hydrogen peroxide and ferric sulfate as inductors of ChL. 0.1 ml of serum, 0.4 ml of phosphate buffer (pH 7.5), 0.4 ml of 0.01M ferric sulfate and 0.2 ml of 2 % hydrogen peroxide were put into cuvette. Luminescence was registered for 40 s.

To estimate the intensity of lipid peroxidation we used following parameters:

- J_{max} is maximum intensity of ChL during the experiment. Value of J_{max} quantifies the level of free radicals, i.e., gives an idea of the potential ability of the blood serum to free radical lipid peroxidation;
- $tan\alpha$ is tangent of maximum slope angle of ChL curve towards time axis. This value characterizes the decay rate of free radical oxidation, i.e., quantifies an effectiveness of the antioxidant system;
- A is an area covered by intensity curve or total light sum. Value of A is inversely proportional to the antioxidant activity of the sample;
- $Z=AJ_{max}^{-1}$ is normalized light sum.

Free radical processes in serum have been studied after exposure of original polystyrene films and fullerene-containing polystyrene films. The mean values of ChL parameters in native serum without film adding were used as controls. 8-10 measurements required for each film were carried out on the same day. The results have been expressed as percentages relative to controls and were given as mean values ± standard deviations. A p-value of 0.05 was chosen as the significance limit.

Also the lipid peroxidation reaction was identified by SF-46 spectrophotometer (Russia) at wave length of 532 nm. Malonic dialdehyde (MDA) was estimated as peroxidation derivate by its complexing with 2-thiobarbituric acid [51]. Total antioxidant reactivity (TAR) was evaluated by measuring the MDA concentration before and after incubation of samples [52].

Figure 18 shows kinetics of chemiluminescence in serum after exposure of original polystyrene and nanocomposite films. Peak of chemiluminescence due to free radical production was in 2 s of reaction. This can be explained by production of active oxygen species (HO_2^*, O_2^*, O_2^-, OH^-). The highest intensity, J_{max}, was registered in case of nanocomposites with 0.01 and 0.03 wt. % of fullerenes.

In Table 8 we represent the main ChL parameters for films prepared by casting of o-xylene solution. It can be seen in case of original polystyrene film the ChL parameters were approximate to controls. Value of J_{max} in case of PS/C_{60}+C_{70} composites is higher than for control serum samples. A light sum, A, was significantly increased only for films with 0.01 and 0.03 wt. % of fullerenes (p<0.05). For nanocomposite containing 1 wt. % of C_{60}+C_{70} no significant changes in value of A was revealed. In addition we found both significant growth in value of tanα and reduction in value of Z in case of all fullerene-containing films (Table 8). So regardless of the fullerenes content the antioxidant activity of PS/C_{60}+C_{70} composites is higher than for original polystyrene.

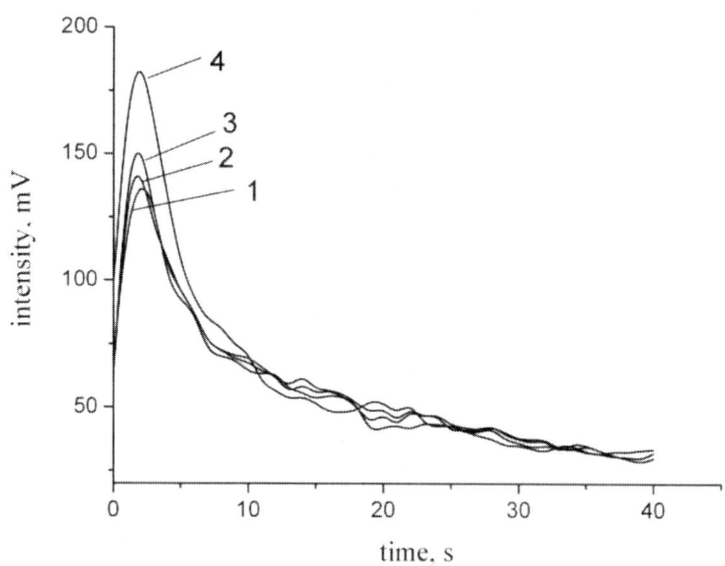

Figure 18. Kinetic chemiluminescence profiles of native blood serum (*1*) and after exposure of studied materials (*2-4*): *2* - PS; *3* – PS/C_{60}+C_{70} (1.0 wt. %); *4* - PS/C_{60}+C_{70} (0.03 wt. %). Solvent: o-xylene.

It seems nanocomposites containing fullerenes were easy to react with oxygen species, preventing lipid peroxidation. We suggest the fullerene molecules can cross the external cellular membrane and they localize to the mitochondrions. Formation of oxygen free radicals occur due to the electrons leakage in the mitochondrial electron transport chain. Therefore, the localization of fullerenes near mitochondrions may contribute to their antioxidant action.

Note similar assumption has been proven for fullerene derivative $C_{61}(CO_2H)_2$ [53]. The authors of ref. [53] researched distribution of the [^{14}C] radioactivity in various cellular compartments and concluded $C_{61}(CO_2H)_2$ molecules localize preferentially to the mitochondrions.

Intensity of lipid peroxidation was also estimated by malodic dialdehyde concentration and total antioxidant reactivity assessed by spectrophotometry (Table 9). We revealed that nanocomposite with 0.03 wt. % of C_{60}+C_{70} increased MDA level in blood serum (p<0.05). The increasing of MDA level indicates the activation of lipid peroxidation and the accumulation of free radicals in the blood serum. But the film specimens with higher fullerene content decreased this parameter. Note the similar trend correlates with dependence of J_{max} value on fullerene concentration in film (Table 8).

It can be seen in Table 9 total antioxidant reactivity was increased in serum samples after exposure of nanocomposites with 0.03, 0.10 and 1.00 wt. % of fullerenes (p<0.05). This proved antioxidant effect of experimental materials. It appears the amount of active centers, which able to effectively capture and inactivate the free radicals, increases with concentrations of fullerenes in composite material.

Table 8. Chemiluminescence parameters in blood serum after exposure of original polystyrene film and fullerene-containing composites (solvent: o-xylene)

Fullerene concentration, wt. %	*A*, %	J_{max}, %	tanα, %	*Z*, %
Controls	100.0	100.0	100.0	100.0
0	96.0±9.0	97.0±7.0	97.5±10.5	99.0±11.5
0.01	121.5±10.5 *	131.5±11.5 *	139.0±27.0 *	92.0±6.5 *
0.03	115.0±8.0 *	125.0±11.0 *	146.0±27.0 *	92.0±7.0 *
0.1	105.0±3.0	110.0±6.0 *	110.0±8.0 *	95.5±5.0
1.0	96.0±12.0	111.0±6.0 *	121.0±7.0 *	86.5±7.0 *

* significant differences compared to control (p<0.05).

Table 9. Lipid peroxidation parameters (MDA, TAR) in blood serum after exposure of original polystyrene film and fullerene-containing composites (solvent: o-xylene)

Fullerene concentration, wt. %	MDA	TAR
Controls	100.0	100.0
0	103.0±23.0	108.0±4.0
0.03	113.0±5.0 *	116.0±5.0 *
0.1	94.0±8.0	110.0±4.5 *
1.0	96.5±10.5	107.0±4.0 *

* significant differences compared to control (p<0.05).

Table 10. Chemiluminescence parameters in blood serum after exposure of original polystyrene film and fullerene-containing composites (solvent: toluene)

Fullerene content, wt. %	A, %	J_{max}, %	tanα, %	Z, %
Controls	100.0	100.0	100.0	100.0
0	98.0±7.5	99.0±7.0	98.5±7.5	100.0±7.0
0.1	101.5±10.5	120.5±14.0 *	100.5±10.5	84.0±8.5 *
0.5	103.0±7.5	101.0±7.5	130.0±14.0 *	102.0±5.0

* significant differences compared to control ($p<0.05$).

It is interesting to reveal the effect of the medium in which the films were fabricated. For this we performed experiments with films prepared by casting of other aromatic compound – toluene.

The main ChL parameters for "toluene" films are given in Table 10. It can be seen that in case of original polystyrene film all values were approximate to controls. But using nanocomposites (0.5 wt. % of $C_{60}+C_{70}$) we found significant increase in value of tanα. It demonstrates the antioxidant activity of researched $PS/C_{60}+C_{70}$ composites.

In addition it can be seen in Tables 8 and 10 at the same concentration of fullerene (0.1 wt. %), the value of tanα is higher for film formed by casting of o-xylene than for "toluene" films. This can be explained by that the solubility of fullerenes in o-xylene is higher than in toluene [54]. It appears that in toluene solution the fullerene molecules are in the clusters form, which does not ensure uniform distribution of the nanoparticles in the composite film during its formation of solution.

Note also we had preliminary experiments with $PS/C_{60}+C_{70}$ composite films that were fabricated by casting of aliphatic compound – chloroform. The results of chemiluminescent analysis and spectrophotometry for serum samples after exposure of both original polystyrene film and composite films regardless of the fullerenes content were approximate to controls. This again emphasizes the significance of the medium in which the films were fabricated.

So our study proved that polystyrene/fullerene nanocomposites have ability to activate lipid peroxidation in blood serum. Moreover the possibility of such activation depends on the composite forming conditions.

Conclusion

The results of studies on the physical and chemical properties as well as biological activity of polystyrene/fullerene composites for development of new functional materials show certain progress in this field.

But in spite of undoubted achievements on the fabrication of polymeric composites containing C_{60}, numerous scientific and applied problems are still remain. The major problem is using fullerenes as modifiers of the polymer accompanied by their aggregation. The aggregation complicates the uniform distribution of nanoparticles in the polymer matrix required for fabrication of materials with the optimum properties. In particular, this phenomenon is observed at the high concentrations C_{60} in the composites. It has to be taken into account at fullerene using.

Many researches concluded that an improvement in physical and chemical properties of composites compared to ones of original polymers is related to structural changes under the effect of introduced fullerene. However, the mechanism of these improvements is still far from clear. Therefore the effect of fullerene doping on phase transition from the glassy state to elastic one, dielectric parameters, supramolecular structures, biological activity etc requires complex analysis. The prolongation of these studies is very important, since the use of the polystyrene/fullerene composites will allow expansion of the applications borders of polystyrene traditionally used in construction industry, home appliances, electronics etc and will allow them to be replaced with fullerene-containing materials.

ACKNOWLEDGMENTS

The study was supported by the Russian Foundation for Basic Research (project no. 12-03-97528-a).

REFERENCES

[1] Uğur, Ş.; Yargi, Ö.; Pekcan, Ö. Conductivity percolation of carbon nanotubes (CNT) in polystyrene (PS) latex film. *Canad. J. Chem.* 2010, 88, 267-276.

[2] Rana, S.; Bhattacharyya, A.; Parveen, S.; Fangueiro, R.; Alagirusamy, R.; Joshi, M. Processing and performance of carbon/epoxy multi-scale composites containing carbon nanofibres and singlewalled carbon nanotubes. *J. Polym. Res.* 2013, 20, 314.

[3] Badamshina, E.R.; Gafurova, M.P. Characteristics of fullerene C_{60}-doped polymers. *Polym. Sci. B.* 2008, 50, 215–225.

[4] Kroto, H.W.; Heat, J.R.; O'Brien, S.C.; Curl, R.F.; Smalley, R.E. C_{60}-Buckminsterfullerene. *Nature.* 1985, 318, 162–163.

[5] Verner, R.F.; Benvegnu, C. (eds). *Handbook on fullerene: synthesis, properties and applications.* Nova Science Publishers, Inc. New York, 2012.

[6] Coetzee, J.F. (ed.). *Recommended methods for purification of solvents and tests for impurities.* Pergamon Press. Oxford, 1982.

[7] Sabban, J.R.; Xu-wu, A.; Chichos, J.S.; Planas Leitão, M.L.; Roux, M.V.; Torres, L.A. Reference materials for calorimetry and differential thermal analysis. *Thermochim. Acta.* 1999, 331, 93-204.

[8] Weng, D.; Lee, H.K.; Levon, K.; Mao, J.; Scrivens, W.A.; Stephens, E.B.; Tour J.M. The influence of Buckminsterfullerenes and their derivatives on polymer properties. *Eur. Polym. J.* 1999, 35, 867-878.

[9] Gladchenko, S.V.; Polotskaya, G.A.; Gribanov, A.V.; Zgonnik, V.N. The study of polystyrene-fullerene solid-phase composites. *Tech. Phys. Russ. J. Appl. Phys.* 2002, 47, 102-106.

[10] Alekseeva, O.V.; Bagrovskaya, N.A.; Kuzmin, S.M.; Noskov, A.V.; Melikhov, I.V.; Rudin, V.N. The influence of fullerene additives on the structure of polystyrene films. *Russ. J. Phys. Chem. A.* 2009, 83, 1170-1175.

[11] Alekseeva, O.V.; Rudin, V.N.; Melikhov, I.V.; Bagrovskaya, N.A.; Kuz'min, S.M.; Noskov, A.V. Kinetics of formation of hierarchical nanostructures in polystyrene films containing fullerene. *Dokl. Phys. Chem.* 2008, 422, 275-278.

[12] Skanavi, G.L. *Fizika dielektrikov (Oblast' slabykh polei) (Physics of dielectrics (Weak fields))*: Gostekhizdat. Moscow, 1949 (in Russian).

[13] Adamopoulos, G.; Heiser, V.; Giovanella, U.; Ould-Saad, S.; van de Wetering, K.L.; Brochon, C.; Zorba, T.; Paraskevopoulos, K.M.; Hadziioannou, G. Electronic transport properties aspects and structure of polymer-fullerene based organic semiconductors for photovoltaic devices. *Thin Solid Films.* 2006, 511-512, 371-376.

[14] Nourdine, A.; Perrin, l.; de Bettignies, R.; Guillerez, S.; Flandin, L.; Alberola, N. Synthesis and characterization of fullerene. *Polymer.* 2011, 52, 6066-6073.

[15] Stankovich, S.; Dikin, D.A.; Dommett, G.H.B.; Kohlhaas, K.M.; Zimney, E.J.; Stach, E.A.; Piner, R.D.; Nguyen, S.B.T.; Ruoff, R.S. Graphene-based composite materials. *Nature.* 2006, 442, 282-286.

[16] Tugov, L.I.; Kostrykina, G.L. *Khimiya i fizika polimerov (Polymer chemistry and physics).* Khimiya. Moscow, 1989 (in Russian).

[17] Sihvola, A. Mixing rules with complex dielectric coefficients. *Subsurf. Sens. Technol. Appl.* 2000, 1, 393-415.

[18] Badamshina, E.; Gafurova, M. Polymeric nanocomposites containing non-covalently bonded fullerene C_{60}: properties and applications. *J. Mater. Chem.* 2012, 22, 9427–9438.

[19] Jurkowska, B.; Jurkowski, B.; Kamrowski, P.; Pesetskii, S.S.; Koval, V.N.; Pinchuk, L.S.; Olkhov, Y.A. Properties of fullerene-containing natural rubber. *J. Appl. Polym. Sci.* 2006, 100, 390-398.

[20] Potalitsin, M.G.; Babenko, A.A.; Alekhin, O.S.; Alekseev, N.I.; Arapov, O.V.; Charykov, N.A.; Nekrasov, K.V.; Gerasimov, V.I.; Semenov, K.N. Caprolons modified with fullerenes and fulleroid materials. *Russ. J. Appl. Chem.* 2006, 79, 306-309.

[21] Olkhov, Y.A.; Jurkowski, B. Effect of fullerenes on the structure and properties of linear and crosslinked polyesterurethane ureas. *J. Appl. Polym. Sci.* 2007, 104, 1431-1442.

[22] Ginzburg, B.M.; Tuichiev, S.; Shukhiev, S. Permittivity of low-concentration c_{60} fullerene solutions in p-xylene. *Tech. Phys. Lett.* 2009, 35, 491-493.

[23] Konarev, D.V.; Lyubovskaya, G.N. Donor–acceptor complexes and radical ionic salts based on fullerenes. *Russ. Chem. Rev.* 1999, 68, 19-38.

[24] Kao, K.C.; Hwang W. *Electrical transport in solids. With particular reference to organic semiconductors.* Pergamon Press. Oxford-New York-Toronto-Sydney-Paris-Frankfurt, 1984.

[25] Krätschmer, W.; Lamb, L.D.; Fostiropoulos, K.; Huffman, D.R. Solid C_{60}: a new form of carbon. *Nature.* 1990, 347, 354-358.

[26] Ginzburg, B.M.; Pozdnyakov, A.O.; Shepelevskij, A.A.; Melenevskaya, E.Y.; Novoselova, A.V.; Shibaev, L.A.; Pozdnyakov, O.F.; Redkov B.P.; Smirnov A.S.; Shiryaeva O.A. Structure of fullerene C_{60} in a poly(methyl methacrylate) matrix. *Polym. Sci. A.* 2004, 46, 169-175.

[27] Guinier, A. *X-ray diffraction: In crystals, imperfect crystals, and amorphous bodies.* Dover Publications. New York, 2001.

[28] N'Dri, K.; Houphouet-Boigny, D.; Jumas, J.-C. Study of first sharp diffraction peak in As_2S_3 glasses by X-ray powder diffraction method. *J. Non-Oxide Glasses.* 2012, 3, 29-37.

[29] Feigin, L.A.; Svergun, D.I. *Structure analysis by small-angle X-ray and neutron scattering*. Plenum Press. New York-London, 1987.

[30] Kirste, V.R.G. Neue vorstellungen über statistische fadenknäuel. *Die Makromolekulare Chemie*. 1967, 101, 91-103 (in German).

[31] Grosberg A.Yu.; Khokhlov, A.R. *Statistical physics of macromolecules*. American Institute of Physics. New York, 1994.

[32] Hare, J.P.; Kroto, H.W.; Taylor, R. Preparation. *Chem. Phys. Lett.* 1991, 177, 394-398.

[33] Honeychuck, R.V.; Cruger, T.W. Behavior of Buckminsterfullerene (C_{60}) in chlorobenzene with regard to Beer's law in the visible and ultraviolet regions. *Anal. Lett.* 1992, 25 1755-1763.

[34] Gun'kin, I.F.; Loginova, N.Yu. Aggregation of fullerene C_{60} in o-dichlorobenzene. *Russ. J. Gen. Chem*. 2006, 76, 1914-1915.

[35] Gun'kin, I.F.; Loginova, N.Yu. Effect of nature of organic solvent on the absorption spectrum of C_{60} fullerene. *Russ. J. Gen. Chem*. 2006, 76, 1911-1913.

[36] Ponomarev, A.N.; Yudovich, M.E.; Nikitin, V.A.; Nikitin, D.V.; Barchenko, V.T.; Sobol', V.N.; Strel'nikov, K.B.; Charykov, N.A.; Myshkov, A.I.; Sergeev, V.I. Some features of analysis of solutions of fullerenes C_{60} and C_{70} by their absorption spectra. *Opt. Spectrosc.* 2000, 88, 195–196.

[37] Yevlampieva, N.P.; Zaitseva, I.I.; Ryumtsev, E.I.; Dmitrieva, T.S.; Melenevskaya, E.Yu. Interaction of polystyrene and fullerene C_{60} in benzene: composition and molecular properties of the product. *Polym. Sci. A*. 2007, 49, 284-291.

[38] Dechant, J.; Danz, R.; Kimmer, W.; Schmolke. R. *Ultraspectroscopische untersuchungen an polymeren*. Akademie Verlag. Berlin, 1972 (in German).

[39] Smith, A.L. *Applied infrared spectroscopy. Fundamentals, techniques, and analytical problem-solving*. Wiley. New York, 1979.

[40] Samonin, V.V.; Slutsker, E.M. The kinetics of adsorption of organic solvent vapors on fullerene blacks. *Russ. J. Phys. Chem. A*. 2005, 79, 87-90.

[41] Samonin, V.V.; Marakulina, E.A. Adsorption properties of fullerene-containing materials. *Russ. J. Phys. Chem. A*. 2002, 76, 787-790.

[42] Li, Y.-H.; Ding, J.; Luan, Z.; Di, Z.; Zhu, Y.; Xu, C.; Wu, D.; Wei, B. Competitive adsorption of Pb^{2+}, Cu^{2+} and Cd^{2+} ions from aqueous solutions by multiwalled carbon nanotubes. *Carbon*. 2003, 41, 2787-2792.

[43] Podosenova, N.G.; Sedov, V.M.; Kuznetsov, A.S.; Knyazev A.S. New sorbents for electron-exchange adsorption of low-density lipoproteins. *Russ. J. Phys. Chem. A*. 1999, 73, 97-100.

[44] Podosenova, N.G.; Sedov, V.M.; Sharonova, L.V.; Drichko N.V. Fullerene effects on the adsorption properties of silica gel with respect to low-density lipoproteides. *Russ. J. Phys. Chem. A*. 2001, 75, 1871-1875.

[45] Saldadze, K.M.; Kopylova-Valova, V.D. *Kompleksoobrazuyushchie ionity (kompleksity) (Complexing Ionites (Complexites))*. Khimiya. Moscow, 1980 (in Russian).

[46] Andreev, I.; Petrukhina, A.; Garmanova, A.; Babakhin, A.; Andreev, S.; Romanova, V.; Troshin, P.; Troshina, O.; DuBuske, L. Penetration of fullerene C_{60} derivatives through biological membranes. *Fuller. Nanotub. Car. N.* 2008, 16, 89-102.

[47] Da Ros, T. Twenty years of promises: fullerene in medicinal chemistry. In: Cataldo, F.; Da Ros, T. (eds) *Carbon Materials: Chemistry and Physics, vol. 1: Medicinal Chemistry and Pharmacological Potential of Fullerenes and Carbon Nanotubes.* Springer, 2008, 1-21.

[48] Piotrovskiy, L.B.; Eropkin, M.Y.; Eropkina, E.M.; Dumpis, M.A.; Kiselev, O.I. Mechanisms of biological activity of fullerenes – relation to aggregate state. *Psychopharmacol. Biol. Narcol.* 2007, 2, 1548-1554 (in Russian).

[49] Lyon, D.Y.; Adams, L.K.; Falkner, J.C.; Alvarez, P.J. Antibacterial activity of fullerene water suspensions: effects of preparation method and particle size. *Environ. Sci. Technol.* 2006, 40, 4360-4366.

[50] Okovitiy, S.V. Clinical pharmacology of antioxidants. *FARMindex: Praktik.* 2003, 5, 85-111 (in Russian).

[51] Ishihara, M. Studies on lipoperoxide of normal pregnant women and patient toxemia of pregnancy. *Clin. Chim. Acta.* 1978, 84, 1-9.

[52] Promyslov, Sh.M.; Demchuk, M.L. A modified procedure for estimation of total antioxidant activity of blood serum. *Voprosy Meditsinskoi Khimii.* 1990, 36, 90-92 (in Russian).

[53] Foley, S.; Crowley, C.; Smaihi, M.; Bonfils, C.; Erlanger, B.F.; Seta, P.; Larroque, C. Cellular localisation of a water-soluble fullerene derivative. *Biochem. Biophys. Res. Commun.* 2002, 294, 116–119.

[54] Zhou, X.; Liu, J.; Jin, Z.; Gu, Z.; Wu, Y.; Sun, Y. Solubility of fullerene C_{60} and C_{70} in toluene, o-xylene and carbon disulfide at various temperatures. *Fullerene Sci. Techn.* 1997, 5, 285-290.

In: Fullerenes
Editor: Shannon B. Ellis

ISBN: 978-1-63321-385-2

Chapter 4

^{13}C NMR IN FULLERENE CHEMISTRY: STRUCTURE/SHIFT RELATIONSHIP AND QUANTUM-CHEMICAL PREDICTIONS

Arthur R. Tulyabaev and Leonard M. Khalilov*
Institute of Petrochemistry and Catalysis,
Russian Academy of Sciences, Ufa, Russia

ABSTRACT

Since the discovery of a fullerene C_{60}, almost thirty years have passed. During this time, a great number of down-to-earth approaches to the synthesis of fullerene derivatives that constitute a new class of organic and organometallic compounds have been described in the worldwide literature. For structure elucidation of new fullerene adducts, a number of state-of-the-art physicochemical means is widely used, among which ^{13}C NMR possesses the most informative value. Using NMR spectra, regioselectivity and the type of addition to a fullerene core, as well as the symmetry set point can be ascertained with high probability. As the literature shows, these characteristics are determined by a number, intensity and position of the resonances in the ^{13}C NMR spectra. As a rule, no exact and precision assignments of the signals to the individual fullerene carbon atom are achieved restricting only by their enumeration. From a different standpoint, the articles that describe how these problems have been solved require special attention. INADEQUATE NMR technique, for which samples with high concentration (200 mg and more) or labeled with the ^{13}C isotope are necessary was successfully applied in this regard. In this chapter, the results of the examination of the relationship between the ^{13}C NMR chemical shift and structure of fullerenes including their higher derivatives are summarized and discussed. These compounds are mono-adducts from [2+n] cycloaddition (where n=1-4), as well as those obtained by diverse radical and nucleophilic addition reactions. It is also an attempt to review a worldwide and own experience in the use of the quantum chemistry approaches to calculate and ascribe the signals in the ^{13}C NMR spectra of fullerene derivatives. We believe that detailed knowledge of the fullerene carbon shielding regularities will create additive rules for

* Tel.: +7(347)2842750, e-mail: khalilovlm@gmail.com.

exact and unequivocal prediction of the ^{13}C NMR chemical shifts, as well as electronic, optical, and magnetic features of fullerene derivatives.

Keywords: Fullerene C_{60}, ^{13}C NMR spectroscopy, INADEQUATE NMR, ^{13}C NMR chemical shift, ^{13}C-enrichment, mono-adducts, bis-adducts, quantum chemistry calculations

1. Introduction

Discovery of a fullerene C_{60} in 1985 facilitated appearance of a new class of organic compounds [1-4]. In the review article (Kroto, Allaf and Balm, 1991), the experimental and theoretical studies on carbon materials, as well as both fullerenes C_{60} and C_{70} are widely discussed [5]. The topological and metric characteristics of polyhedral carbon clusters, quantum chemistry of fullerenes, the mechanism for the formation of fullerenes and endohedral complexes, classification of the carbon forms having regular structure, and chemical properties of carbon clusters have been considered by Sokolov and Stankevich in 1993 [6]. Eletskii and Smirnov have reviewed studies on fullerene structure, the methods for their generation, the processes involving them, and the formation of fullerenes in natural gaseous systems [7]. Wanda Andreoni has focused attention on the results obtained with different computational approaches to the physical chemistry of C_{60} and C_{70}, and their specific derivatives in comparison with experimental data [8]. An issue on spherical fullerene aromaticity has been circumstantially considered by Buhl and Hirsch in 2001 [9,10]. Gomes and Mallion have partially described this problem [11]. Heine has provided a detailed and vivid description of quantum chemistry calculations of NMR parameters for pure fullerenes, endohedral fullerenes, fullerene derivatives, and fullerene dimers as well [12]. A comprehensive review on chirality of soluble C_{60} and C_{70} fullerenes, as well as the higher fullerenes and their exo- and endoderivatives was made by Thilgen and Diederich [13]. Mchedlov-Petrossyan has described a problem regarding the solubility of the fullerene C_{60} in condensed media [14]. There are also some articles on studies of fullerenes reactivity by quantum-chemical methods with the use of diverse reactivity indices [15], and studies on heterophase and liquid-phase oxidation reactions of fullerene with ozone, as well as chemiluminescence generated in these reactions [16]. As expected, there is a great number of the reports that describe different approaches to functionalization of the fullerene core [17-28]. At the same time, there are special articles which are devoted to studies on structure and electronic properties of fullerene C_{60} derivatives. A distinguishing feature of these studies is application of ^{13}C–^{13}C INADEQAUTE NMR technique that requires ^{13}C isotope enriched fullerene samples. This allowed performance of full and unambiguous assignments of the resonances in the ^{13}C NMR spectra of the fullerene derivatives to be discussed. In addition, to prove the structure of fullerene compounds, each synthetic approach was accompanied by the detailed ^{13}C NMR data without special enrichment. Thus, there is enormous actual ^{13}C NMR data flow that requires critical and analytic summarizing. This is an aim of the present chapter. It consists of three parts. The first section covers the experimental data obtained with only precise INADEQUATE NMR technique; the second one involves discussions on ^{13}C NMR features and stereochemical behavior of mono [2+n]-cycloadducts (n=1-4), as well as mono-adducts obtained by various radical and nucleophilic addition reactions without special enrichment; and in the third section, the wide application of the quantum chemistry approaches to ^{13}C NMR chemical shift calculations is considered.

2. ^{13}C-ENRICHED MONO- AND BIS-ADDUCTS OF FULLERENE C_{60}

A molecular carbon form, the fullerene C_{60}, has been obtained (Kroto, 1985) *via* evaporation of graphite under laser irradiation [1-4]. The structure of the derived carbon cluster is made up by 60 carbon atoms that has been identified with the use of the time-of-flight mass-spectrometry technique. It was suggested to be a polyhedron as the truncated icosahedron I_h with the geometry of a soccerball. The proposed molecule should consist of 60 vertices and 32 faces, 12 of which are pentagonal and 20 hexagonal, and each pentagon have not common bonds with hexagon [29, 30]. Such a regularity of mutual arrangement of pentagons and hexagons in C_{60} skeleton has been subsequently formulated as the well-known "Isolated pentagon rule (IPR)". When C_{60} has become available in macroscopic amounts due to the technique developed by the authors [31], the evidence for the C_{60} molecule as a truncated icosahedron has appeared five years later. In that case, D. Bethune and co-workers [32-35] have obtained the C_{60}/C_{70} mixture and have first registered the ^{13}C NMR spectrum of C_{60} at 125 MHz operative magnet frequency in CCl_4 without decoupling. They have assumed that the exact equivalence of all the sixty carbon atoms implied by I_h symmetry and the presence only of both 5,6 and 6,6 bonds can be immediately proved by ^{13}C NMR. The ^{13}C NMR spectrum of such a fullerene contained only a single line with a 142.5 ppm chemical shift that has been assigned to all the 60 equivalent fullerene carbon atoms.

J. Hawkins et al. has first obtained $C_{60}(OsO_4)$(4-*tert*-butylpyridine)$_2$ **1** (Figure 1) in the crystalline form in 1992 [36-38]. Using X-ray technique, C–C bond lengths and dihedral angles of the fullerene core have been discussed. The obtained data were characterized and confirmed by the informative INADEQUATE NMR of the mono-osmylated adduct specially enriched with the ^{13}C isotope. A clear relationship between $^{1}J(^{13}C–^{13}C)$ couplings and C–C bond lengths was observable. The ^{13}C NMR spectrum of this compound revealed five peaks of the coordinated 4-*tert*-butylpyridine fragment and 17 signals of the fullerene core carbon atoms as well (four of them were observed with 2C intensity and the remaining signals with 4C). It clearly indicates that there are two orthogonal mirror planes. Fullerene sphere of compound **1** contains, therefore, 17 types of carbon atoms, from which 13 are represented four times and 4 – two times. The signals with 2C intensity were assigned to carbon atoms that lie in the mirror symmetry planes. The chemical shift dispersion along δ scale was *ca.* 16 ppm.

Another successful example of the combined application of NMR and X-ray techniques has been demonstrated by Barley et al. [39]. Similarly, the relationship between C–C couplings and C–C bond lengths was investigated for compound **2** (Figure 1). The authors observed morphologic distortion of the fullerene core that occurs in the vicinity of the site of addition due to the cyclopropanation. The one-dimensional ^{13}C NMR spectrum of **2** revealed 17 resonances which clearly indicate a C_{2V} set point group symmetry of the molecule. Four of these signals displayed the half the intensity of the other 13 peaks. Signal dispersion along δ scale was 6 ppm, in contrast to **1**.

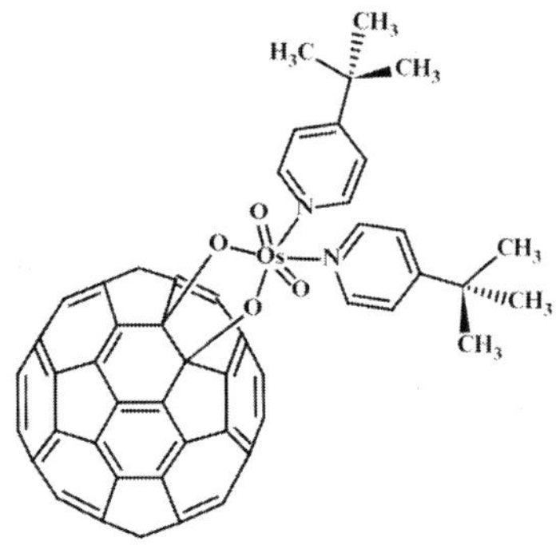

1

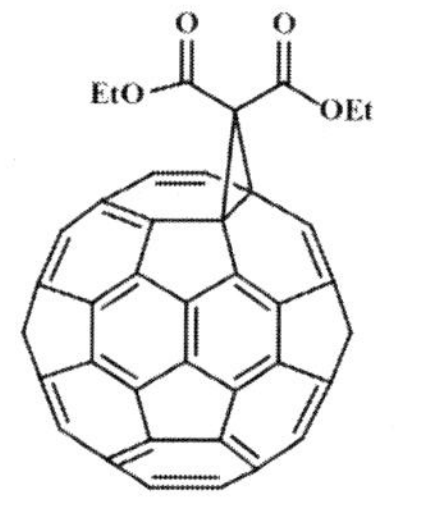

2

Figure 1. Mono-adducts **1** and **2**.

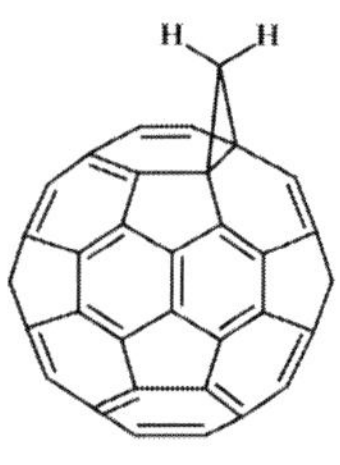

3

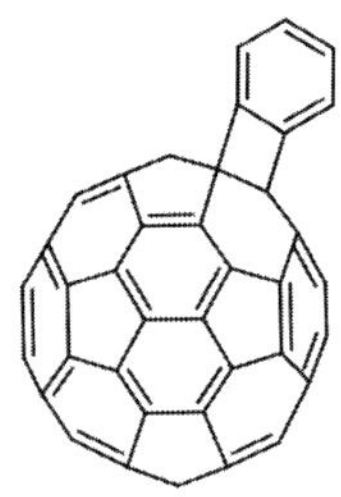

4

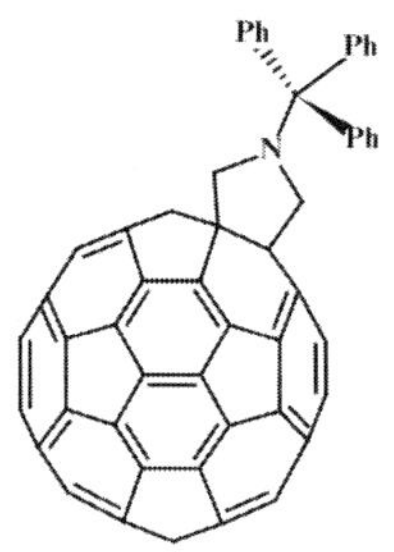

5

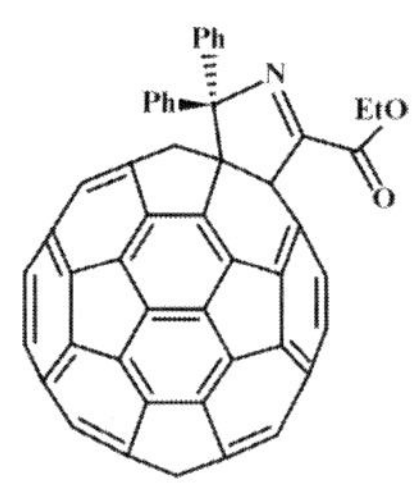

6

Figure 2. Structures of fullerene derivatives **3-6**.

The simplest parent fullerene cyclopropane, 1,2-methanobuckminsterfullerene $C_{61}H_2$ **3**, (Figure 2) has been purely isolated and discussed by Smith III et al. [40,41]. Meier and co-workers obtained later the same fullerocyclopropane using fullerene C_{60} to be enriched with the ^{13}C isotope [42]. This compound was a [2+1]-cycloadduct. The ^{13}C NMR pattern of **2** included 18 signals, one of which was attributed to sp^3 fullerene carbon atoms, while the remaining 17 were assigned to sp^2 species. The other two ^{13}C-labeled fullerenes C_{60}, such as [2+2]-cycloadduct **4** and [2+3]-cycloadduct **5**, were also described. The 17 fullerene core resonances in the ^{13}C NMR spectra of **4** and **5** were noticeable, and those are in agreement with the C_{2V} symmetry point group. The dispersion of the signals of compounds **3-5** was *ca.* 16 ppm.

Ball et al. [43] have performed structural reassignment of compound **6**, because in previous studies [44-47] they incorrectly identified the structure of mono-adduct. The ^{13}C NMR spectrum of **6** that has been recorded at 150 MHz operative frequency revealed 28 peaks with 2C intensity and 2 signals being 1C which were assigned to sp^2 fullerene carbon atoms. In this case, the symmetry of the molecule **6** was found to be C_S. The authors, therefore, concluded that the addition reaction between N-(diphenylmethylene)-glycinate esters and fullerene C_{60} under Bingel conditions gave [60]fullerenyl-dihydropyrrols and not methano[60]fullerenyl iminoesters as it was previously reported.

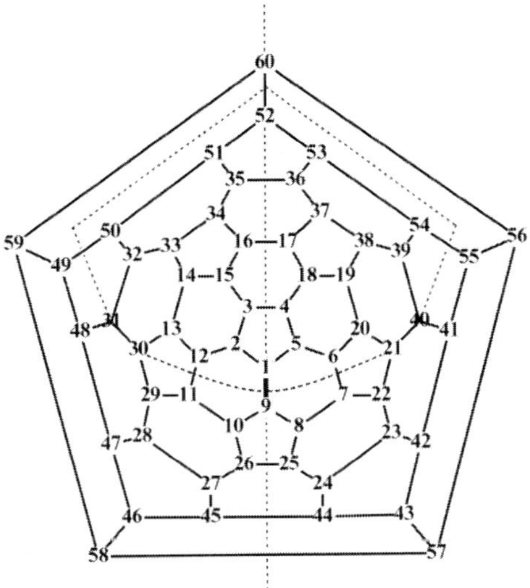

Figure 3. Schlegel diagram of fullerene derivatives **1-5**. The bond between C1 and C9 atoms is shown in bold. The mirror symmetry planes that divide fullerene framework into the four equal moieties are depicted with the dashed lines.

Table 1. ^{13}C NMR chemical shifts for compounds 1-5, in ppm ($CDCl_3$+CS_2, room temperature)

Carbon	1	2	3	4	5	Carbon	6	Carbon	6
1,9	105.38	72.06	71.04[1]	78.75[2]	80.66[3]	1	82.71	32,39	145.81
6,7,11,12	137.02	139.49	136.77	138.51	136.08	2,5	153.06	33,38	144.06
15,18,24,27	139.42	141.22	140.98	140.05	139.78	3,4	145.02	34,37	142.28
16,17,44,45	141.81	143.26	143.15	142.39	141.12	6,12	134.55	35,36	142.65
34,37,43,46	142.32	143.26	142.24	141.85	141.59	7,11	136.53	41,48	146.27
35,36,57,58	142.48	142.48	143.02	144.87	142.20	8,10	148.49	42,47	144.08
31,40	142.55	143.37	143.02	142.47	142.61	9	82.73	43,46	142.25
13,20,22,29	142.75	142.15	142.87	141.62	141.89	13,20	141.71	44,45	141.76
33,38,42,47	144.85	144.16	143.75	144.06	144.09	14,19	144.73	49,55	144.96
49,50,44,45	145.04	144.91	144.76	142.47	144.81	15,18	139.08	51,53	145.67
3,4,25,26	145.76	145.41	145.74	145.84	145.67	21,30	145.25	52	146.79
14,19,23,28	145.77	144.91	144.26	144.94	145.03	50,54	145.28	56,59	145.82
51,53,56,59	145.99	145.47	145.21	145.55	145.61	22,29	141.20	57,58	142.62
21,30	146.10	144.92	144.82	144.79	145.15	23,28	145.35	60	146.93
32,39,41,48	146.32	145.51	145.12	145.60	145.78	24,27	139.61		
52,60	148.41	145.08	144.37	146.31	146.78	25,26	147.49		
2,5,8,10	153.03	145.82	150.10	154.25	154.16	31,40	142.88		

[1] – The published data [40]. CS_2+$CDCl_3$ were used as solvents.

[2] – The published results [49]. The solvents were CS_2+acetone-D_6.

[3] – The published data [50]. $CDCl_3$+CS_2 were as solvents.

In summary, full and unambiguous ^{13}C NMR spectral assignments of compounds **1-6** have been precisely performed with the use of the INADEQUATE NMR technique and ^{13}C-labeled parent fullerene samples. All the aforementioned mono-adducts have essentially the same feature which consists in that the cycloaddition of a fragment to a fullerene core occurred along 6,6-closed bond, *i.e.* between two hexagons. The evidence of this was that in the ^{13}C NMR spectra there exist peaks in the range 70-100 ppm which are typical of the sp^3 fullerene carbon atoms. Another property is that there were 17 sp^2 carbon signals of compounds **1-5**, 13 of which were with 2C intensity, as well as the remaining were 1C.

Let us consider signal distribution along δ NMR scale for compounds **1-5** (Table 1). The fullerene carbon atoms were, in this regard, numbered in accordance with the IUPAC systematic nomenclature [48] (Figure 3). For example, in some articles [37,38], the authors do not use any rules for numbering carbon atoms of the fullerene core.

As can be seen from the Table 1, dispersion range along δ NMR scale for the compounds **1** and **3-5** is *ca.* 17 ppm. This is typical of the C_{60} fullerene molecules with C_{2V} set point group symmetry. The most high-frequency ^{13}C NMR resonances of this range, such as δ 153.03 for **1**, δ 145.82 for **2**, δ 150.10 of **3**, δ 154.25 for **4**, and δ 154.16 for **5** were attributed to C-2/C-5/C-8/C-10 atoms which located in α to a functionalization site. They are one bond away from the C-1/C-9 carbon atoms. It is noteworthy that low-frequency signals, such as δ 137.02 for **1**, δ 139.49 for **2**, δ 136.77 of **3**, δ 138.51 of **4**, and δ 136.08 for **5** were ascribed to β C-6/C-7/C-11/C-12 atoms that locate two bonds away from the site of addition. The second β atoms, C-3/C-4/C-25/C-26, were also two chemical bonds away from the C-1/C-9 and gave rise to middle peaks, such as δ 145.76 for compound **1**, δ 145.41 for **2**, δ 145.74 of **3**, δ 145.84 for **4**, and δ 145.67 of **5**. The mentioned conclusion remarks have been suggested by Meier and co-workers for the series **2–3–4** [42]. In addition, they succeeded in carrying out of some relationships. Those are between the ^{13}C NMR chemical shifts and local strain of the annulated rings, as well as the pyramidalization angle. The correlation between fullerene sp^2 chemical shifts and their properties was unobservable. When analyzing ^{13}C chemical shifts of the vicinity of the addition site and ^{1}H chemical shifts of the attached patterns, the authors, however, concluded that there exists a large paramagnetic ring current circulating around the site of addition. It causes significant deshielding of magnetic nuclei of the attached moieties, so that the NMR signals appear in the high-frequency regions. The similar effects can be observed for compound **6** as well. The molecule **6** is divided into two mirror parts by one symmetry plane passing through C-1, C-9, C-52, and C-60 atoms (Figure 3), because 28 peaks were observed in the ^{13}C NMR spectrum. The high-frequency signals, such as δ 153.06 and 148.49, were assigned to α C-2/C-5 and C-8/C-10 atoms, respectively, whereas the low-frequency resonances, such as δ 134.55 and 136.53, were typical of the C-6/C-12 and C-7/C-11 atoms, respectively. This allowed the authors to establish unequivocally the structure of the fullerene derivative **6**.

The simplest fullerene C_{60} derivatives are hydrofullerenes, or fulleranes, $C_{60}H_2$ **7** and $C_{60}H_6$ **8** (Figure 4) [51-54]. Their structure has been fully characterized by the INADEQAUTE NMR method on ^{13}C-enriched C_{60} samples [55]. These compounds are resulted from reduction of a C=C bond between hexagon patterns of the fullerene core. The detailed description of the hydrogenated C_{60} and C_{70} by means of NMR spectroscopy is presented by Hedenstrom et al. [27]. The 1,16-dialkyl adduct **9** can be freely ascribed to the compounds of the similar type [56-58]. The structure of this first fullerene derivative was assigned from the ^{13}C 2D INADEQUATE NMR spectra by using ^{13}C-labeled material. ^{13}C NMR chemical shifts of the **7**, **8**, and **9** are given in Table 2.

The ^{13}C NMR spectrum of hydrofullerene **7** revealed 17 peaks, one of which, δ 53.00, was assigned to sp^3 fullerene carbon atoms. The other signals that spread out in 16 ppm region are characteristic for the sp^2 fullerene carbon atoms. This is in agreement with C_{2V} symmetry point group of molecule **7**. The similar observations were typical of the aforementioned compounds **1-5**. The C-2/C-5/C-8/C-10 atoms that are α to the functionalization site, i.e., C 1/C 9 (Figure 5), resonated in the high-frequency region at δ 152.69 ppm. The first β carbons, such as C-6/C-7/C-11/C-12 gave a low-frequency peak at δ 136.56, whereas the second β species, such as C-3/C-4/C-25/C-26 gave rise to a signal at δ 148.00.

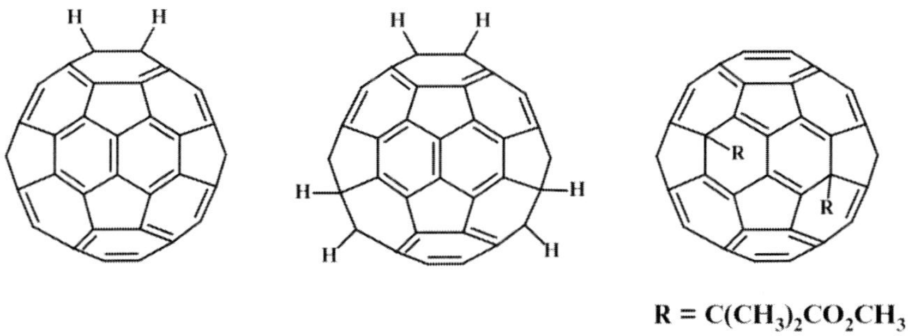

7 **8** **9**

Figure 4. Hydrofullerenes $C_{60}H_2$ **7** and $C_{60}H_6$ **8,** and 1,16-dialkyl derivative **9**.

Table 2. ^{13}C NMR chemical shifts of 7-9, in ppm ($CDCl_3$+CS_2, room temperature)

Carbon	7	Carbon	8	Carbon	9	Carbon	9
1,9	53.00	1,19,36,37,46,58	51.3	1,23	64.8	29,39	144.6
6,7,11,12	136.56	2,8,35,38,47,57	152.8	7,8	135.2	30,54	145.0
15,18,24,27	140.59	3,19,25,28,51,56	150.5	11,40	136.7	20,26	145.2
16,17,44,45	141.81	4,18,26,27,52,60	150.8	12,41	137.6	36,48	145.4
34,37,43,46	142.17	5,10,17,45,53,59	156.9	15,57	138.8	35,49	145.5
35,36,57,58	142.76	6,11,16,44,49,54	141.8	6,25	141.1	10,21	145.5
31,40	143.53	7,12,34,39,43,48	140.7	28,38	141.2	34,59	145.9
13,20,22,29	142.19	13,22,31,33,40,42	143.3	18,45	141.4	33,60	146.6
33,38,42,47	144.94	14,21,23,30,32,41	153.0	16,58	141.6	9,22	146.6
49,50,44,45	145.65	15,20,24,29,50,55	144.9	13,55	142.1	3,43	146.7
3,4,25,26	148.00			17,46	142.5	32,52	147.8
14,19,23,28	145.72			31,53	142.7	50,51	148.6
51,53,56,59	146.54			19,27	144.1	4,44	150.8
21,30	146.33			37,47	144.3	2,42	150.9
32,39,41,48	146.61			14,56	144.5	5,24	160.4
52,60	147.61						
2,5,8,10	152.69						

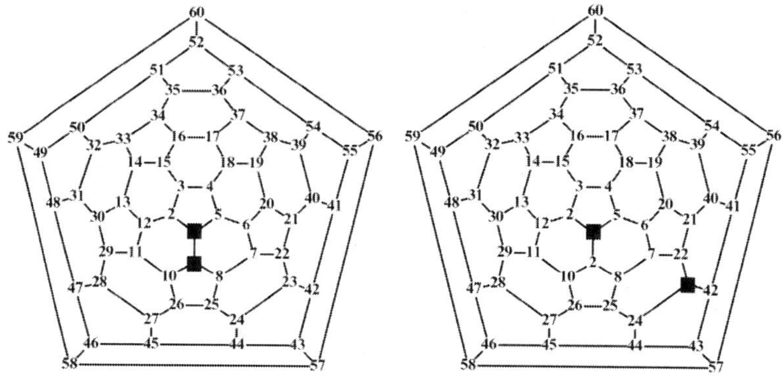

7 9

Figure 5. Schlegel diagrams of **7-9**. The black squares depict carbon atoms attached to the fragments.

The fullerane $C_{60}H_6$ **8** with higher symmetry D_3 is more symmetrical, in contrast to **7**. This is confirmed by the ^{13}C NMR spectrum that revealed 10 peaks of the fullerene carbon atoms. One of them (δ 51.3) characterizes the sp^3 carbon atoms, while the other nine belong to sp^2 species. The most high-frequency signals at 150-160 ppm are attributed to sp^3 carbon atoms of pentagons, whereas the low-frequency resonances at 140-150 ppm were typical of pentagons that contain exclusively sp^2 carbon atoms (Figure 5).

Furthermore, the authors have investigated the relationship between ^{13}C–^{13}C coupling constants and the calculated C–C bond lengths [55]. The used quantum chemistry approach accurately reproduced experimental data. The second dependence was selected between the pyramidalization angle and experimental ^{13}C NMR chemical shifts. It has been concluded that this relationship was not strong, and the pyramidalization angle was not an exclusive crucial factor. Finally, the third relationship was between experimental and theoretical ^{13}C NMR chemical shifts. Once again, the applied quantum chemistry approach reproduced experimental results with high correlation coefficient. As a result, authors have found that ^{13}C chemical shift dispersion of fullerene C_{60} derivative is caused by the paramagnetic ring currents circulating around the site of addition.

The ^{13}C NMR spectrum of compound **9** contained 30 peaks, 29 of them were typical of the sp^2 fullerene carbon atoms, and the last one belonged to sp^3 atoms (Table 2). This observation is in agreement with the fact that C_S group of symmetry is characteristic for this molecule. In addition, using INADEQUATE NMR spectrum, the authors were able to ascertain regioselectivity of addition of alkyl fragments to a fullerene core (Figure 5), so that the structure of **9** has unequivocally been elucidated.

The fullerene core is known to be able attaching more than one fragment due to several reaction sites. For example, addition of a second fragment can occur along some double bonds to form eight different bis-adducts. To distinguish them, Hirsch et al. have proposed description of the spatial arrangement of the added groups [59]. The fragments can be placed in the same hemisphere forming three possible *cis* isomers, or in opposite hemispheres to form four possible *trans* isomers. In addition, the equatorial isomers, such as e_{face} and e_{edge} are also feasible. Burley et al. [47] have discussed two regioisomeric bis-adducts **10** and **11** (Figure 6) with the use of the INADEQUATE NMR technique. The ^{13}C NMR chemical shifts of these compounds are given in Table 3.

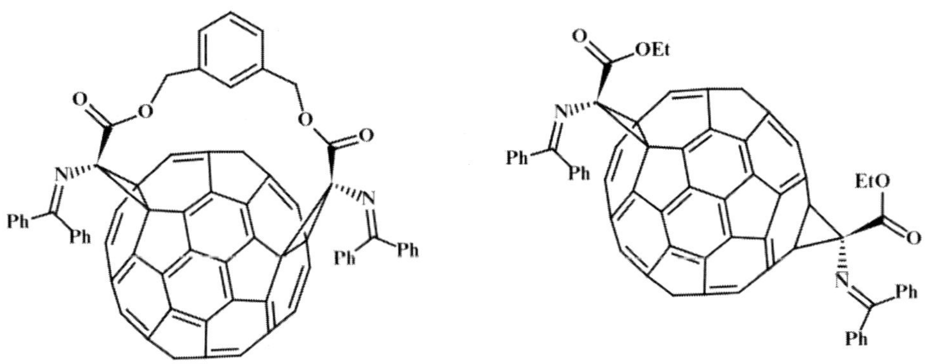

10 **11**

Figure 6. Regioisomeric bis-adducts **10** and **11**.

Table 3. ^{13}C NMR chemical shifts of 10 and 11, in ppm ($CDCl_3$+CS_2)

Carbon	10	Carbon	10	Carbon	11	Carbon	11
1,48	81.4	15,33	139.4	5,59	153.9	16,54	143.8
9,47	81.3	16,34	141.1	2,57	153.3	18,52	143.7
10,28	145.7	17,35	141.3	8,47	151.2	13,42	143.6
11,29	129.0	18,51	140.9	14,41	148.5	35,38	143.2
12,30	136.2	19,52	141.9	25,28	148.5	11,18	142.0
2,31	150.6	20,60	143.4	23,30	145.4	36,37	141.6
3,32	146.5	21,56	148.3	10,45	148.3	15,55	141.3
4,45	146.4	22,57	140.3	3,4	147.6	22,31	140.2
5,49	149.6	23,43	145.3	56,60	147.6	6,49	138.9
6,59	147.1	24,44	138.7	21,32	147.6	24,29	138.2
7,58	136.2	36,37	141.8	19,51	147.2	11,44	136.0
8,46	146.6	38,53	145.5	26,27	145.4	12,43	134.0
25,45	145.3	39,54	145.5	34,39	144.9	1,58	81.9
26,27	143.5	40,55	145.5	17,53	144.7	9,46	81.6
13	142.2	41	150.6	20,50	144.6		
14	148.7	42	145.7	33,40	144.3		

As can be seen from Table 3, the ^{13}C NMR spectrum of **10** consists of 32 peaks of fullerene carbon atoms, 28 of which were with 2C intensity, as well as the other four were with 1C. It corresponds to C_S point group symmetry of the molecule which is divided by one mirror plane. The thirty signals of fullerene carbon atoms of the bis-adduct **11** were observed in the ^{13}C NMR spectrum; it is in agreement with the C_2 point group symmetry. The authors have ascertained the addition sites of the added fragments (Figure 7).

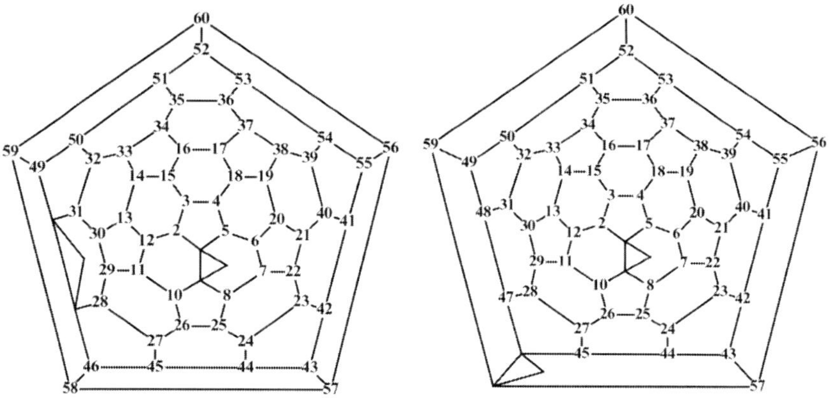

10 **11**

Figure 7. Schlegel diagrams of bis-aducts **10** and **11**. The added moieties are omitted for clarity.

The most high-frequency peak (δ 150.6) of **10** was assigned to C-2/C-31 α-carbons, as well as to C-41. The other α-atoms, such as C-5/C-49 gave the resonance at δ 149.6 ppm. The low-frequency signal at δ 136.2 has been ascribed to C-12/C-30 and C-7/C-58 atoms of β-environment of the site of addition. The α-carbons C-5/C-59, C-2/C-57, and C-8/C-47 of **11** gave rise to resonances at δ 153.9, 153.3, and 151.2, respectively. The atoms C-11/C-44 and C-12/C-43 of β-location gave the peaks at δ 136.0 and 134.0, respectively. In summary, the similar regularities of the nuclear shieldings by electrons for bis-adducts to be discussed are characteristic as for the mono-adducts.

Some germylated C_{60} adducts were also derived and characterized by means of the INADEQUATE NMR technique [60]. Neither in the articles nor in the additional informational, however, ^{13}C NMR assignments for these compounds were reported. We were unable, unfortunately, to discuss the ^{13}C chemical shifts and to compare them with the other fullerene derivatives.

In summary, there are several articles on full description of fullerene core of mono- and bis-C_{60} adducts with the use of the precise ^{13}C–^{13}C INADEQUATE NMR experiment. Some relationships between ^{13}C NMR chemical shifts and the pyramidalization angle, as well as local strain of the annulated ring have been performed. In some cases, this technique was applied to ascertain a regioselectivity of addition of two patterns to the fullerene framework.

3. Mono[2+n]-cycloadducts of [60]Fullerene (n=1-4) without Special ^{13}C Isotope Enrichment

3.1. Mono-adducts from [2+1] Cycloaddition

In the mono-adducts from [2+1] cycloaddition, a fullerene core is able to annulation with three-membered carbo- and heterocycle fragments. Such derivatives are referred to methanofullerenes, fulleroaziridines, and fullerooxyranes. The simplest compound of [2+1] cycloaddition along 6,6-closed bond is methanofullerene **3** described earlier.

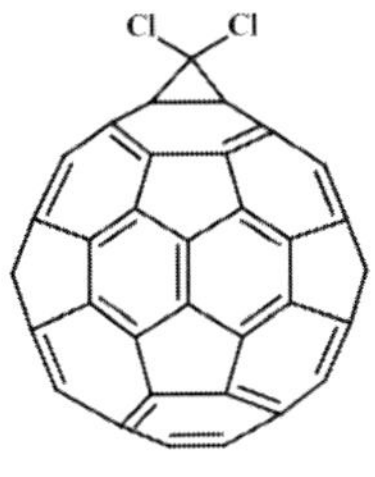

12

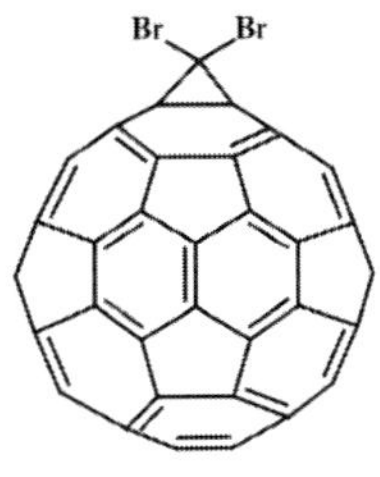

13

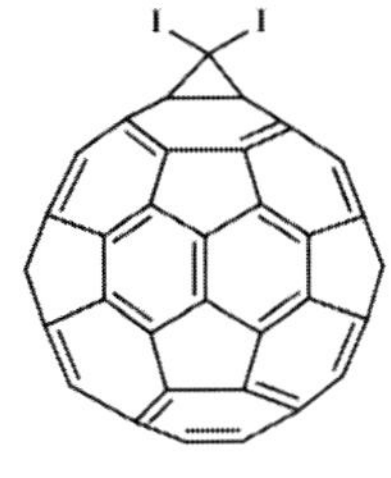

14

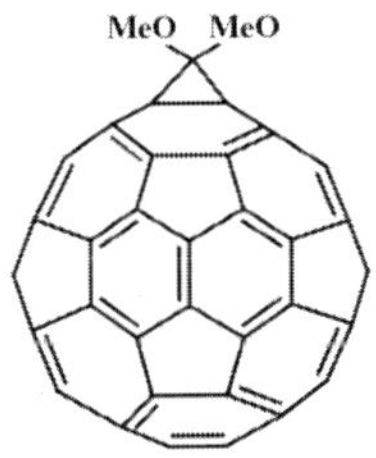

15

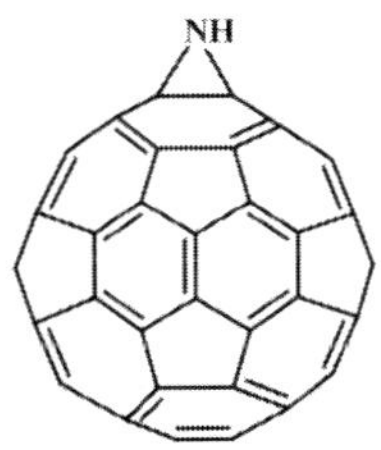

16

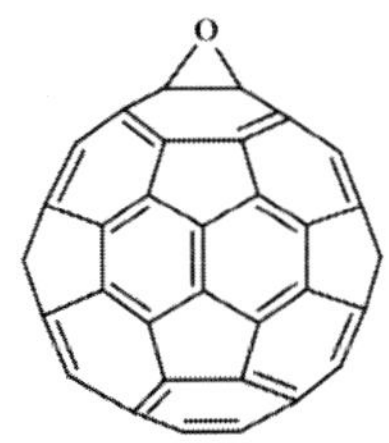

17

Figure 8. The structures of methanofullerenes and the simplest aziridinofullerenes and fullereoxyranes.

The ^{13}C NMR spectrum of this molecule contained 18 signals; one signal was attributed to sp^3 fullerene carbons, as well as the other 17 were typical of sp^2 species. Such a high C_{2V} symmetry is due to two orthogonal mirror planes that divide the fullerene core into four equivalent mirror moieties. The first plane passes through a 6,6-closed fullerene bond, and the second one is perpendicular to it. The similar observations were characteristic for halogenmethanofullerenes **12** (Figure 8) [61-63], **13** [64], and **14** [65,66], 1,2-dihydrofullerenedicarboxylate **15** [67], the simplest aziridinofullerene **16** [68], and fullerooxyrane **17** [69]. It should be pointed out that C_{2V} symmetry of aziridinofullerenes is retained, even there exist asymmetric moieties attached along 6,6-closed ring bond [70,71].

A loss of one symmetry plane leads to C_S or C_2 molecules, so that one-dimensional ^{13}C NMR spectrum of compound **18** (Figure 9) with the attached electron-withdrawing fragment, for example, contained 20 peaks of various intensity in the range of 138-145 ppm. The similar observation was typical of compound **19** that has the same attached moiety. In the range of 134-145 ppm of the ^{13}C NMR spectrum of **19**, 23 signals were characteristic [72]. In **18** and **19**, the symmetry planes pass through the bond of two fullerene hexagons, along which fragment is attached. The molecule **20** with the 5,6-open fullerene bond was also C_S. The thirty signals belonging to the sp^2 fullerene carbons in the range of 134-149 ppm of the ^{13}C NMR spectrum prove this assumption [70]. The same was observed, when 37 resonances of the sp^2 fullerene carbon atoms attributable to the C_2 azafulleroid adduct **21** were derived [73]. The symmetry plane of the two recent molecules passes perpendicular to a 5,6-open bond, so that it divides a fullerene core into two mirror parts.

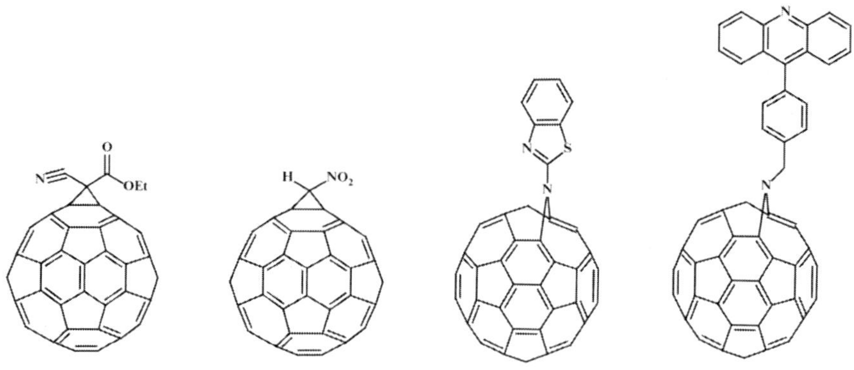

18 19 20 21

Figure 9. 6,6-Closed and 5,6-open adducts with C_S and C_2 point group symmetries.

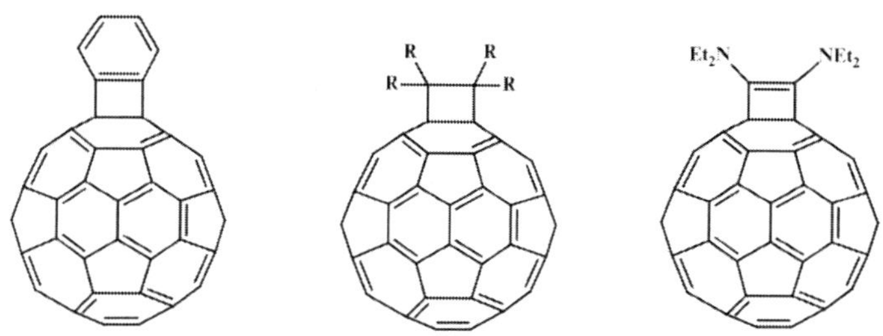

22

23: R = OEt
24: R = OBu

25

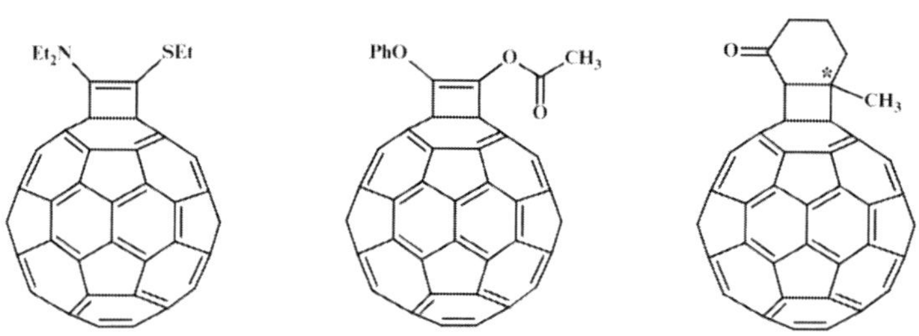

26 **27** **28**

Figure 10. Mono-adducts from [2+2] cycloaddition.

3.2. Mono-adducts from [2+2] Cycloaddition

The annulation of fullerene C_{60} with cyclobutane fragments is also possible to form adducts from [2+2] cycloaddition. These enormous fullerene derivatives are **22** [49], **23** and **24** [74], **25** [75] which are C_{2V}, **26** [75] and **27** [76] with C_S symmetry, as well as C_1 adduct **28** with one chiral centre in the attached moiety [77] (Figure 10).

The ^{13}C NMR spectrum of the mono-adduct **22** contained 16 signals attributable to sp^2 fullerene carbon atoms in the range of 139-156 ppm, of which 12 were with 4C intensity, 3 signals with 2C, and one as 6C. The peak at δ 78.75 has been assigned to the sp^3 species. The same pattern was found for the compounds **23-24**, when three peaks were with 2C intensity, and the rest with 4C. The sp^3 carbon atoms of **23** and **24** gave rise to signals at δ 76.10 and 75.27, respectively. Finally, the ^{13}C NMR spectrum of **25** also revealed 16 types of the sp^2 fullerene carbon atoms at δ 136.9-157.1 ppm, and the sp^3 one at δ 74.9. These observations prove only two orthogonal symmetry planes of **22-25** that divide the fullerene cores into four mirror parts. One of them passes through the 6,6-bond, as well as the second is perpendicular to it.

The annulated cyclobutane fragment with two various substitutions forms fullerene derivatives with one symmetry plane. The ^{13}C NMR spectrum of **26**, for example, contained 31 signals with different intensities at δ 101.5-159.56 which are characteristic for the sp^2 carbon atoms, and two sp^3 ones at δ 76.1 and 76.6 ppm. The similar pattern was found for compound **27**, so that ^{13}C NMR provided 25 lines of sp^2 fullerene carbon atoms at δ 139.27-146.25, one of which was with 6C, 19 were with 2C intensity, three were with 4C, as well as two were as 1C. The sp^3 carbon atoms resonated at δ 76.21 and 76.24 ppm. One symmetry plane bisects mirror-like the molecules **26-27** along 6,6-closed fullerene bond. It is noteworthy that one chiral centre in an attached moiety removes magnetic equivalence almost of the whole fullerene sphere that leads to the molecules with C_1 point group symmetry. The ^{13}C NMR pattern of mono-adduct **28**, for example, has shown 50 signals with various intensity of the sp^2 carbon atoms at δ 135-157 ppm. Such a number of the resonances is determined, in our opinion, by the anisotropic effect of the chiral carbon atom in the attached moiety. The aforementioned symmetry points groups are, therefore, typical of the mono-adducts from [2+2] cycloaddition, as well as for the [2+1] cycloadducts.

3.3. Adducts from [2+3] Cycloaddition

Carbocyclic fragments annulated with fullerene C_{60} can form mono-adducts from [2+3] cycloaddition as well. These compounds are fullerocyclopentanes. One of them is C_{2V} adduct **29** (Figure 11) derived from the reaction between C_{60} and 1,8-dihydronaphthalene [78]. The structure elucidation of this compound has been made based on the ^{13}C NMR data that revealed 16 peaks with different intensities belonging to the sp^2 carbon atoms (δ 135.73-153.59 ppm), and one signals at δ 76.47 of the sp^3 carbon atom. The molecule is divided, in this regard, into four equal parts by two orthogonal mirror planes. The less symmetric C_S adduct, such as fullerocyclopentane **30** [79,80], gave 28 resonances of sp^2 fullerene carbon atoms, and two peaks of sp^3 carbon atoms at 69.91 and 77.08 ppm in the ^{13}C NMR spectrum. In such a case, the molecule is bisected by one mirror plane along 6,6-bond.

The simplest fulleropyrrolidine is mono-adduct **31** (Figure 12) [81,82]. The C_{2V} symmetry of such a compound is confirmed by 16 resonances at δ 136-155 ppm, among which 13 were with 4C intensity, and three peaks with 2C, as well as one peak of sp^3 carbon atoms at δ 71.7 ppm in the ^{13}C NMR spectrum. Fullerene derivative **32** is C_2 molecule [83]. The evidence for this assumption were 29 peaks of the sp^2 fullerene carbon atoms at δ 135-153 ppm of the ^{13}C NMR spectrum, so that two signals were with 4C intensity, the remaining ones were as 2C, and one line at δ 77.16. The C_1 mono-adduct **33** gave the ^{13}C NMR pattern that included 45 signals located at 135-155 ppm, six were as 2C, two were with 3C, and one was as 4C. The sp^3 carbon atoms of the fullerene core resonated at δ 69.54 and 73.24 [83]. One chiral centre of the attached fragment of **34** caused anisochrony almost of the whole fullerene core [84]. In the δ 119.72-156.36 range of the ^{13}C NMR spectrum of **34**, there exist 53 peaks attributable to 58 sp^2 fullerene carbon atoms, whereas two signals at δ 75.28 and 77.15 were typical of the two sp^3 carbon atoms.

An intermediate of 1,3-dipolar cycloaddition, such as fulleropyrazoline **35** (Figure 13), is a C_S molecule, in which the symmetry plane passes through the 6,6-closed ring bond [85]. The ^{13}C NMR spectrum revealed 30 lines with various intensities at δ 135-148 ppm ascribed to sp^2 carbon atoms of the fullerene sphere. The sp^3 carbon atoms gave two signals at δ 82.34 and 92.52. Another fulleropyrazoline **36** with a chiral carbon atom in the attached fragment was characterized by a C_1 symmetry point group [86]. The anisochronous sp^2 carbon atoms of this fullerene derivative resonated as 49 peaks in the range of δ 133-153, among which one was three-fold greater in intensity, seven were with 2C, and the remaining signals were with 1C intensity. The saturated fullerene carbon atoms bound to the attached fragment gave the resonant line at δ 63.73. Fullerotriazoline **37** (Figure 13) can be considered as C_S molecule [87], the ^{13}C NMR spectrum of which contained 32 peaks with different intensities at δ 135-148, and two lines at δ 77.21 and 81.17 ppm.

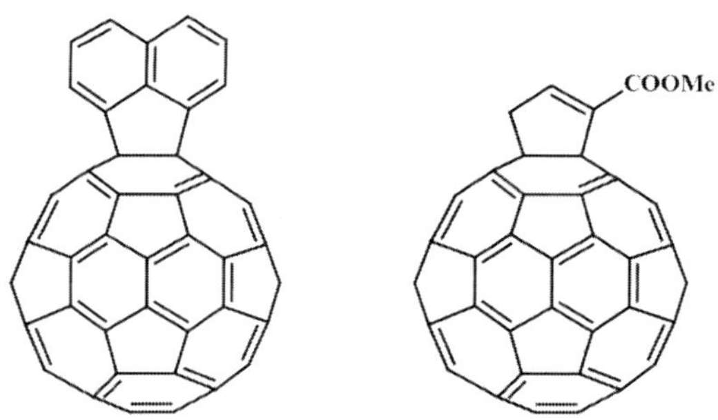

29 **30**

Figure 11. The structures of annulated carbocyclic [2+3] mono-adducts.

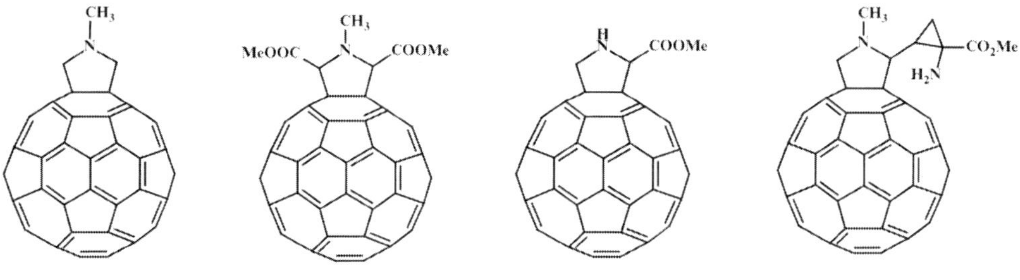

31 **32** **33** **34**

Figure 12. N-pyrrolidino[60]fullerenes.

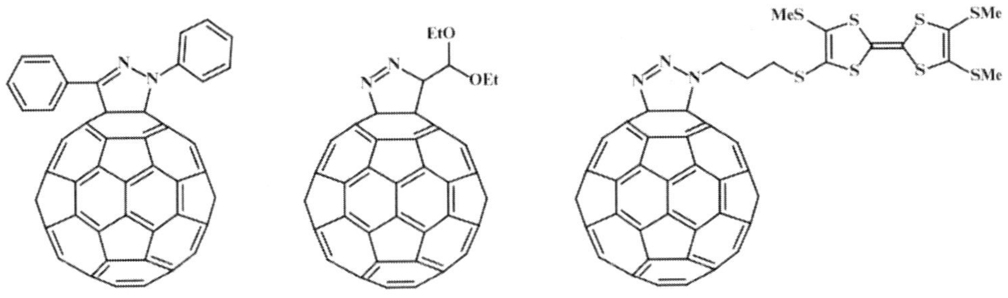

35 36 37

Figure 13. Mono-adducts from [2+3] cycloaddition.

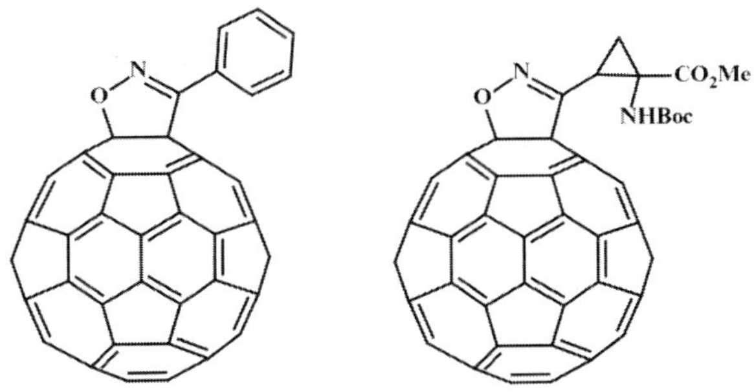

38 39

Figure 14. Fulleroisoxazoline mono-adducts.

Fulleroisoxazoline **38** (Figure 14) with C_S symmetry gave a ^{13}C NMR pattern with 28 signals of the sp^2 fullerene carbon atoms, two of which were with 1C intensity, 24 were with 2C, and two were with 4C [88]. The carbon atoms resonated at δ 146.92 and 147.42 ppm lie in a symmetry plane, since their signals were 1C. All other fullerene carbon atoms have one mirror partner, because their peaks are twice the intensity. The exceptions are those at δ 142.13 and 142.51. The ^{13}C NMR spectrum of chiral mono-adduct **39** consisted in 43 signals of sp^2 carbon atoms at the range of 136.64-155.68 ppm, and two sp^3 peaks at 77.21 and 80.69 ppm [84]. These features are typical of the compounds with the C_1 symmetry point group.

A cycloadduct **40**, in which fullerene core was associated with the saturated five-membered heterocycle fragment (Figure 15), was a molecule with a high C_{2V} symmetry [89]. A confirmation was that there were 16 resonances at δ 136.37-154.88 of the ^{13}C NMR spectrum, which are characteristic of the sp^2 fullerene carbon atoms, and also one resonant line at δ 73.39 attributable to sp^3 carbon atoms. The oxidized derivative **41** is C_S, because its ^{13}C NMR pattern included 27 signals of the sp^2 fullerene core carbon atoms at δ 134.40-154.53, and one resonance at δ 71.26 assigned to sp^3 carbon atoms [89]. Chiral adduct **42** was determined as C_1, because 52 resonances of the sp^2 carbon atoms at δ 136.46-152.22, as well as two peaks at δ 73.19 and 79.35 of the sp^3 species were observable in the ^{13}C NMR spectrum [90]. The high C_{2V} symmetry was typical for the other [2+3] cycloadduct **43** (Figure 15). One signal at δ 65.3 and 16 resonances at δ 136.70-147.16 were derived from the ^{13}C NMR spectrum of this compound in this regard [91].

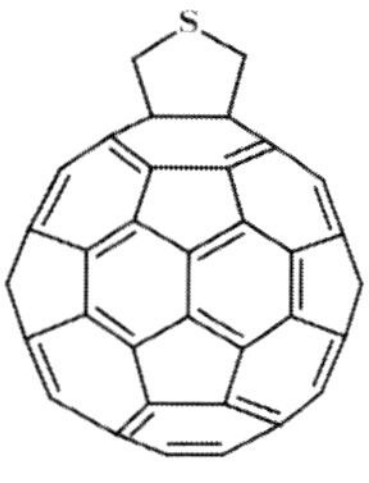

40

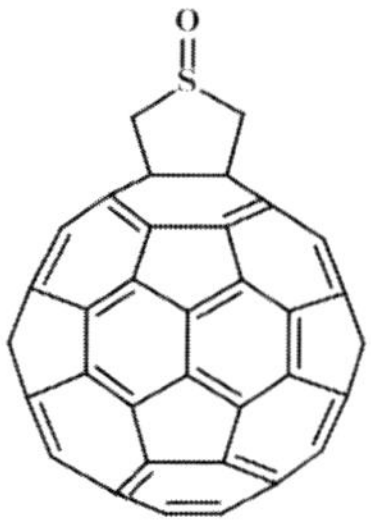

41

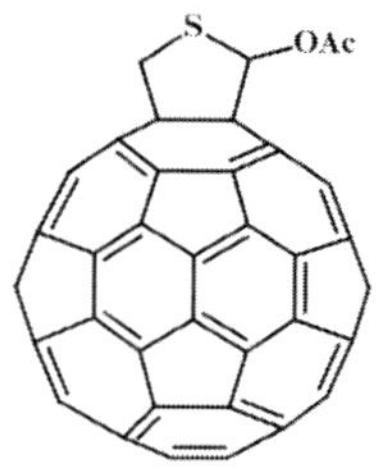

42

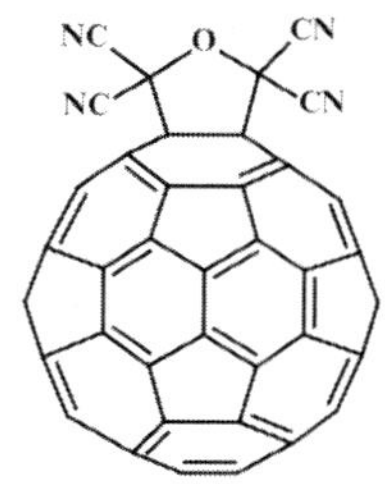

43

Figure 15. Mono-adducts from [2+3] cycloaddition.

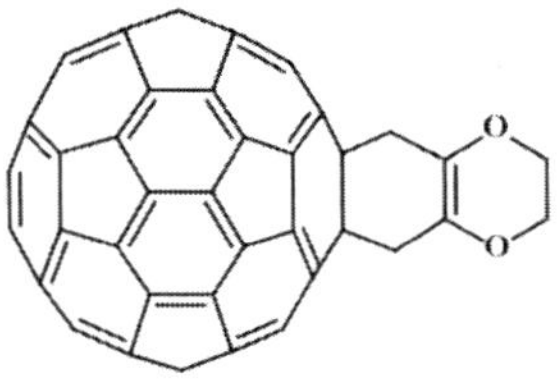

44

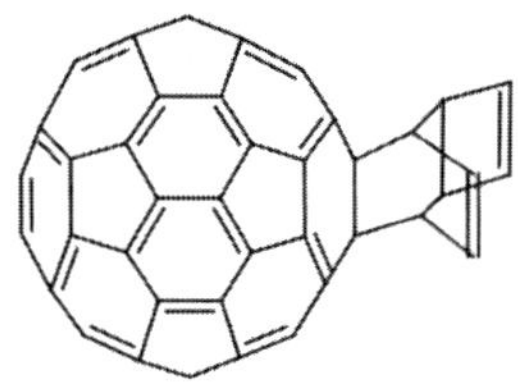

45

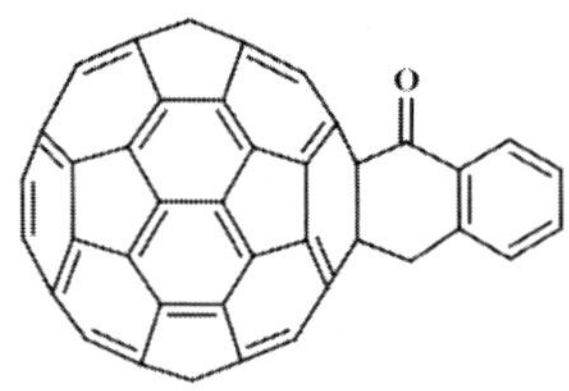

46

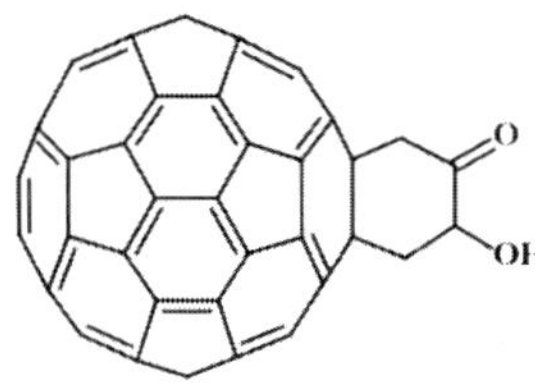

47

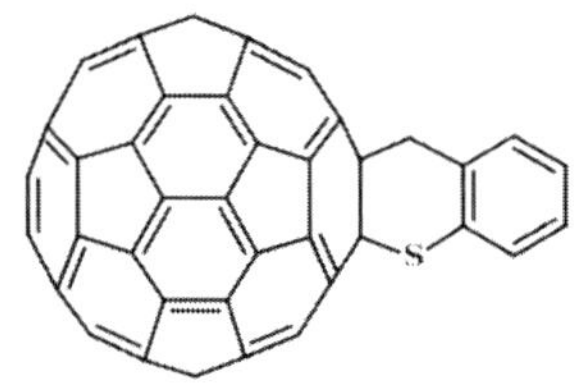

48

Figure 16. [2+4] Cycloadducts.

3.4. Adducts from [2+4] Cycloaddition

C_{2V}-symmetric compound **44** (Figure 16) [92], C_S fullerene derivatives **45** [93] and **46** [94], as well as cycloadducts **47** [92] and **48** [95] with chiral centers in attached moieties are known to be mono-adducts from [2+4] cycloaddition. Their structure has been established with the use of modern physicochemical means. The ^{13}C NMR spectrum of **44**, particularly, contained 16 signals of the sp^2 fullerene carbon atoms at δ 135.29-155.91 and one sp^3 resonance at δ 66.31 ppm. One mirror plane bisects the molecule **45** into equal parts, because the ^{13}C NMR pattern consisted in 31 peaks in the range between δ 136.17 and 155.57 ppm belonging to sp^2 fullerene carbon atoms and one resonant line at δ 69.73 attributable to sp^3 species. The same was characteristic for the cycloadduct **46**, however, the signals of sp^3 carbon atoms have been registered at δ 63.30. The chiral centre in the attached fragment of **47** effects anisotropically the fullerene framework. The confirmation consists in that there were 52 signals of sp^2 carbon atoms of the fullerene core at δ 134.19-154.92 and two resonances at δ 61.49 and 62.00 ascribed to two sp^3 carbon atoms of the same core. Another example of anisotropic influence of chiral carbon on fullerene skeleton was observable for adduct **48**, the ^{13}C NMR pattern of which revealed 50 peaks of sp^2 carbon atoms at δ 134.12-155.58 and two sp^3 carbon atoms at δ 68.37 and 70.79 ppm.

In summary, four set point groups of symmetry, such as C_{2V}, C_S, C_2, and C_1 are characteristic for the mono-adducts from [2+n] cycloaddition (n=1-4), as seen from the reference ^{13}C NMR data. Assignment of the fullerene compounds to one or another symmetry point group is based on a number of signals, their location along δ scale and relative intensities. The relationship between symmetry type of molecule and symmetry of fragment to be attached to a fullerene core is clearly observed. In most cases, addition of moieties to the 6,6-closed ring bonds of the fullerene core is characteristic for such adducts. There is lack of examples describing cycloadditions to the 5,6-ring bonds. It is probably because the closed mono-adducts are more stable thermodynamically than their 5,6-open isomers.

4. Quantum Chemistry Approaches to Calculate ^{13}C NMR Chemical Shifts of Fullerene Derivatives

A current trend in fullerene organic chemistry is that quantum chemistry approaches are widespread for calculating both geometric parameters and the most diverse physicochemical properties of the molecules [96,97]. The latter are tensors of magnetic shielding with the use of ab initio and density functional methods. The calculation of second-order properties of large molecules, such as C_{60}, has been demonstrated, for example, for the static dipole polarizabilities and nuclear magnetic shielding using SCF level of theory on workstation computers [98]. The algorithmic advances were shown together with illustrative application to fullerenes C_{60} and C_{70}. The authors have ascertained that the absolute carbon shielding calculated by the GIAO-CPHF formalism at the MP2/DZP equilibrium geometry was C_{60}:σ=44.5 ppm. This result is in excellent agreement with the experimental data. The authors proposed calculation of ^{13}C chemical shifts of fullerene molecules for each site *I* from the following equation

$$\delta_{TMS}(I) = \delta_{TMS}(C_{60}) + \sigma(C_{60}) - \sigma(I),$$

where $\delta_{TMS}(C_{60})$ is 142.68 ppm, $\sigma(C_{60})$ and $\sigma(I)$ are the shieldings in the scaled valence-electron IGLO-DFTB model [99]. It is noteworthy that the chemical shift of C_{60} which equal to 142.68 ppm is usually assumed to be a reference signal in the experimental spectra of higher fullerenes. In addition, the relationship between the ^{13}C chemical shifts and π orbital axis vector, or the pyramidalization angle, has been provided. Sun and Kertesz have performed calculations of total energies and ^{13}C NMR chemical shifts for all isolated-pentagon-rule (IPR) C_{60} and higher fullerenes in terms of density functional theory (B3LYP/6-31G(d)) [100]. The NMR chemical shifts were evaluated employing the GIAO approach. The common patterns of the ^{13}C NMR spectra to be calculated were in good agreement with the experimental data. The authors have concluded that the signals above δ 140 are in better agreement with the experiment than those below 140 ppm, which are commonly overestimated by 1-2 ppm. In another article, the same authors have calculated the relative energies and the ^{13}C chemical shifts for the IPR C_{82} isomers at the same level of theory [101]. They have found at the first time C_2 isomer of C_{82} by comparison of experimental and calculated ^{13}C NMR chemical shifts. The structures of all the 19 IPR isomers of C_{86} have been calculated at B3LYP/6-31G level of theory [102]. The better agreement between theoretical and experimental ^{13}C NMR data was observed for C_2 and C_S isomers which have been predicted as the most stable among the other discussed species. Sternfeld et al. have achieved combined experimental and theoretical study on numerous higher fullerene anions with different sizes and shapes using GIAO B3LYP/6-31G(d) method [103]. They have concluded that the carbon atoms of all the multiply charged fullerene anions were deshielded relative to those of the neutral fullerenes.

A complete assignment of all the 31 ^{13}C NMR peaks of fullerene carbons of C_S monoazafullerenes has been provided for the first time by systematic comparison of theoretical and experimental data at DFT level, such as GIAO-B3LYP/6-31G(d)/ B3LYP/6-31G(d) [104].

The striking similarities of the signal patterns of the sp^2 fullerene carbons for two different heterofullerenes with both aryl and alkyl attached moieties were observable. As seen from the comparison, there was a good agreement between theoretical and experimental data, so that $\Delta\delta$ for the sp^2 fullerene carbons was 0.03-1.35 ppm, however for the sp^3 species it was *ca.* 10 ppm.

Later, the attempt to investigate the ^{13}C NMR chemical shifts and hyperfine coupling constants of fullerenols $C_{60}(OH)_{1-3}$ using DFT has been made [105]. To calculate chemical shifts, the GIAO approach at the B3LYP level with 6-31G(d,p) basis set was used. It was shown that one hydroxyl group attached to C_{60} causes preferentially the localized electron state at the first neighboring carbon site that belongs to a pentagon ring. The two hydroxyl moieties lead to reduction in the chemical shifts of the vicinal hydroxylated sites as compared to $C_{60}OH$. The ^{13}C NMR chemical shifts of $C_{60}(OH)_2$ reveal that opposite symmetrical sites are considerably shifted downfield. In $C_{60}(OH)_3$, there was obtained a significant reduction of spin density and negligible hyperfine constant couplings at the neighbor hydroxyl groups.

Laikov has developed quantum chemistry package "PRIRODA" in 2000 [106-108], in which GIAO method [109] at PBE/3ζ level [110] has been realized. It was shown that this method describes the structure and IR-spectra of fullerene C_{60} and C_{70} and their derivatives with high probability [111,112]. Using this approach, we have performed calculations of ^1H and ^{13}C NMR chemical shifts of relatively simple organic compounds [113], as well as some fullerene derivatives [114]. A good agreement between theoretical and experimental data has been observed.

The confidential intervals for proton chemical shifts of simple organic compounds was, for example, $\delta_{H,calc}\pm0.35$ ppm, and for carbon chemical shifts was $\delta_{C,calc}\pm6.05$ ppm. For the carbon atoms of fullerene derivatives, for which experimental ^{13}C NMR assignments were performed by means of INADEQUATE NMR method, a systematic error at GIAO PBE/3ζ level was 5 ppm. The better agreement between the theoretical ^{13}C chemical shifts and experimental data was observed for the sp^2 fullerene carbons. To assign experimental the ^{13}C NMR spectra of spiro [2+1] cycloadducts **49-56** (Figure 17), we have carried out quantum chemistry calculations [115].

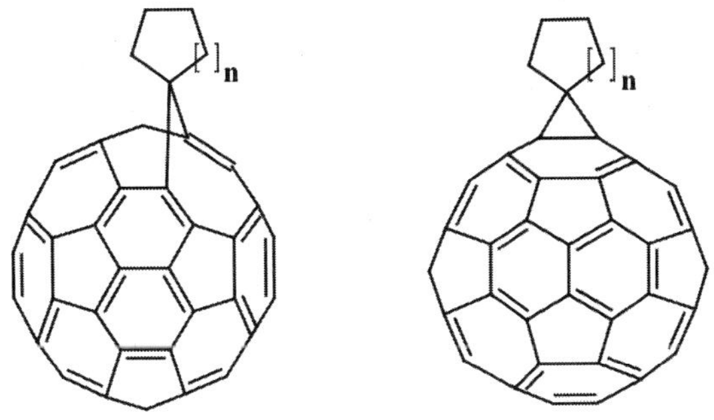

49: n=1 **51:** n=3 **53:** n=1 **55:** n=3
50: n=2 **52:** n=4 **54:** n=2 **56:** n=4

Figure 17. [2+1] Spiro cycloadducts of C_{60}.

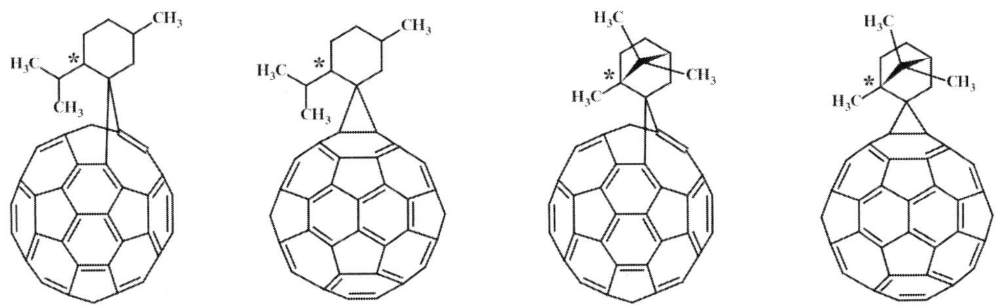

57 **58** **59** **60**

Figure 18. Mono-adducts **57-60**. Chiral centers are depicted with asterisks.

The ^{13}C NMR spectra of homofullerenes **49-52** revealed 31 peaks of sp^2 fullerene carbon atoms at δ 135-148 ppm. The relative intensity of four signals was 1C, one peak as 4C, and of the remaining ones was 2C. It corresponds to a C_S symmetry point group. The most high-frequency signal, such as δ 147.56, was typical of the C-3/C-4 atoms (see systematic numeration in Figure 3) that bound to spiro carbon atom of attached moiety due to calculations. The most low-frequency resonance, δ 135.06, was characteristic of the C-14/C-19 carbon atoms which localized in β to a functionalization site as well. The peaks with 1C intensity have been ascribed to C-1, C-9, C-52, and C-60 atoms that lie in a symmetry plane. The ^{13}C NMR spectra of methanofullerenes **53-56** contained 16 signals at δ 137-148 ppm conformed to sp^2 carbon atoms and one resonance at δ 70-80 ppm assigned to the sp^3 fullerene carbon atoms. The 13 signals were with 4C intensity, as well as three were as 2C. It is in agreement with a C_{2V} set point of symmetry of these molecules. The high-frequency peaks at δ 148 ppm were attributed to C-2/C-5/C-8/C-10 of α to a functionalization site. The most low-frequency signals at δ 137 belonged to C-6/C-7/C-11/C-12 of β environment to the same functionalization site. The four resonances with 2C intensity were typical for the C-1/C-9, C-31/C-40, C-52/C-60, and C-21/C-30 atoms that localized in the symmetry planes. In summary, the similar regularities were observable for carbocyclic homo- and methanofullerenes **49-56**, as for the C_{2V} [2+1] cycloadducts, **3**, **12-17**, discussed above.

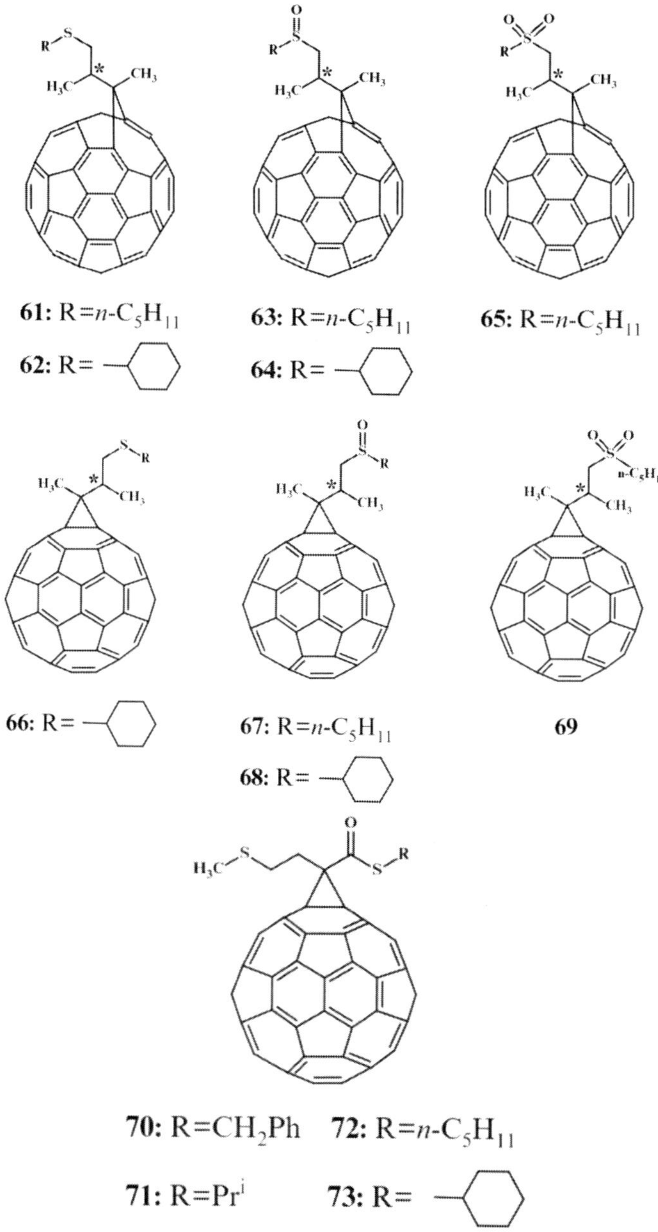

Figure 19. Sulfur homofullerenes and methanofullerenes. The chiral centers are depicted with asterisks.

Chiral mono[2+1]-cycloadducts **57-60** are more complicated example for consideration (Figure 18) [116].

52 Peaks of sp^2 fullerene carbon atoms at δ 132-147 ppm were derived from the ^{13}C NMR spectrum of adduct **57**. This allowed us to conclude that molecule **57** is assigned to a C_1 set point group of symmetry. Anisochrony of the fullerene core was, probably, caused by chiral centre of the attached moiety. The most low-frequency resonances, such as δ 132.94 and 134.77, were typical of C-14 and C-19, respectively, according to calculations. These

carbon atoms were located in β to a functionalization site. The most high-frequency signals, δ 147.55 and 147.19, were characteristic of C-33/C-38 that associated with C-14/C-19 and located in γ-environment to the functionalization site. The methanofullerene **58** which has been obtained by thermal isomerization of **57** is the same C_1 molecule. 37 Peaks at 136-151 ppm confirmed this assumption. The most high-frequency signals, such as δ 148.88, 149.11, 151.05, and 151.35 attributable to fullerene C-10, C-8, C-5, and C-2 atoms, respectively, located in α to the functionalization site, whereas the low-frequency resonances, δ 136.60, 136.73, 138.02, and 138.09 were ascribed to β-carbon atoms C-12, C-7, C-6, and C 11, respectively. The ^{13}C NMR pattern of C_1 homofullerene **59** revealed 48 peaks of sp^2 diastereotopic fullerene carbon atoms. The C-33 and C-38 atoms had high-frequency peaks δ 147.77 and 147.88, respectively, while C-14 and C-19 resonated as low-frequency signals at δ 132.99 and 133.97 ppm. The isomer **60** has been derived with the same method, as methanofullerene **58**. 45 Peaks at δ 136-151 region of the sp^2 fullerene core carbon atoms were observable in the ^{13}C NMR spectrum of **60**. The similar shieldings of α and β fullerene carbon atoms were obtained, as for compound **58**.

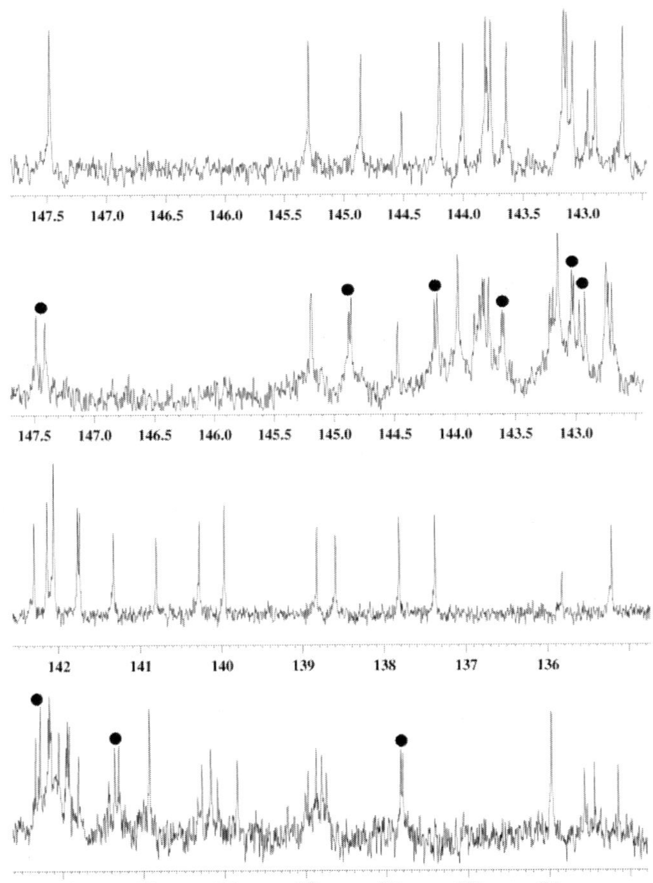

Figure 20. Comparison of the ^{13}C NMR spectra between the reference fullerene derivative **50** (the upper spectrum) and homofullerene **61** (the lower spectrum). Diastereotopic pairs are depicted as black balls.

Table 4. ^{13}C NMR chemical shifts of diastereotopic pairs of sp^2 fullerene carbon atoms for 61-65, in ppm, $CDCl_3+CS_2$. The magnitudes $\Delta\delta_{dias}$ are in parentheses, in ppm

50	61	62	63	64	65
135.06 (C-14/C-19)				135.02/135.07 (0.05)	135.10/135.16 (0.06)
135.89 (C-7/C-11)		136.00/136.04 (0.04)			
137.82 (C-8/C-10)	137.81/137.83 (0.02)		137.92/137.96 (0.08)	137.91/137.97 (0.06)	
138.79 (C-6/C-12)		138.84/138.89 (0.05)			
138.98 (C-2/C-5)				139.02/139.09 (0.07)	138.98/139.07 (0.09)
139.55 (C-21/C-30)			139.89/139.92 (0.03)	139.18/139.25 (0.07)	139.14/139.36 (0.12)
140.18 (C-35/C-36)		140.10/140.19 (0.09)	141.07/141.17 (0.10)	139.89/139.92 (0.03)	140.41/140.43 (0.02)
141.53 (C-31/C-40)	141.31/141.36 (0.05)				141.31/141.45 (0.14)
141.97 (C-13/C-20)		141.88/141.92 (0.04)			
142.22 (C-41/C48 or C-57/C58)		142.13/142.17 (0.04)	142.15/142.18 (0.03)		
142.30 (C-43/C-46)	142.27/142.30 (0.03)	142.29/142.33 (0.04)			142.37/142.54 (0.17)
142.76 (C-25/C-26)		142.75/142.79 (0.04)	142.76/142.80 (0.04)	142.70/142.76 (0.06)	
142.96 (C-56/C-59)	142.92/142.97 (0.05)	142.93/142.98 (0.05)			
143.08 (C-34/C-37)	143.03/143.07 (0.04)			143.02/143.09 (0.07)	
143.54 (C-24/C-27)	143.59/143.61 (0.02)				
143.70 (C-49/C-55)					143.70/143.73 (0.03)
144.00 (C-32/C-39)			143.99/144.04 (0.05)	143.98/144.03 (0.05)	
144.20 (C-32/C-39)	144.15/144.17 (0.02)	144.16/144.18 (0.02)	144.20/144.25 (0.05)	144.20/144.25 (0.05)	144.25/144.29 (0.04)
144.86 (C-22/C-29)	144.86/144.88 (0.02)				
145.25 (C-33/C-38)		145.21/145.24 (0.03)			145.07/145.20 (0.13)
147.56 (C-3/C-4)	147.41/147.48 (0.07)	147.42/147.50 (0.08)	147.31/147.36 (0.05)	147.31/147.36 (0.05)	147.26/147.31 (0.05)

The ^{13}C NMR chemical shifts and diastereotopic splittings of sulfur homofullerenes **61-65** (Figure 19) and methanofullerenes **66-73** have been predicted with the use of GIAO PBE/3ζ approach [117].

To estimate diastereotopic splittings of the signals of ^{13}C NMR spectra for homofullerenes **61-65**, we used the reference fullerene derivative **50** [115], as well as compound **56** [115] was applied for the methanofullerenes **66-73**. The key point of estimation was comparison between reference and discussed compounds in the same regions of the spectra. If there is a diastereitopic splitting, the magnitude of this parameter is calculated from formula $\Delta\delta_{dias} = |\delta_1 - \delta_2|$, where δ_1 and δ_2 are ^{13}C chemical shifts of symmetric carbon atoms that resonate as separate signals. The clearly distinguishable diastereotopic carbons are, for example, C-3/C-4, because, according to calculation, they appeared as splitted peaks at δ 147.41 and 147.48. In such a case, the magnitude $\Delta\delta_{dias}$ is 0.07 ppm. An example of such a comparison is given in Figure 20 and Table 4.

To estimate a diastereotopicity of fullerene carbon atoms for methanofullerenes **66-73**, compound **56** is suggested as a reference one. Thus, in the ^{13}C NMR spectrum of **56**, for example, a resonance at δ 148.74 was observed which can be opposed to the four signals, such as δ 147.42, 147.58, 148.60, and 148.62 ppm assigned to diastereotopic carbon atoms of compound **66**. In such a case, diastereotopic differences $\Delta\delta_{dias}$ are assumed to be 1.32, 0.16, 1.02, and 0.02 ppm, respectively. In accordance with quantum chemistry calculations, these signals of **66** were attributed to α C-2/C-5/C-8/C-10 atoms.

Somewhat different pattern was observed in the case of methanofullerenes **70-73**. The ^{13}C NMR spectra of all these compounds contained 24 signals, 18 of which were with 2C intensity, three were as 3C, and one was with 5C. The signal at δ 137.60 of **56** can be compared with two splitted resonances, δ 136.87 and 138.11 for **70,** δ 137.59 and 138.06 for **71**, δ 137.60 and 138.08 for **72**, and δ 137.57 and 138.06 for **73**. These peaks are typical of C-6/C-7/C-11/C-12 atoms. In addition, the resonance δ 140.99 is assigned to C-15/C-18/C-24/C-27 carbons, while δ 143.77 and 145.75 were characteristic for C-33/C-38/C-42/C-47 and C-3/C-4/C-25/C-26, respectively. The other signals at δ 142.0-142.5 and 142.5-146.0 overlapped one another and require to be distinguished by the use, for example, the precise INADEQUATE NMR technique.

In summary, the quantum chemical approaches allows prediction of geometric parameters and physicochemical properties of pure fullerenes and their derivatives. The GIAO PBE/3ζ approach was shown to be the most reproductive method for modeling ^{1}H and ^{13}C NMR chemical shifts of relatively simple organic compounds and prediction of diastereotopic splittings of C_{2V}, C_S, and chiral C_1 spiro cycloadducts of fullerene C_{60}.

CONCLUSION

Currently, ^{13}C NMR spectroscopy is nowadays the most universal and informative technique that applied for structure elucidation and determination of physicochemical properties of the novel fullerene derivatives. The precise ^{13}C NMR assignments for mono- and bis-adducts fullerene C_{60} were achieved by means of INADEQUATE NMR technique on ^{13}C-enriched samples. The dispersion trend of the ^{13}C resonances along δ NMR scale has been shown for the some fullerene derivatives; this can be used for strong characteristic of the novel fullerene adducts. The general point groups of symmetry for mono-adducts from [2+n] cycloaddition (n=1-4) were ascertained based on ^{13}C NMR data. This reflects stereochemical behavior of such molecules in space. Using quantum chemistry methods one can be predicted the major structural and physicochemical properties of the novel fullerene derivatives with high probability and in relatively reasonable time. This is, therefore, a powerful arsenal of the methods to describe the structure and properties of such unusual organic molecules.

ACKNOWLEDGMENT

This work was supported by the Russian Scientific Fund (project number 14-13-00296).

REFERENCES

[1] Smalley, R. E. *Angew. Chem. Int. Ed.* 1997, 36, 1594–1601.
[2] Kroto, H. *Angew. Chem. Int. Ed.* 1997, 36, 1578–1593.
[3] Curl, R. F. *Angew. Chem. Int. Ed.* 1997, 36, 1566-1576.
[4] Kroto, H. W.; Heath, J. R.; O'Brien, S. C.; Curl, R. F.; Smalley, R. E. *Nature*. 1985, 318, 162-163.
[5] Kroto, H. W.; Allaf, A. W.; Balm, S. P. *Chem. Rev.* 1991, 91, 1213-1235.
[6] Sokolov, V. I.; Stankevich, I. V. *Russ. Chem. Rev.* 1993, 62, 419-435.
[7] Eletskii, A. V.; Smirnov, B. M. *Phys. Usp.* 1995, 38, 935-964.
[8] Andreoni, W. *Annu. Rev. Phys. Chem.* 1998, 49, 405-439.
[9] Buhl, M.; Hirsch, A. *Chem. Rev.* 2001, 101, 1153-1183.
[10] Chen, Z.; King, R. B. *Chem. Rev.* 2005, 105, 3613-3642.
[11] Gomes, J. A. N. F.; Mallion, R. B. *Chem. Rev.* 2001, 101, 1349-1383.
[12] Heine, T. Fullerenes. Calculation of NMR and EPR Parameters: Theory and Applications; WILEY-VCH Verlag GmbH & Co. KGaA: Weinheim, 2004; Vol. 25, pp. 409-420.
[13] Thilgen, C.; Diederich, F. *Chem. Rev.* 2006, 106, 5049-5135.
[14] Mchedlov-Petrossyan, N. O. *Chem. Rev.* 2013, 113, 5149-5193.
[15] Sabirov, D. S.; Bulgakov, R. G.; Khursan, S. L. ARKIVOC. 2011, 2011, part (viii), 200-224.
[16] Bulgakov, R. G.; Sabirov, D. S.; Dzhemilev, U. M. *Russ. Chem. Bull. Int. Ed.* 2013, 62, 302-324.
[17] Diederich, F.; Isaacs, L.; Philp, D. *Chem. Soc. Rev.* 1994, 243-255.
[18] Kepert, D. L.; Clare, B. W. *Coord. Chem. Rev.* 1996, 155, 1-33.
[19] Hirsch, A. Fullerenes and Related Structures. *Top. Curr. Chem*; Springer Berlin Heidelberg: Berlin, 1999; Vol. 199, pp. 1-248.
[20] Yurovskaya, M. A.; Trushkov, I. V. *Russ. Chem. Bull. Int. Ed.* 2002, 51, 367-443.
[21] Thilgen, C.; Gosse, I.; Diederich, F. *Top. Stereochem.* 2003, 23, 1-124.
[22] Sinyashin, O. G.; Romanova I. P. *Vestnik RFFI* (in Russian). RFBR Information Bulletin "*RFBR Journal*". 2004, 29-65.
[23] Mateo-Alonso, A.; Bonifazi D.; Prato M. Functionalization and applications of [60]fullerene. Carbon Nanotechnology. Recent Developments in Chemistry, Physics, Materials Science and Device Applications; Elsevier B.V.: Amsterdam, 2006; Vol. 7, pp. 155-189.

[24] Reinov, M. V.; Yurovskaya, M. A. *Russ. Chem. Rev.* 2007, 76, 715-730.
[25] Troshin, P. A.; Lyubovskaya R. N. *Russ. Chem. Rev.* 2008, 77, 305-349.
[26] Tuktarov, A. R.; Dzhemilev, U. M. *Russ. Chem. Rev.* 2010, 79, 585-610.
[27] Iglesias-Groth, S.; Cataldo, F. Fulleranes. The Hydrogenated Fullerenes. Carbon Materials: Chemistry and Physics; Springer Science+Business Media B.V.: New-York, 2010; Vol. pp. 1-277.
[28] Yamada, M.; Akasaka, T.; Nagase, S. *Chem. Rev.* 2013, 113, 7209-7264.
[29] Osawa, E. *Kagaku (Kyoto).* 1970, 25, 854-863 (in Japanese).
[30] Bochvar, D. A.; Gal'pern, E. G. *Proc. Acad. Sci. USSR.* 1973, 209, 239-241.
[31] Kratschmer, W.; Fostiropoulos, K.; Huffman, D. R. *Chem. Phys. Lett.* 1990, 170, 167-170.
[32] Johnson, R. D.; Meijer, G.; Bethune D. S. *J. Am. Chem. Soc.* 1990, 112, 8983-8984.
[33] Johnson, R. D.; Bethune, D. S.; Yannoni, C. S. *Acc. Chem. Res.* 1992, 25, 169-175.
[34] Yannoni, C. S.; Bernier, P. P.; Bethune, D. S.; Meijer, G.; Salem, J. R. *J. Am. Chem. Soc.* 1991, 113, 3190-3192.
[35] Johnson, R. D.; Yannoni, C. S.; Salem, J. R.; Meijer, G.; Bethune, D. S. One- and two-dimensional NMR studies. C_{60} and C_{70} in solution and in the solid state. Fullerenes; ACS: Washington, DC, 1992; Vol. 481, pp. 107-116.
[36] Hawkins, J. M.; Meyer, A.; Loren, S.; Nunlist R. *J. Am. Chem. Soc.* 1991, 113, 9394-9395.
[37] Hawkins, J. M.; Loren, S.; Meyer, A.; Nunlist, R. *J. Am. Chem. Soc.* 1991, 113, 7770-7771.
[38] Hawkins, J. M. *Acc. Chem. Res.* 1992, 25, 150-156.
[39] Burley, G. A.; Keller, P. A.; Pyne, S. G.; Ball, G. E. *Magn. Reson. Chem.* 2001, 39, 466-470.
[40] Smith, A. B. III; Strongin, R. M.; Brard, L.; Furst, G. T.; Romanow, W. J.; Owens, K. G.; Goldschmidt, R. J.; King, R. C. *J. Am. Chem. Soc.* 1995, 117, 5492-5502.
[41] Smith, A. B. III; Strongin, R. M.; Brard, L.; Furst, G. T.; Romanow, W. J.; Owens, K. G.; King, R. C. *J. Am. Chem. Soc.* 1993, 115, 5829-5830.
[42] Meier, M. S.; Spielmann, H. P.; Bergosh, R. G.; Tetreau, M. C. *J. Org. Chem.* 2003, 68, 7867-7870.
[43] Ball, G. E.; Burley, G. A.; Chaker, L.; Hawkins, B. C.; Williams, J. R.; Keller, P. A.; Pyne, S. G. *J. Org. Chem.* 2005, 70, 8572-8574.
[44] Burley, G. A.; Keller, P. A.; Pyne, S. G.; Ball, G. E. *Chem. Commun.* 1998, 2539-2540.
[45] Burley, G. A.; Keller, P. A.; Pyne, S. G.; Ball, G. E. *Chem. Commun.* 2000, 1717-1718.
[46] Burley, G. A.; Keller, P. A.; Pyne, S. G.; Ball, G. E. *Chem. Commun.* 2001, 563-564.
[47] Burley, G. A.; Keller, P. A.; Pyne, S. G.; Ball, G. E. *J. Org. Chem.* 2002, 67, 8316-8330.
[48] Powell, W. H.; Cozzi, F.; Moss, G. P.; Thilgen, C.; Hwu, R. J.-R.; Yerin, A. *Pure Appl. Chem.* 2002, 74, 629-695.
[49] Hoke II, S. H.; Molstad, J.; Dilettato, D.; Jay, M. J.; Carlson, D.; Kahr, B.; Cooks, R. G. *J. Org. Chem.* 1992, 57, 5069-5071.
[50] Maggini, M.; Scorrano, G.; Prato, M. *J. Am. Chem. Soc.* 1993, 115, 9798-9799.
[51] Meier, M. S.; Weedon, B. R.; Spielmann, H. P. *J. Am. Chem. Soc.* 1996, 118, 11682-11683.

[52] Bergosh, R. G.; Meier, M. S.; Laske Cooke J. A.; Spielmann, H. P.; Weedon, B. R. *J. Org. Chem.* 1997, 62, 7667–7672.
[53] Spielmann, H. P.; Wang, G.-W.; Meier, M. S.; Weedon, B. R. *J. Org. Chem.* 1998, 63, 9865–9871.
[54] Weedon, B. R.; Haddon, R. C.; Spielmann, H. P.; Meier, M. S. *J. Am. Chem. Soc.* 1999, 121, 335–340.
[55] Meier, M. S.; Spielmann, H. P.; Bergosh, R. G.; Haddon, R. C. *J. Am. Chem. Soc.* 2002, 124, 8090-8094.
[56] Ford, W. T.; Nishioka, T.; McCleskey, S. C. *Macromolecules.* 2000, 33, 2413–2423.
[57] Ford, W. T.; Nishioka, T.; Qiu, F.; D'Souza, F.; Choi, J.-p.; Kutner, W.; Noworyta, K. *J. Org. Chem.* 1999, 64, 6257–6262.
[58] Ford, W. T.; Nishioka, T.; Qiu F. *J. Org. Chem.* 2000, 65, 5780-5784.
[59] Hirsch, A.; Lamparth, I.; Karfunkel, H. R. *Angew. Chem. Int. Ed. Engl.* 1994, 33, 437-438.
[60] Akasaka, T.; Maeda, Y.; Wakahara, T.; Mizushima, T.; Ando, W.; Walchli, M.; Suzuki, T.; Kobayashi, K.; Nagase, S.; Kako, M.; Nakadaira, Y.; Fujitsuka, M.; Ito, O.; Sasaki, Y.; Yamamoto, K.; Erata, T. *Org. Lett.* 2000, 2, 2671-2674.
[61] Tsuda, M.; Ishida, T.; Nogami, T.; Kurono, S.; Ohashi, M. *Tetrahedron Lett.* 1993, 34, 6911-6912.
[62] Nogamia, T.; Tsuda, M.; Ishida, T. *Fullerene Sci. Technol.* 1993, 1, 275-285.
[63] Ishida, T.; Furudate, T.; Nogami, T.; Kubota, M.; Hirano, T.; Ohashi, M. *Fullerene Sci. Technol.* 1995, 3, 399-409.
[64] Benito, A. M.; Darwish, A. D.; Kroto, H. W.; Meidine, M. F.; Taylor, R.; Walton, D. R. M. *Tetrahedron Lett.* 1996, 37, 1085-1086.
[65] Yinghuai, Z.; Bahnmueller, S.; Chibun, C.; Carpenter, K.; Hosmane, N. S.; Maguire, J. A. *Tetrahedron Lett.* 2003, 44, 5473-5476.
[66] Yinghuai, Z. *J. Phys. Chem. Solids.* 2004, 65, 349-353.
[67] Isaacs, L.; Diederich, F. *Helv. Chim. Acta.* 1993, 76, 2454-2464.
[68] Averdung, J.; Luftmann, H.; Mattay, J.; Claus, K.-U.; Abraham, W. *Tetrahedron Lett.* 1995, 36, 2957-2958.
[69] Creegan, K. M.; Robbins, J. L.; Robbins, W. K.; Millar, J. M.; Sherwood, R. D.; Tindall, P. J.; Cox, D. M. *J. Am. Chem. Soc.* 1992, 114, 1103-1105.
[70] Jagerovic, N.; Elguero, J.; Aubagnac, J.-L. *Tetrahedron.* 1996, 52, 6733-6738.
[71] Gonzalez, S.; Martin, N.; Swartz, A.; Guldi, D. M. *Org. Lett.* 2003, 5, 557-560.
[72] Keshavarz-K., M.; Knight, B.; Haddon, R. C.; Wudl, F. *Tetrahedron.* 1996, 52, 5149-5159.
[73] Yamakoshi, Y. N.; Yagami, T.; Sueyoshi, S.; Miyata, N. *J. Org. Chem.* 1996, 61, 7236-7237.
[74] Zhang, X.; Fan, A.; Foote, C. S. *J. Org. Chem.* 1996, 61, 5456-5461.
[75] Zhang, X.; Foote, C. S. *J. Am. Chem. Soc.* 1995, 117, 4271-4275.
[76] Matsui, S.; Kinbara, K.; Saigo, K. *Tetrahedron Lett.* 1999, 40, 899-902.
[77] Schuster, D. I.; Cao, J.; Kaprinidis, N.; Wu, Y.; Jensen, A. W.; Lu, Q.; Wang, H.; Wilson, S. R. *J. Am. Chem. Soc.* 1996, 118, 5639-5647.
[78] Hoke II, S. H.; Molstad, J.; Yang, S.-S.; Carlson, D.; Kahr, B. *J. Org. Chem.* 1994, 59, 3230-3231.

[79] Shu, L.-H.; Sun, W.-Q.; Zhang, D.-W.; Wu, S.-H.; Wu, H.-M.; Xu, J.-F.; Lao, X.-F. *Chem. Commun.* 1997, 79-80.
[80] O'Donovan, B. F.; Hitchcock, P. B.; Meidine, M. F.; Kroto, H. W.; Taylor, R.; Walton, D. R. M. *Chem. Commun.* 1997, 81-82.
[81] Maggini, M.; Scorrano, G.; Prato, M. *J. Am. Chem. Soc.* 1993, 115, 9798-9799.
[82] Prato, M.; Maggini, M. *Acc. Chem. Res.* 1998, 31, 519-526.
[83] Gan, L.; Zhou, D.; Luo, C.; Tan, H.; Huang, C.; Lü, M.; Pan, J.; Wu, Y. *J. Org. Chem.* 1996, 61, 1954-1961.
[84] Illescas, B.; Rifé, J.; Ortuño, R. M.; Martín, N. *J. Org. Chem.* 2000, 65, 6246-6248.
[85] Matsubara, Y.; Tada, H.; Nagase, S.; Yoshida, Z.-i. *J. Org. Chem.* 1995, 60, 5372-5373.
[86] Schick, G.; Hirsch, A. *Tetrahedron.* 1998, 54, 4283-4296.
[87] Guldi, D. M.; González, S.; Martín, N.; Antón, A.; Garín, J.; Orduna, J. *J. Org. Chem.* 2000, 65, 1978-1983.
[88] Meier, M. S.; Poplawska, M. *Tetrahedron.* 1996, 52, 5043-5052.
[89] Ishida, H.; Ohno, M. *Tetrahedron Lett.* 1999, 40, 1543-1546.
[90] Ishida, H.; Itoha, K.; Ohno, M. *Tetrahedron.* 2001, 57, 1737-1747.
[91] Jagerovic, N.; Elguero, J.; Aubagnac, J.-L. *J. Chem. Soc. Perkin Trans. 1.* 1996, 499-499.
[92] Torres-García, G.; Mattay, J. *Tetrahedron.* 1996, 52, 5421-5426.
[93] Ishida, H.; Komori, K.; Itoha, K.; Ohnob, M. *Tetrahedron Lett.* 2000, 41, 9839-9842.
[94] O-kawa, K.; Nakamura, Y.; Nishimura, J. *Tetrahedron Lett.* 2000, 41, 3103-3106.
[95] Ohno, M.; Kojima, S.; Shirakawa, Y.; Eguchi, S. *Tetrahedron Lett.* 1995, 36, 6899-6902.
[96] Sabirov, D. S. Polarizability of C_{60}/C_{70} Fullerene [2+1]- and [1+1]-Adducts: A DFT-Prognosis. Density Functional Theory: Principles, Applications and Analysis; Nova Science Publishers, Inc.: New York, 2013; Vol. pp. 147-170.
[97] Helgaker, T.; Coriani, S.; Jørgensen, P.; Kristensen, K.; Olsen, J.; Ruud, K. *Chem. Rev.* 2012, 112, 543-631.
[98] Hfiser, M.; Ahlrichs, R.; Baron, H. P.; Weis, P.; Horn, H. *Theor. Chim. Acta.* 1992, 83, 455-470.
[99] Heine, T.; Seifert, G.; Fowler, P. W.; Zerbetto, F. *J. Phys. Chem. A.* 1999, 103, 8738-8746.
[100] Sun, G.; Kertesz, M. *J. Phys. Chem. A.* 2000, 104, 7398-7403.
[101] Sun, G.; Kertesz, M. *J. Phys. Chem. A.* 2001, 105, 5468-5472.
[102] Sun, G.; Kertesz, M. *Chem. Phys.* 2002, 276, 107-114.
[103] Sternfeld, T.; Thilgen, C.; Chen, Z.; Siefken, S.; Schleyer, P. v. R.; Thiel, W.; Diederich, F.; Rabinovitz, M. *J. Org. Chem.* 2003, 68, 4850-4854.
[104] Hauke, F.; Chen, Z.-F.; Hirsch, A. *Polish J. Chem.* 2007, 81, 971-983.
[105] Fileti, E. E.; Rivelino, R. *Chem. Phys. Lett.* 2009, 467, 339-343.
[106] Laikov, D. N. The development of an economic approach to calculate molecules by means of Density Functional Theory and its implementation to solve the complex chemical researches, *Ph.D. Dissertation*, Moscow State University, 2000.
[107] Laikov, D. N. *Chem. Phys. Lett.* 1997, 281, 151-156.
[108] Laikov, D. N.; Ustynyuk, Y. A. *Russ. Chem. Bull. Int. Ed.* 2005, 54, 820-826.
[109] Wolff, S. K.; Ziegler, T. *J. Phys. Chem.* 1998, 109, 895-905.
[110] Perdew, J. P.; Burke, K.; Ernzerhof, M. *Phys. Rev. Lett.* 1996, 77, 3865-3868.

[111] Shestakov, A. F. *Russ. J. Gen. Chem.* 2008, 78, 811-821.
[112] Sabirov, D. S.; Garipova, R. R.; Bulgakov. R. G. *Chem. Phys. Lett.* 2012, 523, 92-97.
[113] Pankratyev, E. Y.; Tulyabaev, A. R.; Khalilov, L. M. *J. Comput. Chem.* 2011, 32, 1993-1997.
[114] Tulyabaev, A. R.; Khalilov, L. M. *Comput. Theor. Chem.* 2011, 976, 12-18.
[115] Khalilov, L. M.; Tulyabaev, A. R.; Yanybin, V. M.; Tuktarov, A. R. *Magn. Reson. Chem.* 2011, 49, 378-384.
[116] Khalilov, L. M.; Tulyabaev, A. R.; Tuktarov, A. R. *Magn. Reson. Chem.* 2011, 49, 768-774.
[117] Tulyabaev, A. R.; Tuktarov, A. R.; Khalilov, L. M. *Magn. Reson. Chem.* 2014, 52, 3-9.

In: Fullerenes
Editor: Shannon B. Ellis
ISBN: 978-1-63321-385-2

Chapter 5

ORGANIC AND INORGANIC AZIDES IN FULLERENE CHEMISTRY: ACHIEVEMENTS AND TRENDS

Airat R. Tuktarov**, *Arslan R. Akhmetov* and *Usein M. Dzhemilev

Institute of Petrochemistry and Catalysis of Russian Academy of Sciences, Ufa, Bashkortostan, Russian Federation

ABSTRACT

In this review, the research results on the application of organic and inorganic azides in fullerene chemistry are summarized and systematized.

The published data on the chemistry of intermolecular cycloaddition of organic azides to fullerenes C_{60} in terms of thermal and photochemical reactions to give triazolinofullerenes, 5,6-azahomofullerenes, 5,6- and 6,6-aziridinofullerenes are presented in detail.

The first reperesentatives of azidofullerenes were synthesized by the reactions between fullerenes and inorganic azides, namely HN_3, IN_3, BrN_3, metal azides and organometallic azides.

The mechanisms for the formation of new nitrogen-containing fullerene derivatives are discussed.

Keywords: Fullerenes; organic azides; hydrazoic acid; halogen azides; metal azides; organometallic azides; addition; cycloaddition; triazolinofullerenes; 5,6-aziridinofullerenes; 5,6-azahomofullerenes; 6,6-aziridinofullerenes; azidofullerenes

* e-mail: tuktarovar@gmail.com.

1. Introduction

Since the discovery of fullerenes, nearly 30 years have passed. During this relatively short period, the investigations related to various aspects of the synthesis, properties and applications of carbon clusters have rapidly developed as evidenced by more than 40 000 publications to date.

Almost immediately, it has become clear that this unusual and nontraditional form of carbon may appear to be principally new material with unique properties. However, the first attempts to expand the possible boundaries of fullerene applications in different industrial fields as well as in the laboratory practice have failed owing to the problems associated with poor solubility of fullerenes in organic solvents.

Further investigation efforts were preferably aimed at developing preparative procedures that allow the introduction of different substituents, including functional and organometallic groups, into the fullerene molecule, thus improving the solubility of fullerene derivatives in organic and aqueous media as well as giving them required properties for further application.

Many scientists, both in our country and abroad, have succeeded in their efforts. As a resut, the sufficiently selective methods for the introduction of functional groups, including heterocyclic, into the fullerene molecule have been developed by using electrophilic and nucleophilic reagents.

The preparative methods such as the Prato reaction [1], the Bingel-Hirsch [2, 3] cycloaddition reaction of diazo compounds [4], hydroamination, hydrometallation and cyclometallation [5] of carbon clusters have gained the most popularity and practical importance in synthetic fullerene chemistry.

These works are widely cited in the world scientific literature and summarized in numerous articles, books and patents.

In recent years, the investigations of Kazan School of Organic Chemists (Russia) under the leadership of academician Sinyashin O.G. into the reaction of C_{60} fullerene with organic azides have also gained popularity and fame.

Recently, the authors of this review have implemented the reactions between C_{60} fullerenes and inorganic and organometallic azides.

As a result, the literature has accumulated extensive material devoted to the study of the reactions between C_{60} fullerenes and organic, inorganic or organometallic azides, unfortunately, published in the disparate articles, patents and reviews that require analysis and summerizing in order to identify prospects for the development of this exciting and promising research field, namely, chemistry of nitrogen-containing fullerenes.

2. Reactions of Fullerenes with Organic Azides

2.1. Thermal and Photochemical Cycloadditions of Organic Azides to C_{60} and C_{70} Fullerenes

Thermal cycloadditions between organic azides and carbon clusters occur in two main directions, namely, synchronous [2+1]-cycloaddition of nitrenes, generated *in situ* during the thermal decomposition reaction of the corresponding azides, to fullerene affording aziridinofullerenes **1** and 1,3-dipolar cycloaddition of organic azides followed by extrusion of nitrogen from the triazoline-containing fullerene leading to a mixture of [2+1]-opened and [2+1]-closed isomers. By theory, this reaction can afford four isomeric fullerene monoadducts **1–4** with [6,6]- and [5,6]-opened, or [6,6]- and [5,6]-closed carbon spheres.

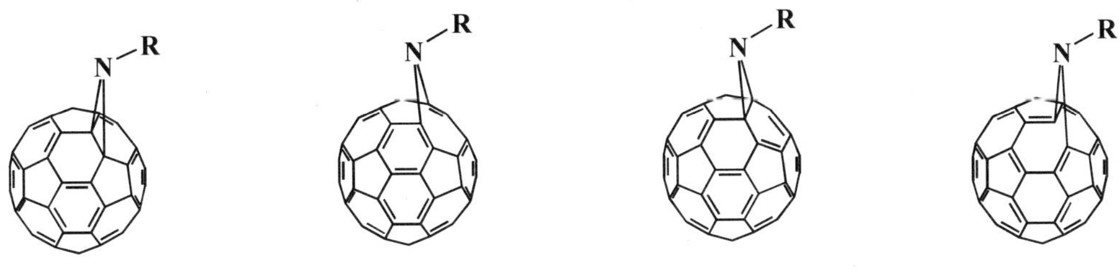

1 6,6 closed

2 5,6-opened

3 5,6-closed

4 6,6-opened

In contrast to diazo compounds [4], the [2+1]-cycloaddition reaction between organic azides and C_{60} affords [5,6]-closed isomers [6, 7].

Thus, under conditions of the reaction C_{60} with isocyanurato-substituted azides, the authors of [6, 7] have first synthesized [5,6]-aziridinofullerenes **5** (Scheme 1) which were quantitatively converted to corresponding [5,6]-azahomo- **6** and [6,6]-aziridinofullerenes **7**.

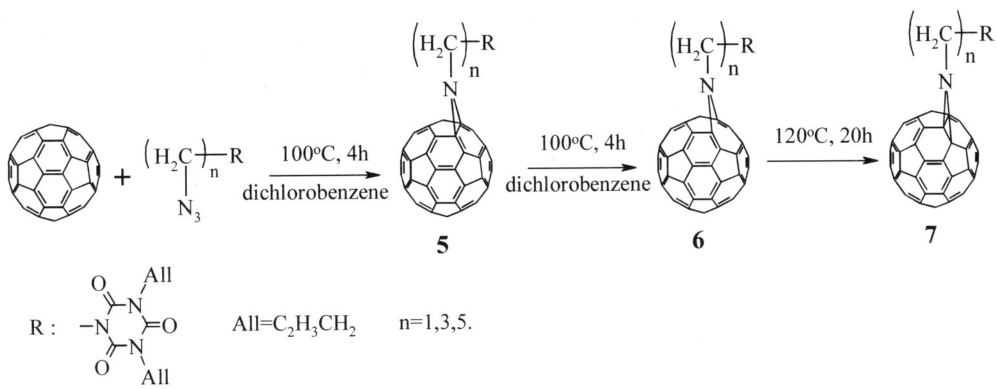

Scheme 1.

It is noteworthy that most thermal reactions of C_{60} with organic azides irrespective of temperature proceed via [3+2]-cycloaddition to give initially fullerotriazolines **8**, which in some cases were isolated in individual form determining their structures by X-ray christallography [8-12]. The subsequent extrusion of the nitrogen molecule from triazoline derivative **8** leads to 5,6-azahomofullerenes **9** and 6,6-aziridinofullerenes **10**, whose structures were confirmed by NMR spectral methods including those on ^{15}N nuclei. Typically, the yields of 6,6-closed [2+1]-cycloadducts in these reactions do not exceed 1–3%.

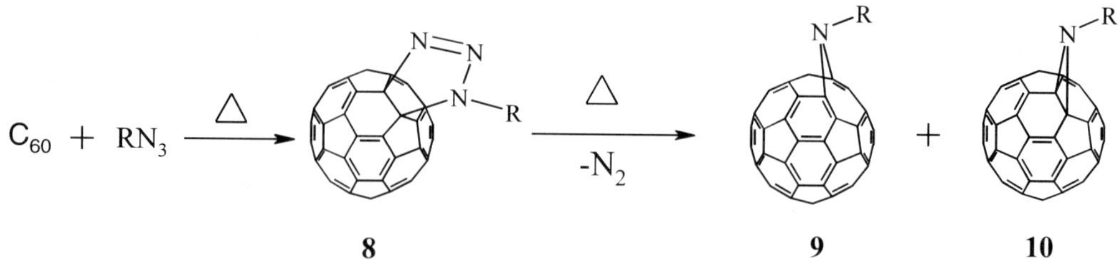

Scheme 2.

Using quantum-chemical calculations of the model reaction between methyl azide and C_{60} fullerene it was shown that the addition of methyl azide to fullerene initially occurs across a [6,6]-double bond junction yielding triazoline intermediate with a energy barrier of about 20 kcal/mol and exothermicity of ca. 2 kcal/mol at B3LYP/6-31G*//AM1 level of theory [12]. The thermal loss of the nitrogen molecule occurs successively. In the first stage, the cleavage of the N–N single bond precedes the breaking of the N–C bond with the total activation energy of 45 kcal/mol. The extrusion of N_2 occurs simultaneously with the formation of the new N–C bond in the second stage when the steric factor of the leaving N_2 molecule contributes to the addition of nitrene to [5,6] junction. The adducts being formed across the [6,6] junction were negligable.

A good example of the synthesis of stable [2+3]-cycloadducts from organic azides and C_{60} fullerenes is the interaction between C_{60} and the tetrathiafulvalene-based azides affording individual triazolines **11**, which are stable at room temperature (Scheme 3) [12]. While heating compound **11** up to 120°C in 1,2-dichlorobenzene, a mixture of [6,6]-aziridinofullerenes **12** (30 %) and [5,6]-azahomofullerenes **13** (20 %) exhibiting properties of amphoteic redox-system, has been obtained.

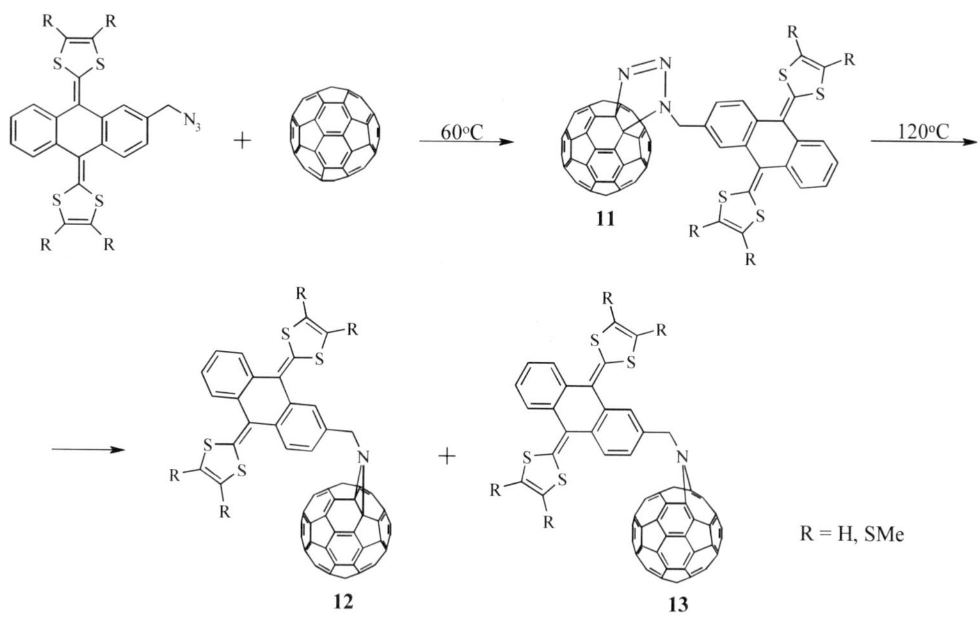

Scheme 3.

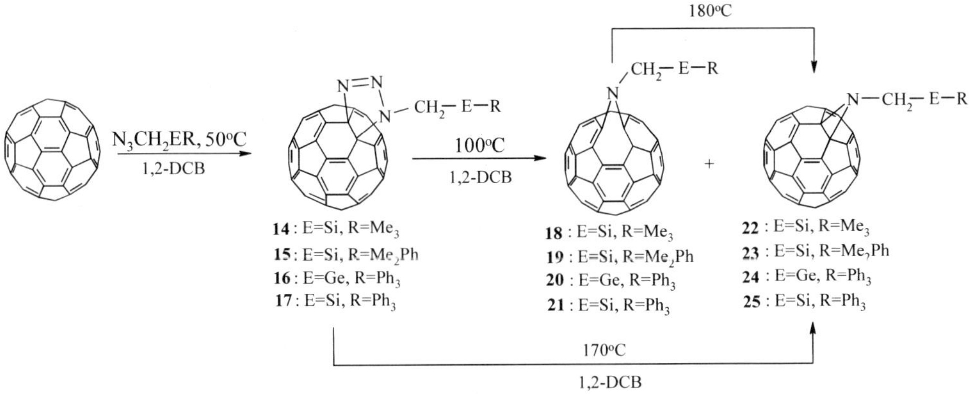

Scheme 4.

When the thermal decomposition proceeds via the predominant formation of [5,6]-azahomofullerenes, in that case, under treatment of these compounds with trifluoromethanesulfonic acid or boron trifluoride etherate, exclusively [6,6]-closed [2+1]-cycloadducts **27** were obtained in quantitative yields [14].

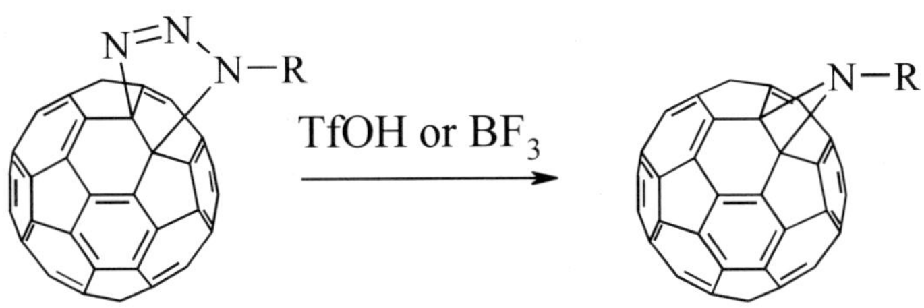

26 **27**

R = CH_3, n-C_4H_9, n-C_8H_{17}, $(CH_2)_3CO_2Me$, CH_2Ph, Ph

Scheme 5.

Another alternative method for decomposing triazolinofullerenes was suggested in the work [15]. The authors have found that when irradiated with light of the wavelength of 300 nm, [3+2]-cycloadducts **28** converted into corresponding [6,6]-closed aziridines **30**, which can be also obtained by isomerization of [5,6]-azahomofullerenes under photochemical conditions. The authors have found that when irradiated by light of the wavelength of 300 nm, [3+2]-cycloadducts were transformed to corresponding [6,6]-closed aziridines **30**, which can be also prepared by isomerization of [5,6]-azahomofullerenes **29** under photochemical conditions (Scheme 6).

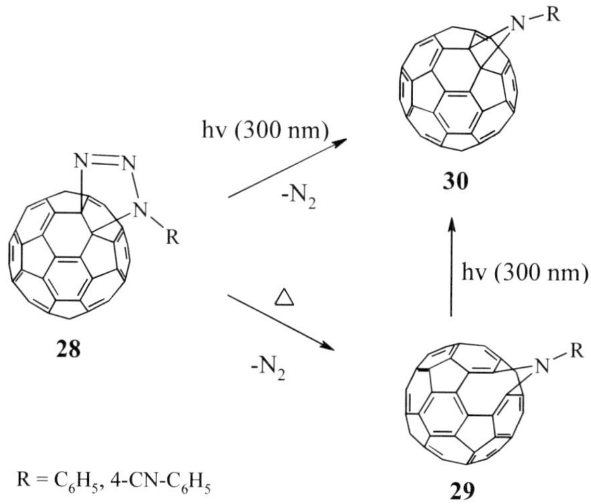

Scheme 6.

Later on, the authors of [16, 17] have studied the effect of structure of the substituent at the nitrogen atom in the [5,6]-open cycloadduct **31** on the rate of its photorearrangement into corresponding [6,6]-aziridinofullerenes **32** (Scheme 6). The required time for completion of the reaction was varied in a series of fullerenes **a: b: c: d** as follows: 1440: 1:> 2160: 360.

Thus, more than a 2,000-fold difference between the photochemical rearrangement rates for the fastest (b) and the slowest (c) processes exists. A 3000-fold increase in photochemical rearrangement rate, from the slowest (e) to the fastest (g), was caused by the introduction of the methyl substituent in various positions of the phenyl ring [18]. Photolysis of compounds **a–d** in solutions saturated by air was much slowly as in an inert atmosphere indicating the intermediate triplet state of the original azafulleroid.

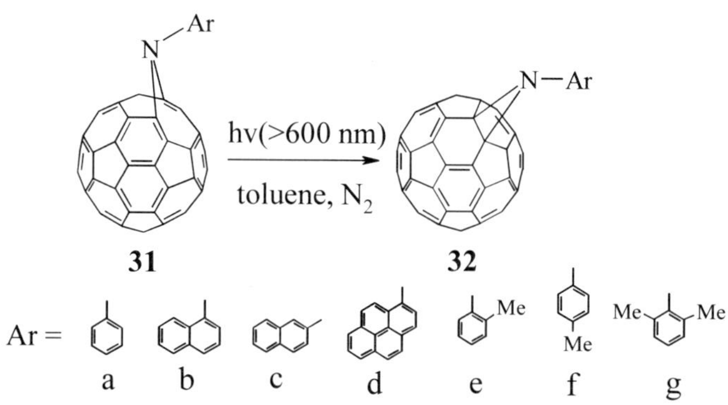

Scheme 7.

In furtherance of the above-mentioned works, there was shown that photolysis of [5,6] azahomofullerene **33** in saturated with oxygen 1,2-dichlorobenzene, provides the formation of ketolactame **34** as the promising material for manufacturing solar cells [19].

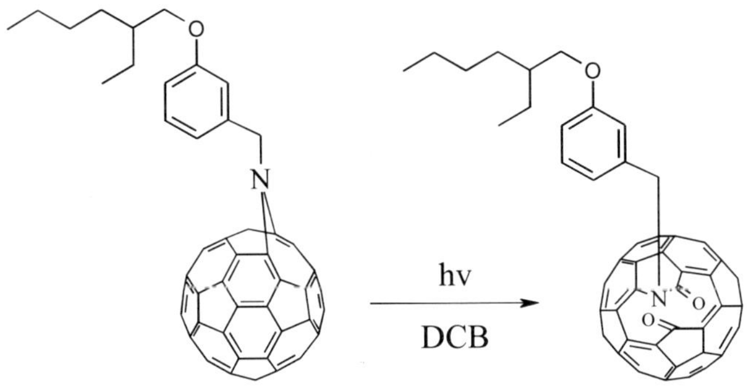

33 **34**

Scheme 8.

The interesting results have been obtained by a group of authors of [20, 21]. They showed that [6,6]-aziridinofullerenes **35a–в**, in turn, can undergo inverse catalytic isomerization to corresponding [5,6]-azahomofullerenes 36a–d using chloroamine and molecular sieves (MS4Å) (Scheme 9).

Thus, the elaborated methods for isomerization of [5,6]-azahomofullerenes to corresponding aziridines, and vise versa, allow to sufficiently extend the possible application areas for N-heterocyclic [2+1]-adducts of fullerenes, in particular, using them as perspective molecular switches.

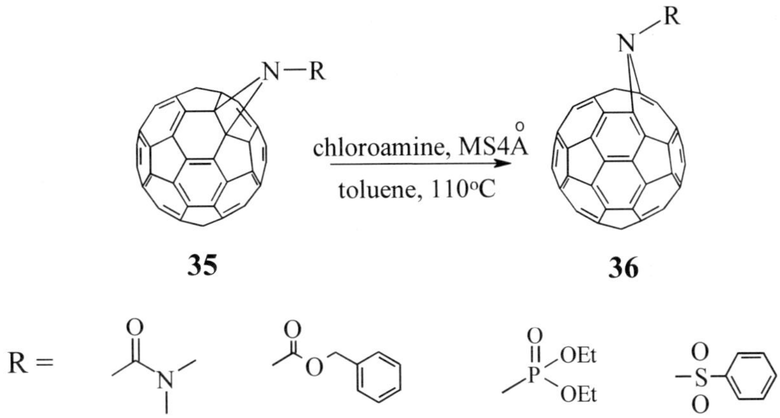

Scheme 9.

Subsequent experiments showed that functionally substituted azides can be involved in the reaction with C_{60} fullerene. For example, $N_3CO_2{}^tBu$ reacted with C_{60} in 1-chloronaphthalene at 60°C affording stable fullerotriazoline **37** (Scheme 10). Refluxing of this triazoline in toluene led to a mixture of 5,6-open **38** (azahomofullerene) and [6,6]-closed **39** (aziridinifullerene) adducts in a 16:3 ratio respectively [22].

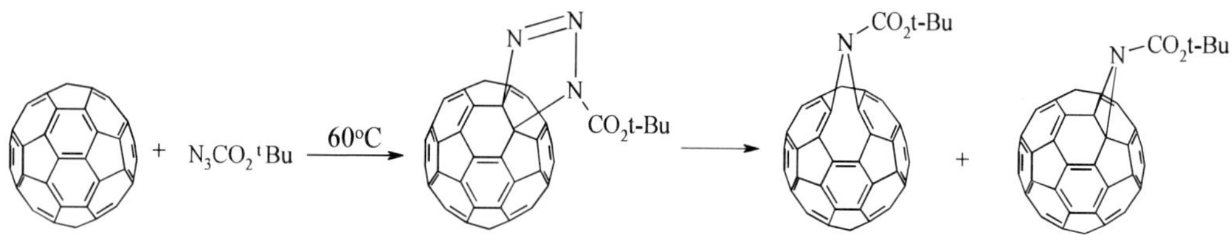

37 38 39

Scheme 10.

Treatment of compound **38** with trifluoroacetic acid at room temperature led to elimination of the ester group resulting in [5,6]-azahomofullerene **40** [23], which is of interest as a precursor for the preparation of different functionally substituted fullerenes.

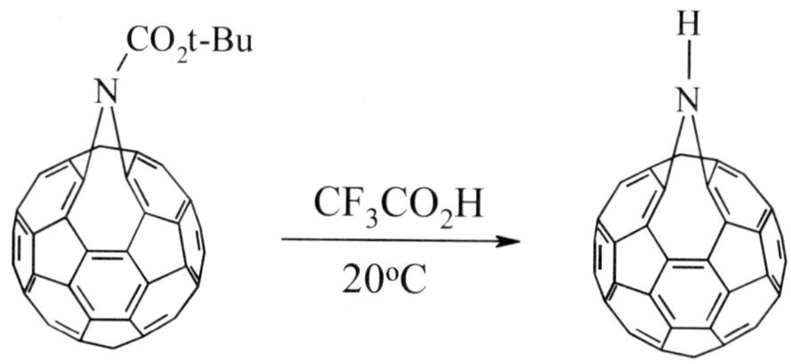

38 40

Scheme 11.

The addition of *tert*-butylazidoformate to C_{60} in 1,1,2,2,-tetrachloroethane at 147°C occurs by the nitrene-transfer mechanism, that is, with the N_2 elimination followed by fast (within a few minutes) formation *N-tert*-butoxycarbonylaziridino[2',3':1,2][60]fullerene **39** (Scheme 12) [24].

The same result could also be achieved by the generation of nitrene under milder conditions, that is, by base-induced α-elimination of O-4-nitrophenyl-*tert*-butylsulfonylhy droxamic acid from compound **41** in the presence of C_{60} under phase-transfer conditions at room temperature. Subsequent elimination of CO_2 from 39 led to unsubstituted aziridinofullerene **42** in the form of stable and easily isolable solid.

The elimination of the *tert*-butoxycarbonyl group from cycloadduct **39**, as shown in [20], can be done by chromatography of 39 in chloroform on neutral silica (pH 7). This methodology for deblocking the N atom of *N*-aziridine ring was also used in the case of phosphorous-containing aziridinofullerene **43** [21, 25].

Using similar schemes, a series of functionally substituted fullerenes **44** and **45** have been obtained, wherein the [6,6]-closed adducts represented a major part of this mixture (Scheme 12) [26, 27].

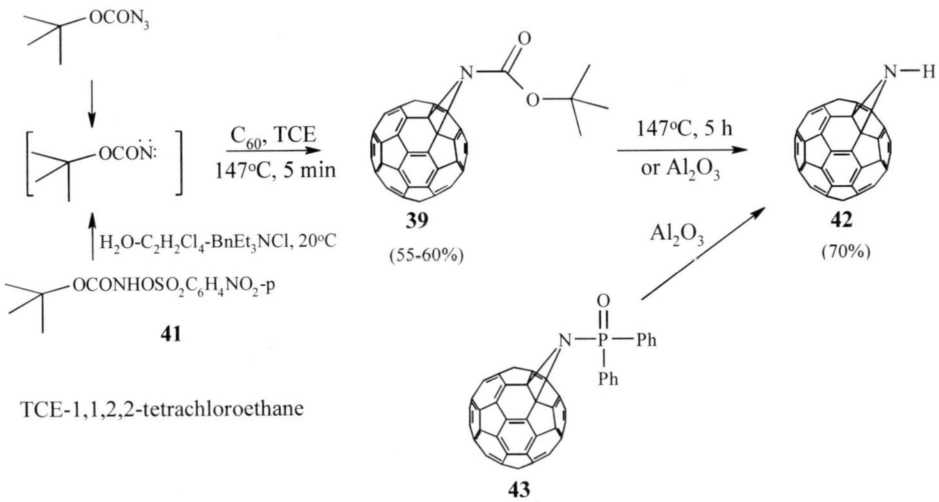

Scheme 12.

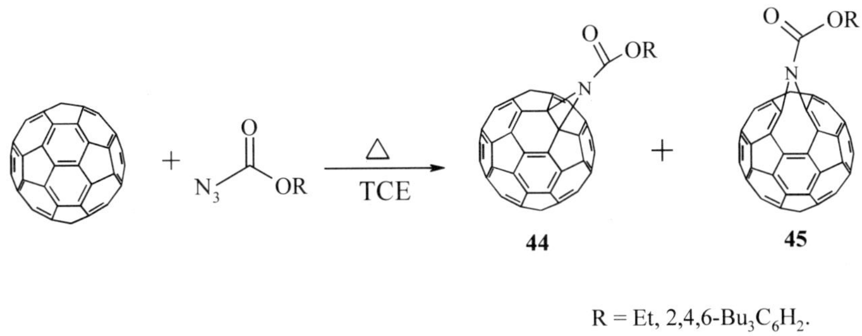

Scheme 13.

Some of directional transformations of aziridinofullerenes **42** were studied in the work [25]. Thus, the water-soluble fullerenes **46** (urethanes, amides, sulfonamides) have been synthesized through acylation of **42** [28].

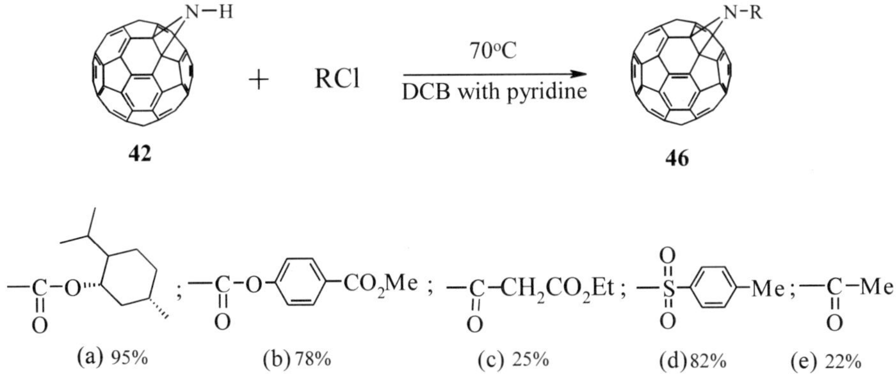

Scheme 14.

Lately, there appeared an increased research interest in production of semiconductors on the basis of fullerene derivatives, including those containing nitrogen atoms [29]. The presence of heteroatom in the molecule has a positive effect on the electronic properties of C_{60}, thus greatly improving the quality of devices constructed. In this regard, [5,6]-azahomofullerenes **47**, are of special interest, (Scheme 15), since they can retain unique electronic structure of original carbon cluster [31–33].

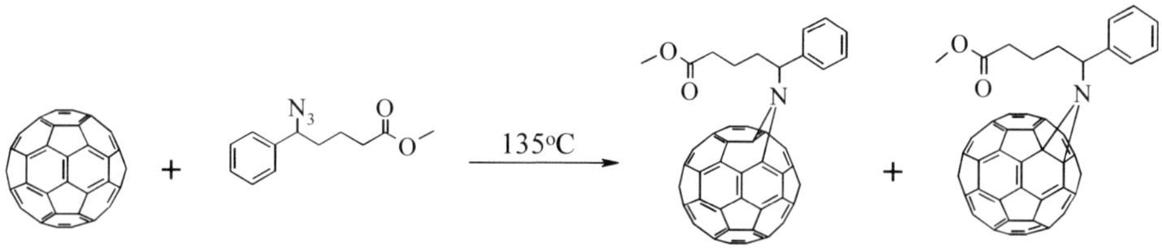

Scheme 15.

At the same time, there was found that the introduction of four electron-donating groups into the molecule of **48** contributes to improving photovoltaic properties of [5,6]-azahomofullerene (adduct **50**) as the perspective material for solar panels [34].

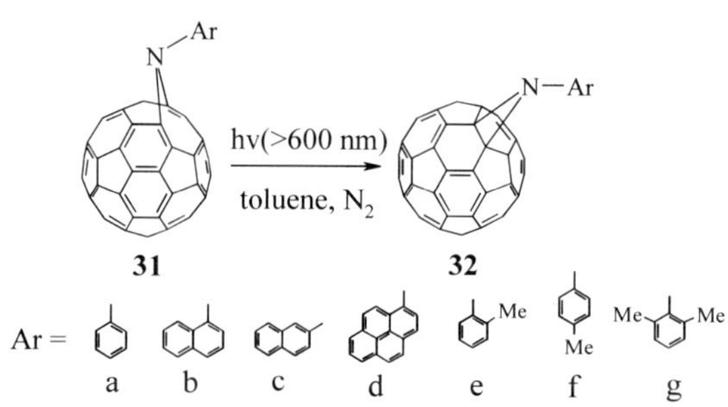

Scheme 16.

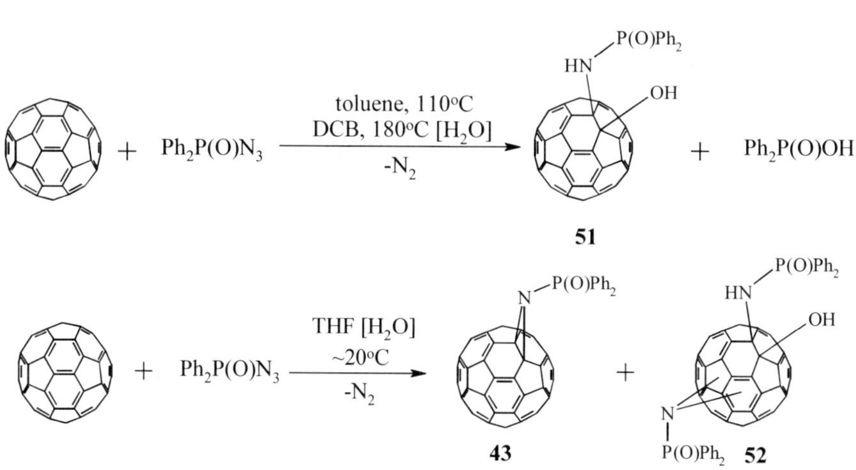

Scheme 17.

Of special interest and practical value are the phosphorylated derivatives of C_{60}. The first message on the synthesis of heterocyclic phosphorous-containing fullerenes appeared in the works [35, 36] devoted to the chemical reaction between diphenylphosphinic azide and C_{60} fullerene. The detailed study of this reaction showed that if using toluene or 1,2-dichlorobenzene, acyclic fullerene adducts **51** were the end products. In the presence of THF as the solvent, this reaction will produce the cyclic compound **43** (Scheme 17).

Because of the poor solubility, the structures for acyclic and cyclic adducts **43, 51** and **52** synthesized have been determined only by using IR spectroscopy and mass spectrometry.

Phosphorylated C_{60} fullerenes appear to be the most promising for production of solar energy photoconverters, but its low solubility do not allow extending the influence of the phosphorylated moiety on the fullerene core thus improving its electrochemical reducing ability.

To increase the solubility of the target phosphorus-containing fullerene adducts a methylene spacer was inserted in the molecule of the original azide between the phosphoryl and azido groups [37]. As a result of interactions between such azides and C_{60} fullerene, a hard separable mixture of adducts has been identified.

In the absence of positive results on the selective cycloaddition between C_{60} fullerenes and phosphorus-containing azides with the methylene groups, the authors of [38] have introduced hydrophilic substituents into the molecule of the original azides. The experiments have shown that C_{60} reacts with diethylazido phosphonate also giving rise to a hard separable mixture of [5,6]-azahomofullerenes **53** and [6,6]-aziridinofullerenes **54** (Scheme 18).

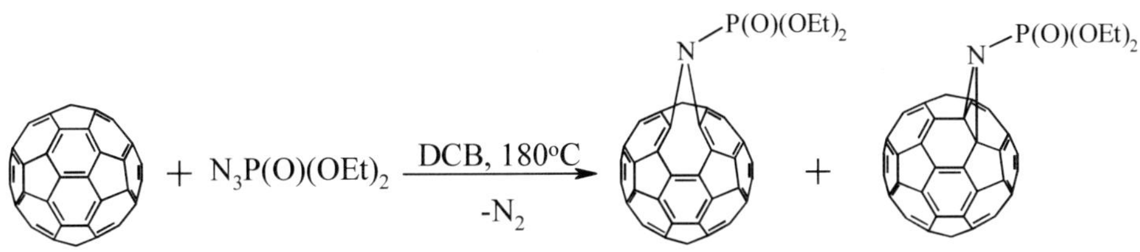

53 **54**

Scheme 18.

However, when using dibutyl- or diphenylazido phosphonates in the reaction with C_{60} [37], the major reaction products were [5,6]-azahomofullerenes **55** and **56**. Their yields were achieved 20–25%, while the content of [6,6]-aziridinofullerenes **57** and **58** did not exceed 1–3 % (Scheme 18). Reducing the reaction temperature to 130°C, the proportion of 6,6-closed adducts was increased up to 5–7 %.

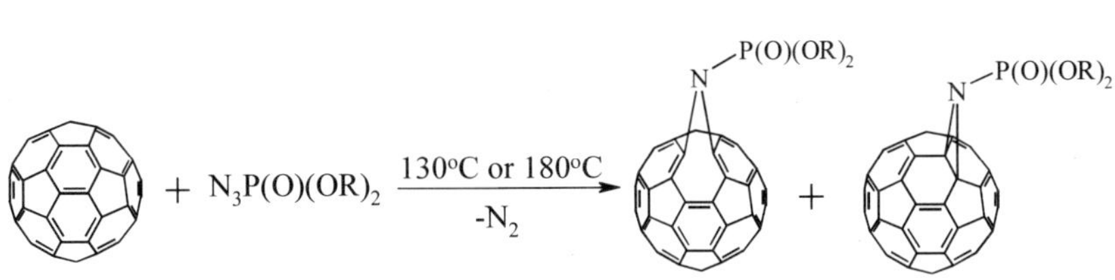

55, 56 **57, 58**

R=Bu(**55, 57**), R=Ph(**56, 58**)

Scheme 19.

Later, it was found that the reaction of sulfonyl azides with C_{60} fullerenes led to a mixture of open-cage adducts **59** and closed-cage adducts **60** in the case of the alkyl derivatives of the said azides. But only [5,6]-open azahomofullerenes **59c,d** were the reaction products while using aryl sulfonyl azides (Scheme 20) [39]. At the same time, there were shown that all synthesized open cycloadducts **59** can undergo phorochemical rearrangement to corresponding [6,6]-aziridinofullerenes **60**. In addition, adduct **60b** possesses the unigue ability to the inverse thermal rearrangement into [5,6]-azahomofullerene **59b** [40] that makes them promising compounds in production of molecular switches.

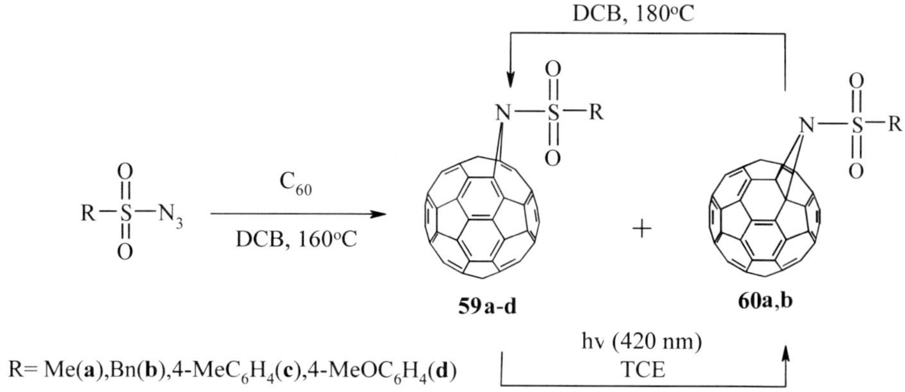

Scheme 20.

The selective synthesis of [5,6]-azahomofullerenes **61 a–c** from carbon clusters and sulfonyl azides occurs under microwave irradiation [41]. In that case, the reaction time was reduced from 2 hours to 20 minutes as compared to the thermal process (Scheme 21).

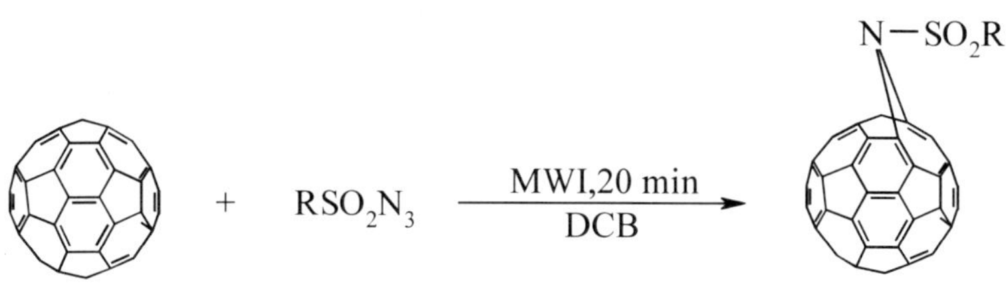

MWI-microwave irradiation **61**

R = **a:** $ICF_2CF_2OCF_2CF_2$; **b:** $HCF_2CF_2OCF_2CF_2$; **c:** C_4H_9

Scheme 21.

Unlike C_{60} fullerene which contains 30 equivalent double bonds, the molecule of C_{70} fullerene (symmetry group D_{5h}), wherein only a portion of hexagons have adjacent hexagons, there exist eight different types of junctions that theoretically may give a great number of possible isomeric adducts from cycloaddition reactions. However, C_{70} and other higher fullerenes show remarkable selectivity under formation of monoadducts and bisadducts [42, 43–45], as evidenced by numerous calculations [46–48] and experimental [49, 50] results.

Thus, the reaction sulfonyl azides and C_{70} under thermal conditions (160°C) led to regioisomeric 5,6-azahomofullerenes **62** (60 %) and **63** (40 %) as being isomeric adducts of azide cycloaddition across the C(2)–C(3) and C(7)–C(8) bonds (Scheme 22) [39].

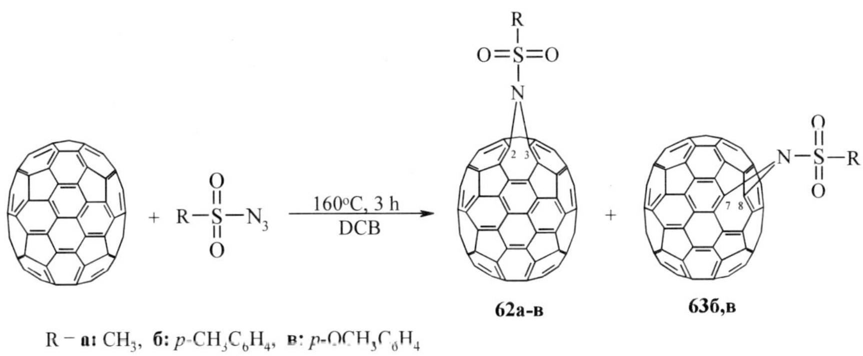

Scheme 22.

Similarly, the reaction between C_{70} and ethylene glycol methyl ether azide afforded two regioisomeric azahomo[70]fullerenes **64** and **65** [51].

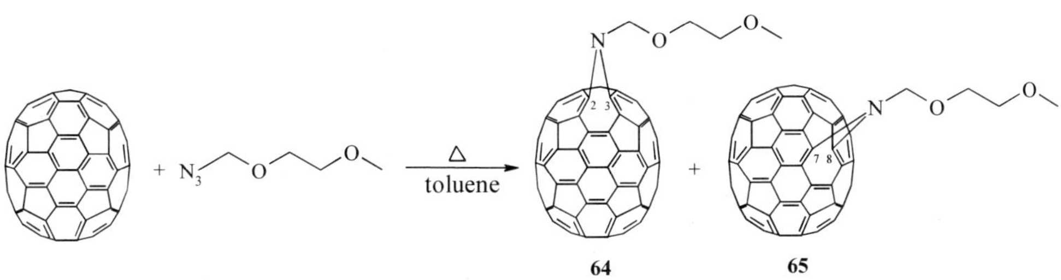

Scheme 23.

The above examples show that cycloaddition of organic azides to C_{60} or C_{70} fullerenes is considered as one of the popular procedures for functionalization of fullerene including those for the synthesis of important carbon clusters. Access to a wide range of original azides is the main advantage of this method [52–54]. Using this approach, azahomofullerenes **66** having the acrydine residue which is capable of intercalation into the DNA double helix have been synthesized (Scheme 24) [52]. Insertion of catechine as a hydrophilic residue into the molecule of C_{60} allowed synthesizing compounds, which can produce singlet oxygen during photooxidation process under visible or UV irradiation. It was established that the aziridine residue in the C_{60} molecule is responsible for the visible-light activity, while the hydrophilic groups enhance mobilization of cimetidine from an aqueous phase onto the surface of catalytic species [55].

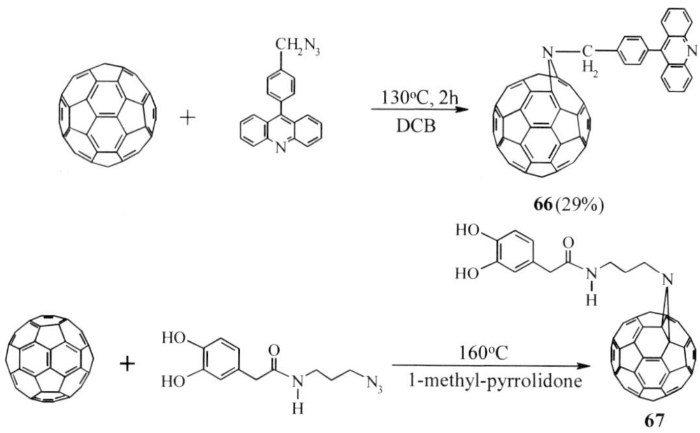

Scheme 24.

The ability of fullerene derivativess to self-assembly on different surfaces makes it possible to use them as semiconductors, insulators and high-temperature superconductors. Thus, the cycloaddition reaction between oligopyridine-based azides and C_{60} led to derivatives with bi- (**68**) and tripyridyl **69** groups, being used for self-assembly on the surface of gold nanoparticles (Scheme 25) [56]. It was revealed that nonlinear absorption and nonlinear refraction of the obtained monoadducts **68** and **69** depend on the interactions between the gold surface and the stabilizer based on these compounds. The results of scanning tunneling microscopy showed that the fullerene derivatives (type **68**) spontaneously form self-assembled monolayers on the surface of gold particles with an average radius of about 10 nm [57].

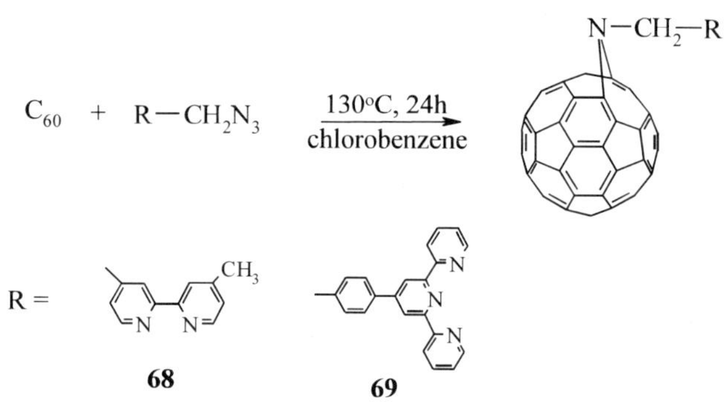

Scheme 25.

In recent years, fullerene derivatives with chromophoric function have attracted special attention because of their ability to intramolecular energy transfer from the addend to the fullerene molecule. In this regard, C_{60} fullerene as acceptor was bonded with azido-1,8-naphthyl imide in refluxing 1,2-dichlorobenzene or under microwave irradiation (MWI) that was much more effective. As a result, [6,6]-aziridinofullerene **70** has been obtained in approximately 60 % yield (Scheme 26) [58].

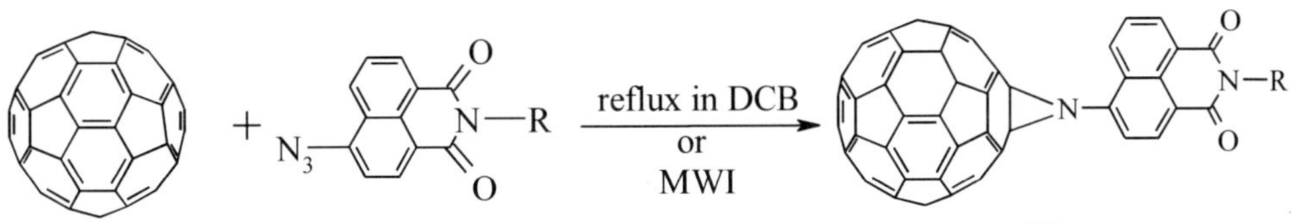

70

$R = C_8H_{17}, C_{12}H_{25}, C_{16}H_{33}, C_2H_5OC_6H_4, CH_3C_6H_4, C_6H_5$

Scheme 26.

Functionalized conducting polymers are an interesting class of materials which can specifically interact with photons. This property can be used to create solid-state devices capable of photoinduced charge transfer and electropolymerization. Therefore, in order to synthesize such conducting polymers, the authors of [59] have synthesized monomer **71** with the bithienyl fragment. Unfortunately, the resulting material did not possess the property of donor–acceptor electron transfer.

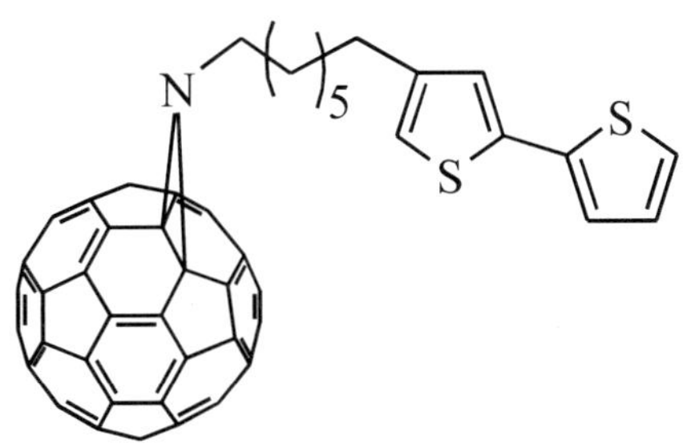

71

To design water-soluble fullerenes **72–74** per-O-acetyl glycosyl azides have been incorporated into C_{60} fullerene. The choice of these compounds was conditioned by the fact that the sugar moiety does not only increase water solubility of the molecule, but also plays the significant role in the interactions between these compounds and cell membranes. The research into the photochemical properties of [5,6]-azahomofullerenes **73** and **74** revealed their ability to convert triplet oxygen into singlet form (Scheme 27) [61]. The most efficient generators of singlet oxygen were the compounds bearing the residue of acetic acid, as compared to those having unsaturated hydroxyl groups.

The alternative method of producing water soluble fullerenes includes the reaction between fullerene and glycosyl azides under ultrasonic conditions [62]. But these transformations occurred during 2 days, and the yields of the desired [5,6]-azahomofullerenes, in that case, did not exceed 6%.

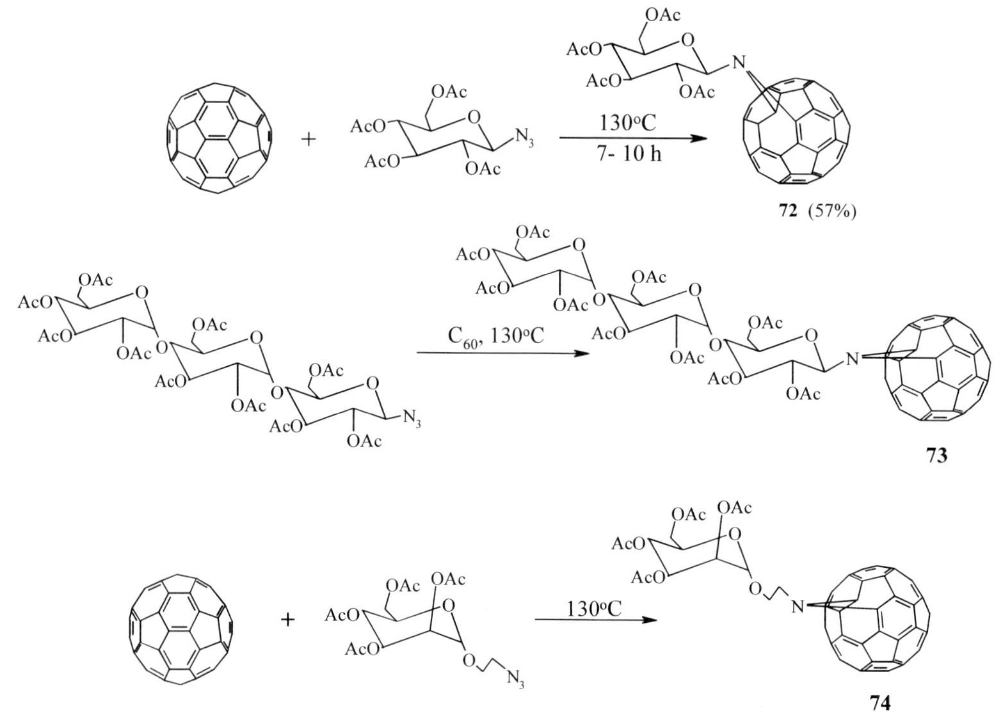

Scheme 27.

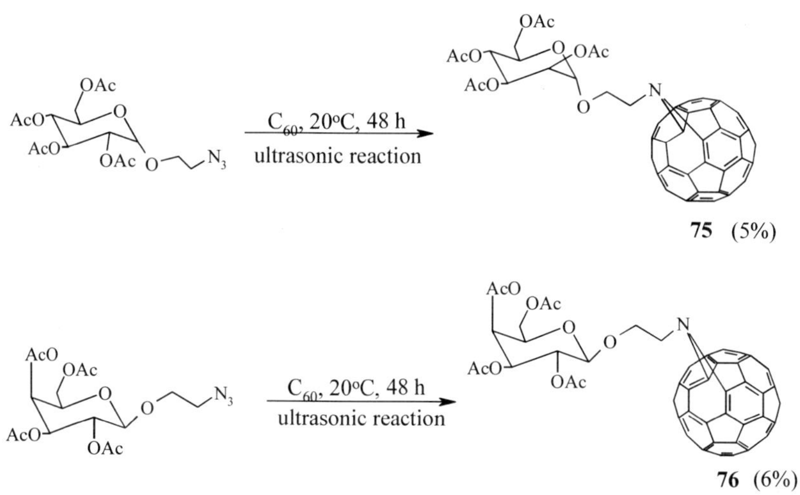

Scheme 28.

Along to glycozyl azides, C_{60} fullerene can undergo the cycloaddition reaction with linear polysaccharides, which were isolated from the fungi *Aureobasidium pullulans* as water soluble non-ionic polymers. Thus, acylated azide entered into reaction with C_{60} in refluxing chlorobenzene to afford appropriate [5,6]-azahomofullerene **77**, hydrolysis of which led to water-soluble derivative **78** (Scheme 29) as perspective material for medicine [63].

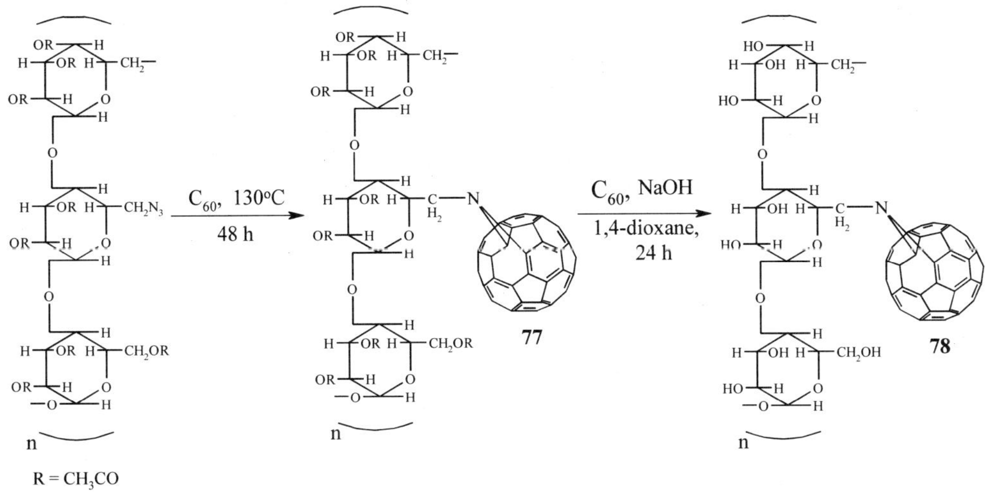

Scheme 29.

Polyphenylene vinylene and its derivatives (PPV-1 **79**, PPV-2 **80** and PPV-3 **81**) being covalently bonded to C_{60} fullerene exhibit unusual electrochemical and electronic properties (Scheme 30). The composite films made on their basis are efficient photocells for opthoelectronic devices, light emitting diodes (LED) and solar batteries [64, 65].

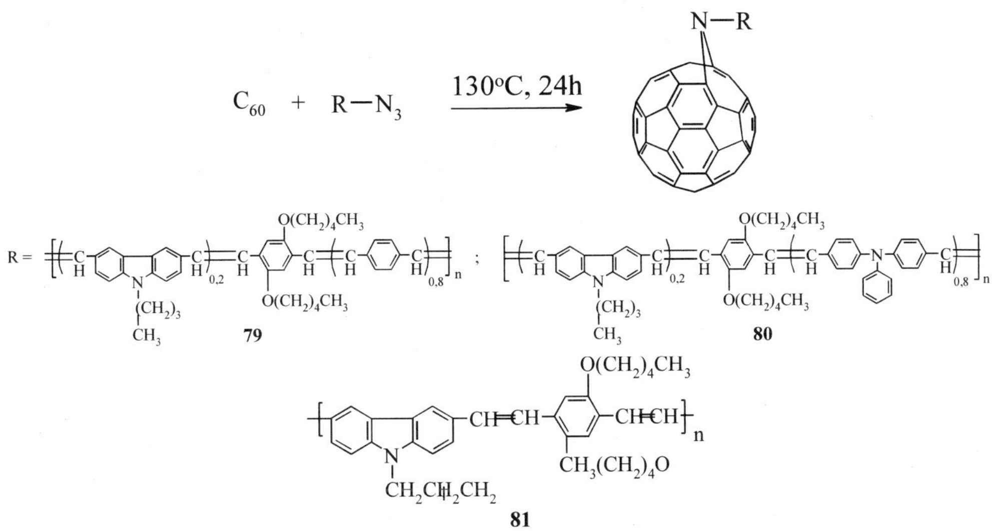

Scheme 30.

A short hydrophilic chain that covalently bonded to the hydrophobic C_{60} molecule in **56** allows the formation of stable, well-ordered fullerene monolayers [66–68]. It was found that, upon addition of Ag cations, adduct **82** changes its absorption spectrum due to direct interaction of Ag^+ with C=C double bond on the fullerene surface in the position of the organic addend attachment [69]. This system serves as a novel methodology to change spectroscopic and redox properties of C_{60}.

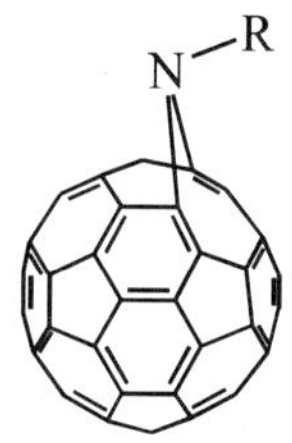

82: R = $CH_2CH_2OCH_2CH_2OCH_2CH_2OCH_3$
83: R = $CH_2OCH_2CH_2OCH_3$

Compound **83**, under photolytic conditions in dichlorobenzene saturated with pure oxygen, formed racemic ketolactam **84** as being the first representative of open-cage fullerenes [70], which under acidic conditions in the gas phase can lose 2-methoxyethanol to give heterodimer **85** [71, 72]. When treating **84** with sulfur in the presence of tetrakis(dimethylamino)ethane and *p*-toluenesulfonic acid, the corresponding sulfur-containing derivatives **86** and **87** have been obtained (Scheme 31) [73]. Removal of N-MEM protecting group led to the first open-cage fullerene derivative **87** without organic addend on the rim of the orifice.

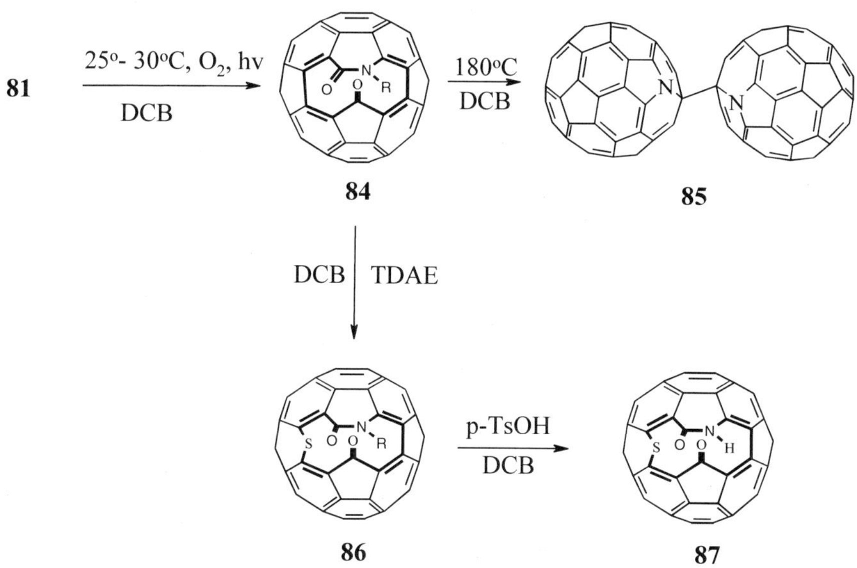

Scheme 31.

Thermal cycloaddition of organic azides to fullerenes was effectively used to design the unique conformationally flexible porphyrin hybrid containing the molecule of [5,6]-azahomofullerene **88** in which the intramolecular interactions were strengthened due to complexation with K+ and Na+ cations [74].

This approach to the synthesis of functionally substituted fullerenes allowed to produce, for the first time, [6,6]-aziridinofullerenes **89** and **90**, containing *N*-ferrocene and *N*-porphyrine directly of the fullerene core (Scheme 33). Compound 89 was established to be

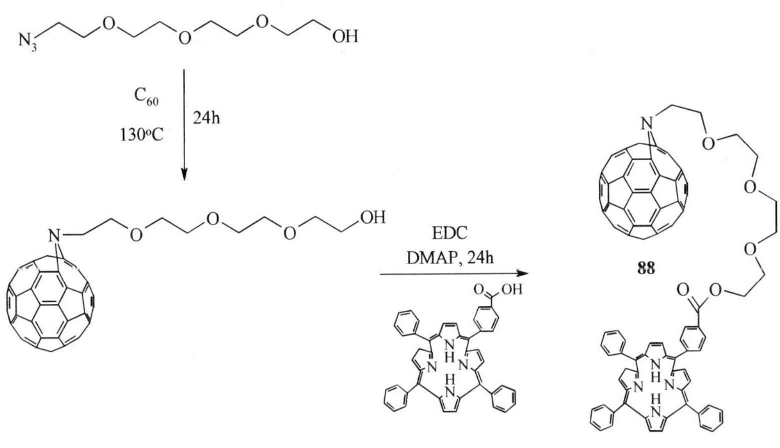

Scheme 32.

the major product in the reaction of ferrocenyl azide with C_{60}, whereas the reaction of the latter with porphyrine azide gave an equal (1:1) mixture of [5,6]-azahomofullerene **90** and [6,6]-aziridinofullerene **91**. Electrochemical measuments showed that chromophores attached to [5,6]-open isomers are more easily oxidizable than those attached to[6,6]-closed adducts. This was due to the nitrogen lone-electron pair and the degree of its delocalization in [6,6]-aziridinofullerene **91** that is higher than in [5,6]-azahomofullerene **90** [75].

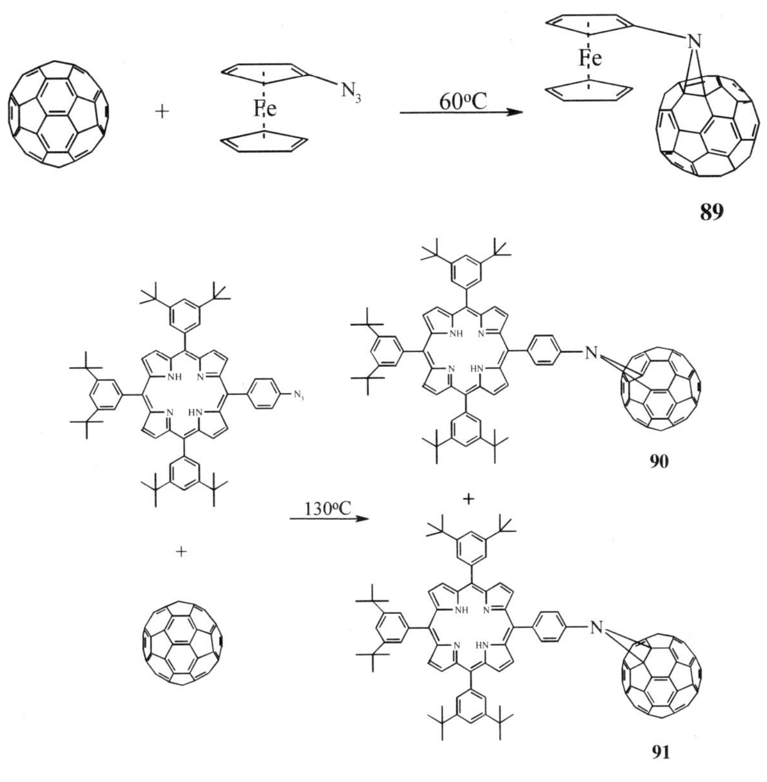

Scheme 33.

Aziridinofullerene **92** bearing fragment of 2,6-diacetylaminopyridine is able to form a supramolecular system **93** with uracil derivatives via three hydrogen bonds (Scheme 34) [76]. Such supramolecular systems are of increasing interest for designing three-dimensional nanomaterials in the form of photovoltaic devices. It is possible to obtain fullerene based supramolecular systems involving four hydrogen bonds in the absence of uracil derivative when [6,6]-aziridinofullerene **92** self assembles into a dimer structure **94** [77].

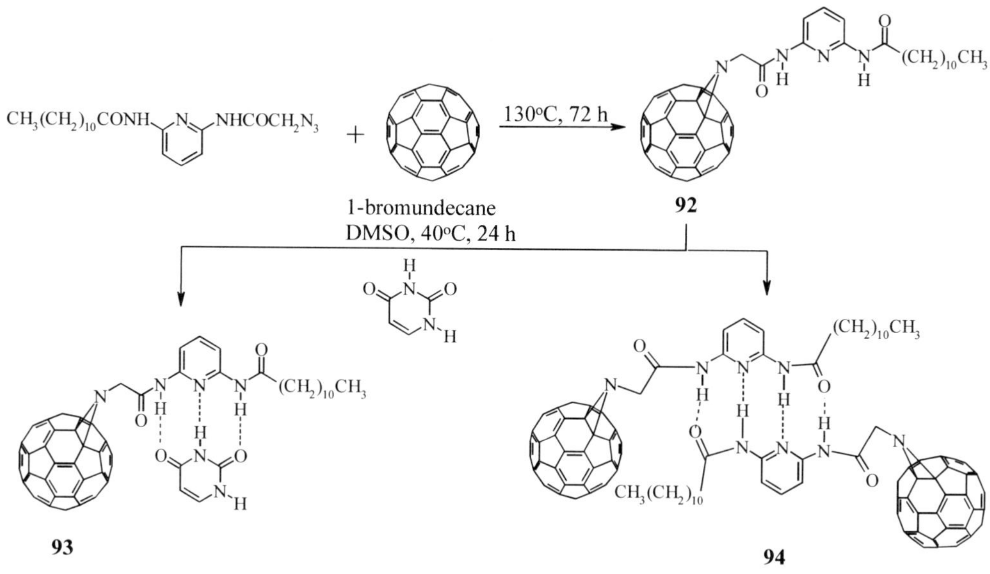

Scheme 34.

The reactions with nitropyrimidine-substituted azides usually provide the formation of [5,6]-azahomofullerenes **95** [78] and [6,6]-aziridinofullerenes **96** (Scheme 35) [79], electrochemical reduction of which occurs more easily unlike unmodified C_{60}.

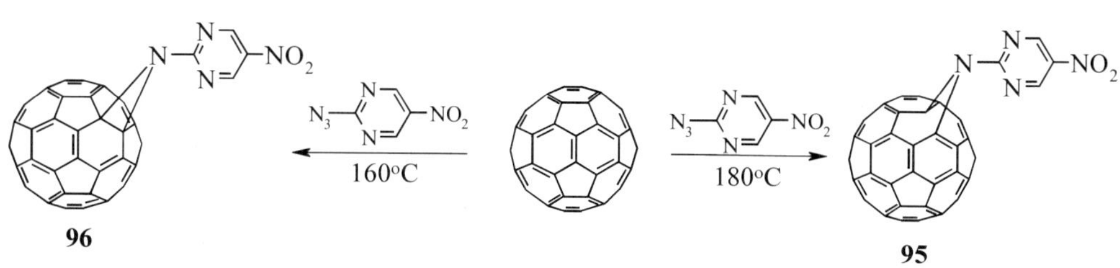

Scheme 35.

The decrease in temperature to 100 °C in the above reaction favored the formation of novel types of fullerene derivatives such as fullero[1;2;4;5]imidazo[1,2-b]pyrimidines **97** [79], that was attributed to the successive transformations of the intermediate molecules after nitrogen extrusion [80].

The reaction C_{60} and 2-azidopyrimidine having the phenyl substituents as electron donors at the positions 4 and 6 provides the formation of classic [5,6]-azahomofullerenes **98** and [6,6]-aziridinofullerenes **99** (Scheme 37) [81].

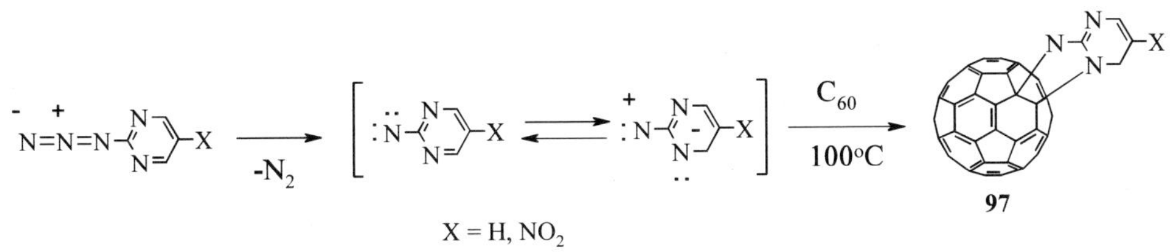

97

Scheme 36.

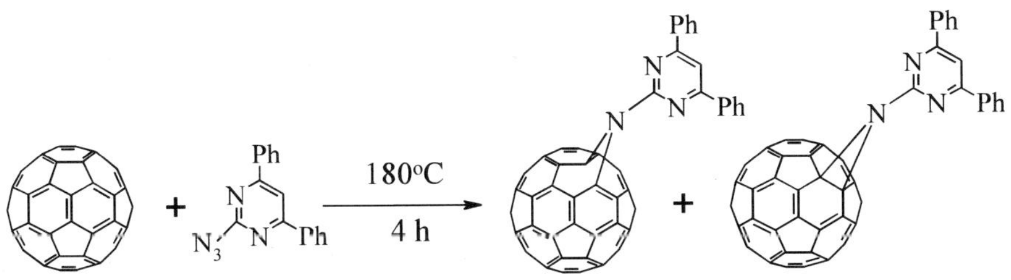

98 99

Scheme 37.

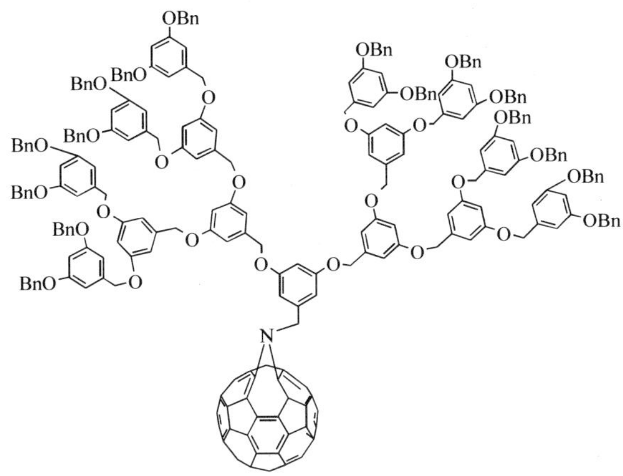

100

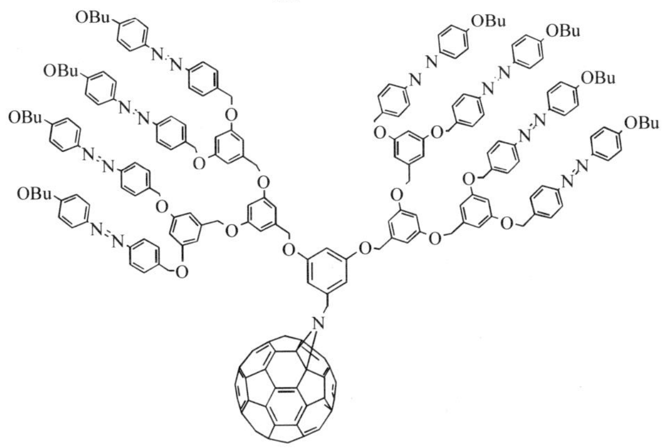

101

Several publications [82, 83] describe the successful synthesis of novel C_{60}-based dendrimer polyfunctional molecules **100** and **101** using the azide method. A growing interest to such dendritic fullerenes is conditioned by their globular three-dimensional structure, which allows encapsulation, and an opportunity to prepare monodisperse materials with known molecular weight for electronics. Thus, dendritic fullerene 101 contains 8 photoisomerizable azobenzene groups in the periphery.

Hybrid systems for molecular recognition in various devices or chemical processes constitute another promising application of multifunctional fullerenes. For example, derivatives of fullerenes with peracylated α-(n = 6), β-(n = 7) and γ-(n = 8) cyclodextrins (CD) **102, 103** and **105** were able to exhibit the properties of the catalysts with hydrophobic cavities [84], while the ability of photoluminescence quenching of rhodamine B as a "guest" was detected for of β-CD-C_{60} conugates.

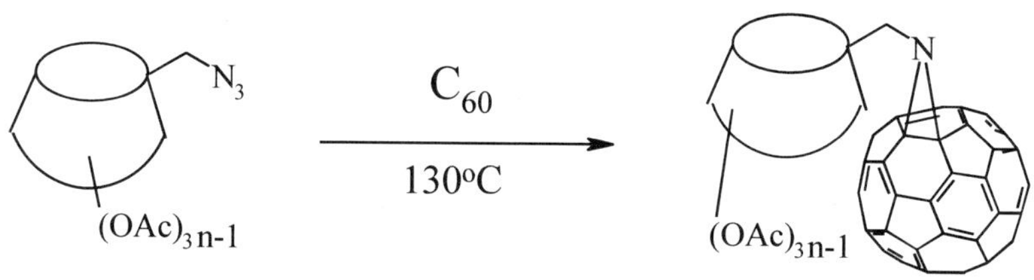

102

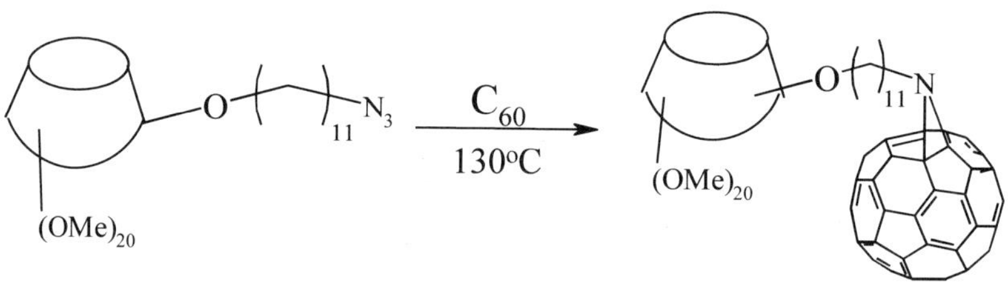

103

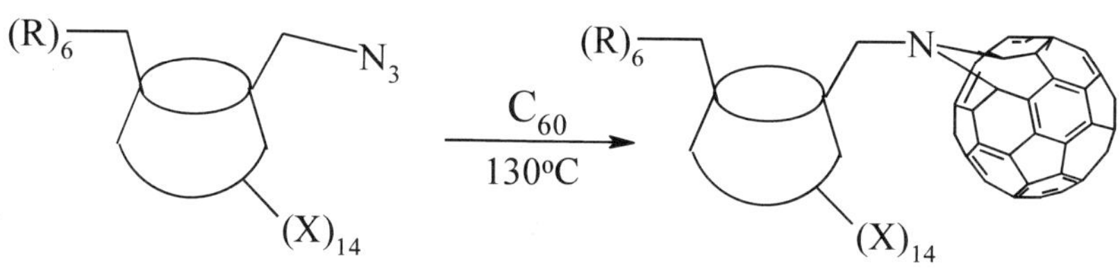

104 R = MeO; X = MeO
105 R = AcO; X = AcO
106 R = TBDMSO; X =OH

104-106

The authors of [87] have studied the effect of the initial structure of aryl and heteroaryl azides on the direction of the cycloaddition to C_{60} fullerene. They concluded that the reaction with aromatic azides led to aziridinofullerenes **107**, while individual azahomofullerenes **108** were the end adducts in the reaction with heterocyclic azides (Scheme 38).

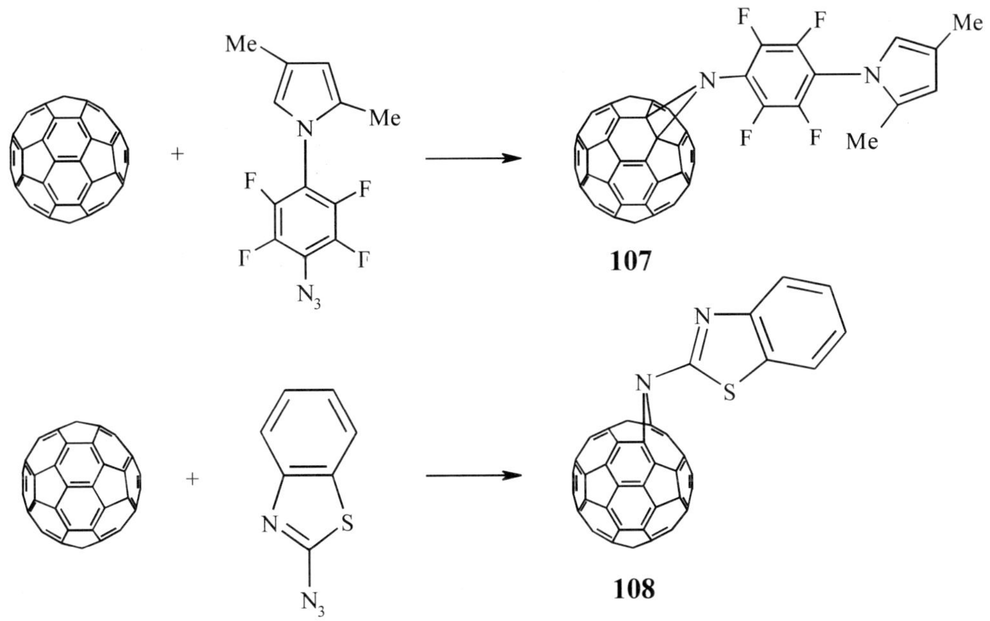

Scheme 38.

By using the azide method, nitrofluorophenyl **109** and fluoroquinoxaline derivatives of C_{60} fullerene **110** and **111** having a high electron affinity were also synthesized (Scheme 39) [88].

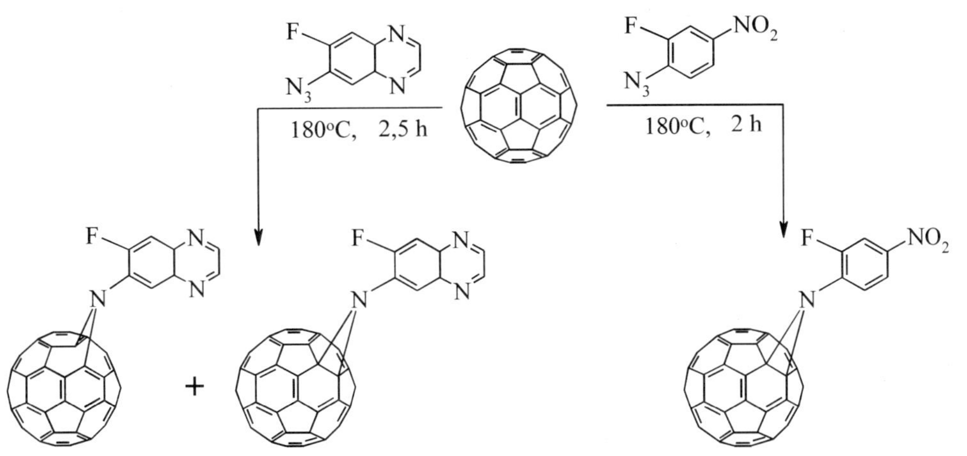

Scheme 39.

The Rh-complexes containing optically active derivatives of [5,6]-azahomofullerenes **112** and **113** in the coordination sphere [89], have been used as the catalysts for asymmetric hydrogenation of prochiral olefins, whereas novel efficient catalysts for oxidation of various sulfides to sulfoxides and sulfones were obtained upon interaction of copper complexes and

112 (Scheme 40) [90]. In turn, compounds **112** and **113** were synthesized from C_{60} fullerene and the chiral azide **114** in refluxing chlobenzene.

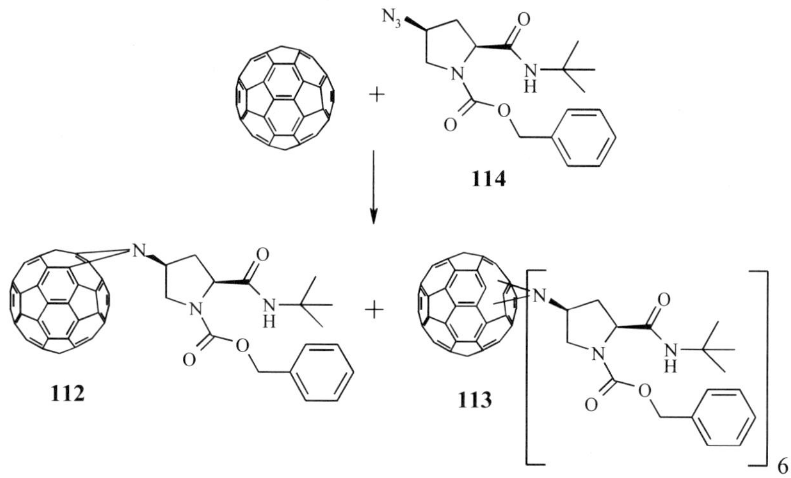

Scheme 40.

In continuation of the abovementioned investigations, there was undertaken the reduction of *N*–alkoxycarbonylaziridino[2',3':1,2][60]fullerene **115** in glacial acetic acid in the presence of Zn or Mg that suprizingly resulted in expansion of the fullerene core (Scheme 41) [91]. As it turned out, under these conditions, the reversable reductive cleavage of the bridgehead C–C bond in **115** provides a first example of bridged fulleroid **116** having an 6,6-open bond. The subsequent removal of *tert*-butoxycarbonyl protecting group (R = *t*-Bu) in **116** by treating with 50% aquous CF_3CO_2H allows the convenient synthesis of $C_{60}H_2NH$ **117** as the parent member of a new class of fullerenes. For several minutes, under the action of 1,4-diazabicyclo[2.2.2]octane as a base, fullerene **117** can transform to fulleroaziridine 30.

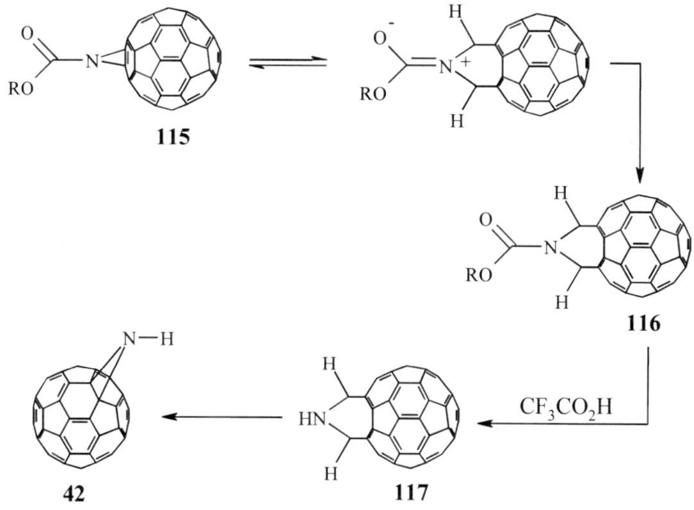

Scheme 41.

Unlike thermal cycloaddition reactions of organic azides to C_{60} fullerene affording azahomofullerenes, aziridinofullerenes or triazolinofullerenes, the same reactions under photolytic conditions gave exclusively 6,6-aziridinofullerenes. As the authors suggested [92], the mechanism for the formation of [6,6]-closed cycloadducts, in this case, involves the step of forming nitrene intermediates as evidenced by the formation of only one isomer.

Thus, the photochemical reaction between aryl azides and C_{60} in dichloroethane or tetrachloromethane afforded corresponding aziridines **118** capable of quantitative rearrangement to fullerooxazoles **119** under thermal or photochemical conditions (Scheme 42).

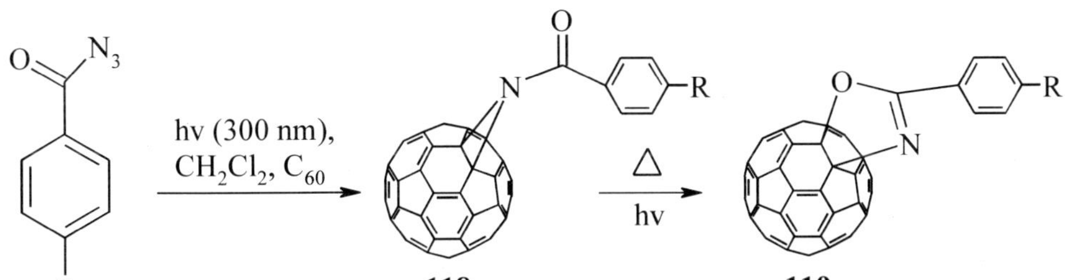

118 **119**

R = H (90%), Br (97%), OMe (92%), CN (90%)

Scheme 42.

This method has been successfully applied to the synthesis of perfluorophenyl derivatives of C_{60} **120** (Scheme 43). These fullerenes constitute a promising class of reagents as photoaffinity labels to crosslink proteins, polymers, and for immobilization of enzymes.

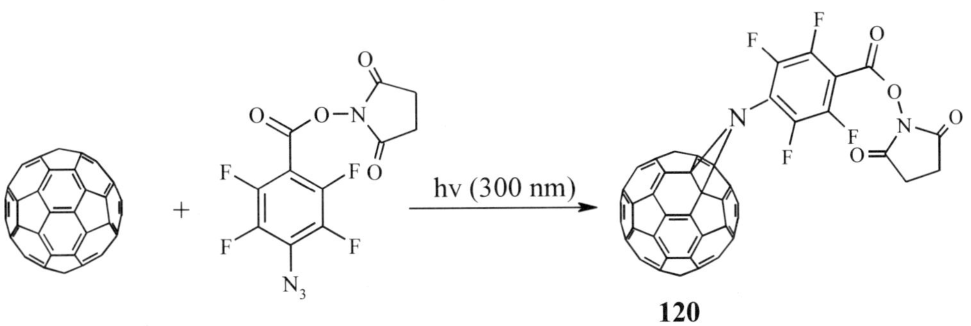

120

Scheme 43.

2.2. Organic Azides in the Synthesis of Bis-Adducts of C_{60} Fullerenes

Currently, bis-adducts of C_{60} fullerene with organic azides are mainly synthesized via two pathways. The first one involves compounds having two azide groups, which react with fullerene giving bis-adducts. The second pathway suggests successive transformations of azahomofullerenes or aziridinofullerenes already synthesized.

Typically, the double addition of different substrates to C_{60} fullerene is characterized by forming a mixture of regioisomers that are difficult to separate. The addition of the second substituent only by [6,6]-junction of monoadduct may afford 9 regioisomers (Figure 1). The number of theoretically possible isomers increases sharply if considering that cycloadditions of organic azides to carbon clusters are accompanied by the formation of adducts involving [6,6]- and also [5,6]- bonds.

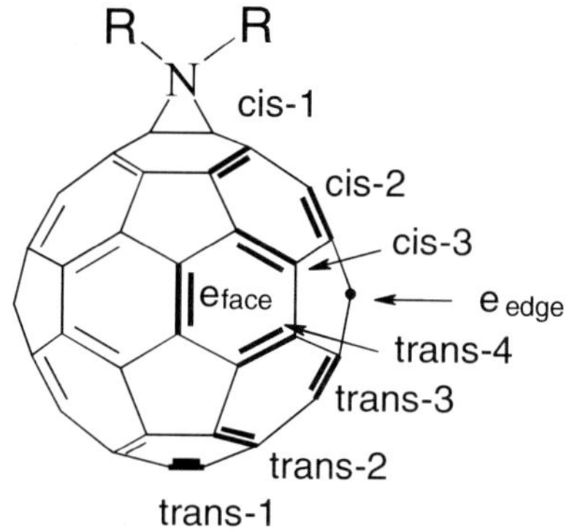

Figure 1. Possible areas for attachment of the second azide molecule to aziridinofullerene.

Meanwhile, the C_{60} fullerene exhibits surprising selectivity in reactions with organic azides, as evidenced by numerous theoretical and experimental data dealing with the preparation of bis-adducts.

Thus, the authors of [94] have demonstrated for the first time that the addition of bis-azides to fullerene can afford bis-azahomofullerene (doubly bridged fulleroid) **121** (~65% yield) involving two [5,6]-bonds of the same pentagon fragment in the original C_{60} molecule (Scheme 44).

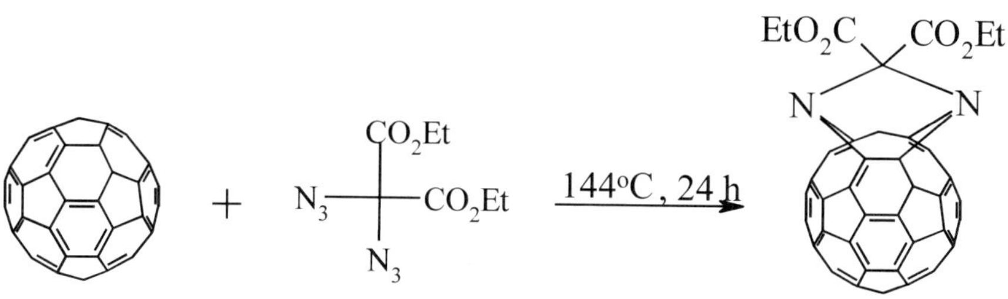

121

Scheme 44.

It was found that α,ω-diazides regioselectively bind with C_{60} fullerene to give corresponding bis-adducts **122** [95] and **123** [96]. Compound **123** has a structure of crown ethers with excellent metal binding capacity. The authors suppose that similar complexes can be used as the new superconducting materials.

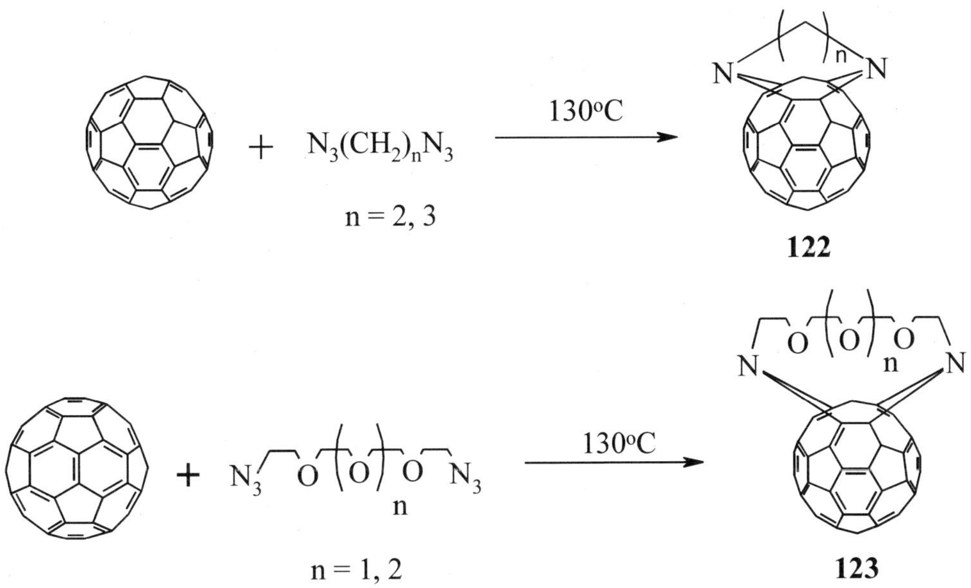

Scheme 45.

The reaction C_{60} and chiral 1,4-di(*tert*-alkoxy)-2,3-bis (azido)butanes was found to produce corresponding optically active bis(aza)fulleroids (2R,3R)-**124a**, (2S,3S)-**124a**, (2R,3R)-**124b**, and (2S,3S)-**124b** in 54, 49, 56 and 51% yield respectively (Scheme 46) [97].

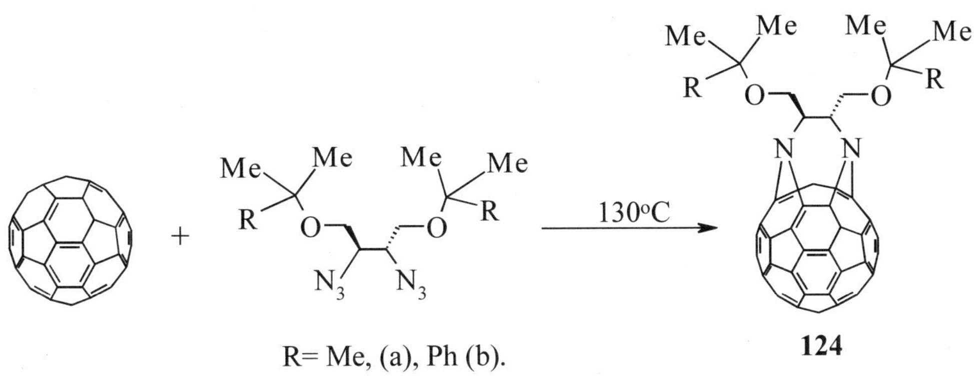

Scheme 46.

Similarly 2,2-dibenzyl-1, 3-diazidopropane entered into reaction with C_{60} fullerene furnishing, together with a mixture of bis-aziridinofullerene **125** and bis-azahomofullerene **126**, regioisomeric mixture of bis-triazolinofullerenes **127** and **128**, thermolysis of which resulted in a mixture of **125** and **126** with quantitave yields [98].

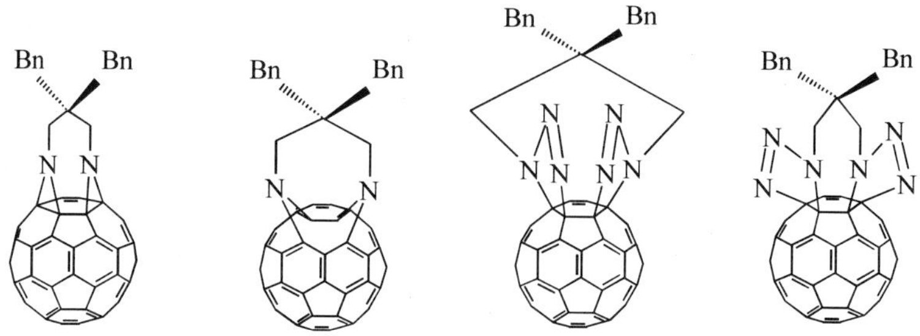

125 **126** **127** **128**

In continuation of these studies, the authors of [99] have synthesized novel bis-adducts **129** and **130** (Scheme 47). They found that under the influence of daylight compound **129** can undergo quantitative rearrangement to more polar adduct **130**, in which the nitrogen atoms are attached to [5,6]- and [6,6]- bonds. The reversible rapid isomerization of **130** to **129** has been observed under thermal reaction conditions.

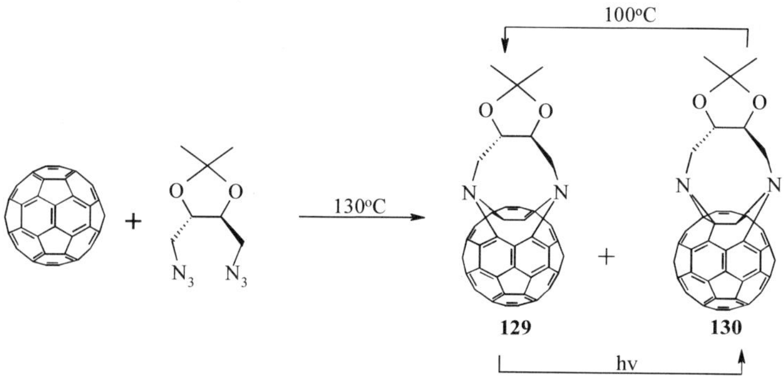

Scheme 47.

The similar structure of bis(aza)fulleroids has been suggested for fullerene derivatives of calix[4]arene (e.g., compounds **131** and **132**), which are capable of forming exohedral complexes with cations such as Li^+, Na^+ and Ag^+ [100, 58].

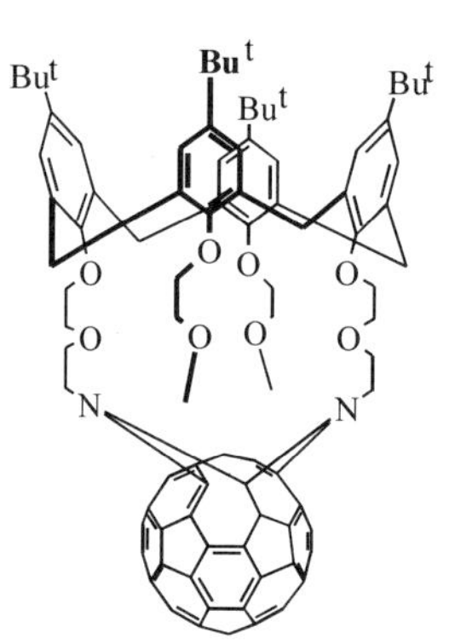

131

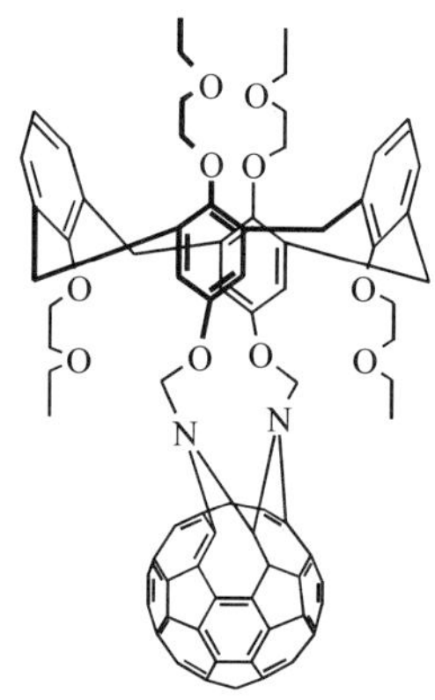

132

Later, the authors of [101] have shown that azahomofullerene bis-adducts can be regioselectively obtained also by using monoazides. Thus, the interaction between isocyanurato-substituted azides and C_{60} fullerene led to the regioselective formation of various azahomofullerene adducts depending upon the bulkness and structure of organic substituents. For example, compound **133** having isocyanate moieties with methoxycarbonylethyl substituents on nitrogens represents the described above bis-adduct, in which the attached fragments are not linked.

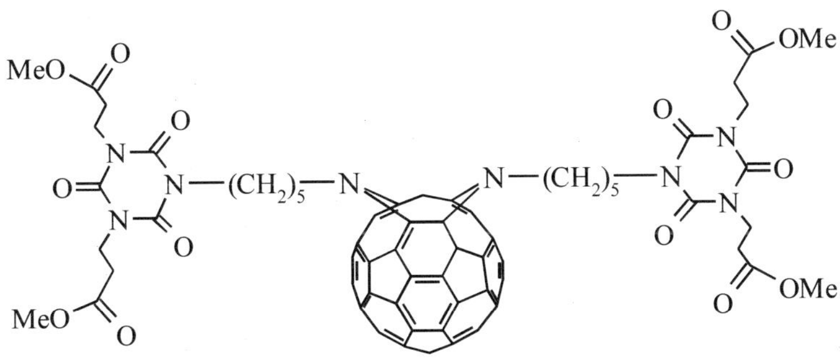

133

Changes in the structure of the substitutent R at the nitrogen atoms in the isocyanurate substituted azides unexpectedly resulted in the formation of bisadducts of other structure **134**. In this case, the length of the methylene chain between azido group and heterocycle does not practically influence the course and direction of the reaction under discussion. The authors of the work [102] have postulated the structure of adducts **134 a,b** considering the ^{13}C NMR signal at δ_C 160 ppm in their spectra, which is typical for carbons bonded to two atoms of nitrogen. Bis-azafulleroid **135** having similar structure was described in the work [103].

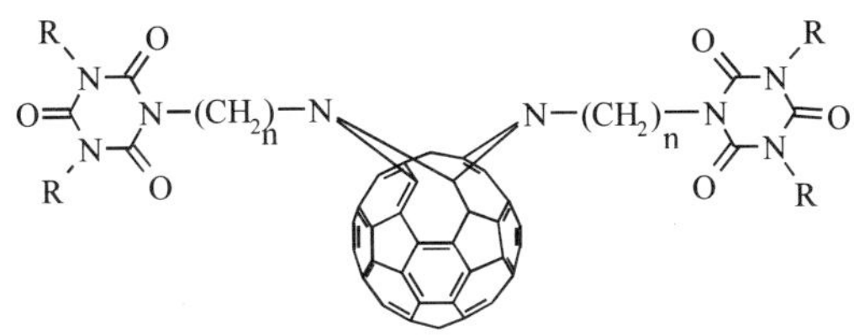

134 a - г

a: n=2, R= ; b: n=3, R=

c: n=5 R= ; d: n=5, R= $(CH_2)_2CN$

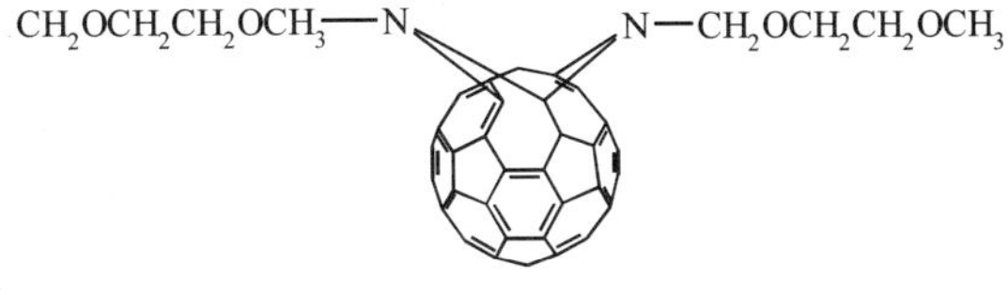

135

Bis-adducts were also detected during the reaction between C_{60} and methyl 5-azido-5-phenylpentanoate in refluxing *o*-dichlorobenzene [104]. Electrochemical properties of the obtained adducts were significantly higher than that of their analogues through conservation of π-electronic system and density in solid state making this new material less flowable in the solar cells and similar devices [105].

The postulated possible routes to variuos azahomofullerene bis-adducts **136** and **157–159** [10, 68] are shown below (Scheme 48).

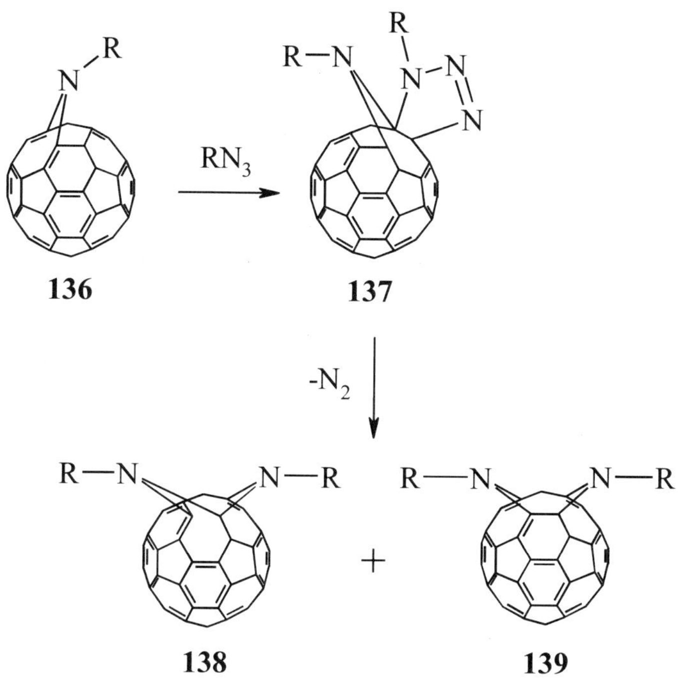

Scheme 48.

Functionalization of C_{60} with ethyl or *tert*-butyl azidoformates resulted in a mixture of regioisomeric [6,6]-closed bis(imino)fullerenes, in which *cis-1* isomeric bisadducts **140** as the first example of fullerene derivatives with open *trans*-annular [6,6]-bonds have been detected [23].

Relatively regioselective synthesis of compound **140** based on monoiminofullerene has been implemented through the formation stage of the intermediate **142**. The latter compound was presumably formed as a result of [3+2]-cycloaddition between azide and **141**, when the negatively polarized (R)N atom joins to the positively polarized atom C(4) atom of the neighboring non-functionalized [6,6]-bond (Scheme 49).

This polarization of bonds in the carbon cage of the molecules **141** was confirmed using calculation semiempirical methods, e.g., AM1. As shown, thermal extrusion of the nitrogen molecule leads to biradical intermediates **143**. Localization of the first addend on the same six-membered ring of the fullerene core favors the predominant formation of 1,2;3,4-bis(imino)[60]fullerenes. The subsequent retro-Diels-Alder reaction (2π–2σ-isomerization) gave the desired adducts **140**.

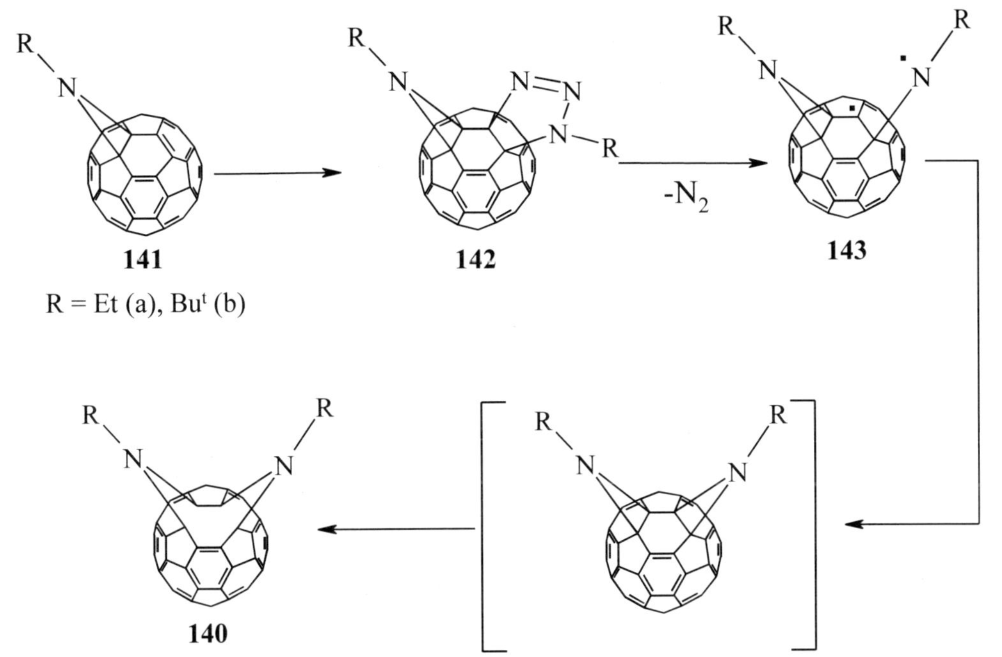

Scheme 49.

The formation of bisadduct with open transannular [6,6]-bonds was also observed under the interaction between C_{60} and the excess of 2-azido-5-nitropyrimidine [106]. The authors noted that electrochemical reduction of the latter adduct proceeds as compared to non-modified C_{60}.

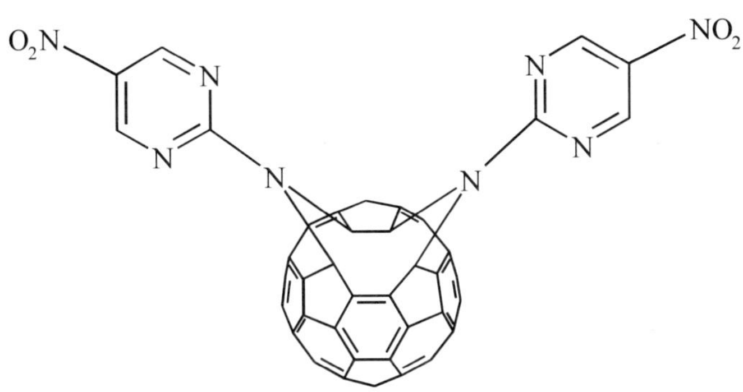

144

Similar results have been obtained by a group of authors of [107]. While investigating the reaction of C_{60} with a *tert*-butyl ester of 11-azido-3,6,9- trioxaundecane acid, they demonstrated the formation of three types of the end products depending upon the reaction temperature, namely, [6,6]-closed monoadduct **145**, [5,6]-open monoadduct **146** or [5,6]-open bis-adduct **147**, being the end product in the cycloaddition reaction between [5,6]-open fullerene monoadduct **146** and another azide molecule (Scheme 50).

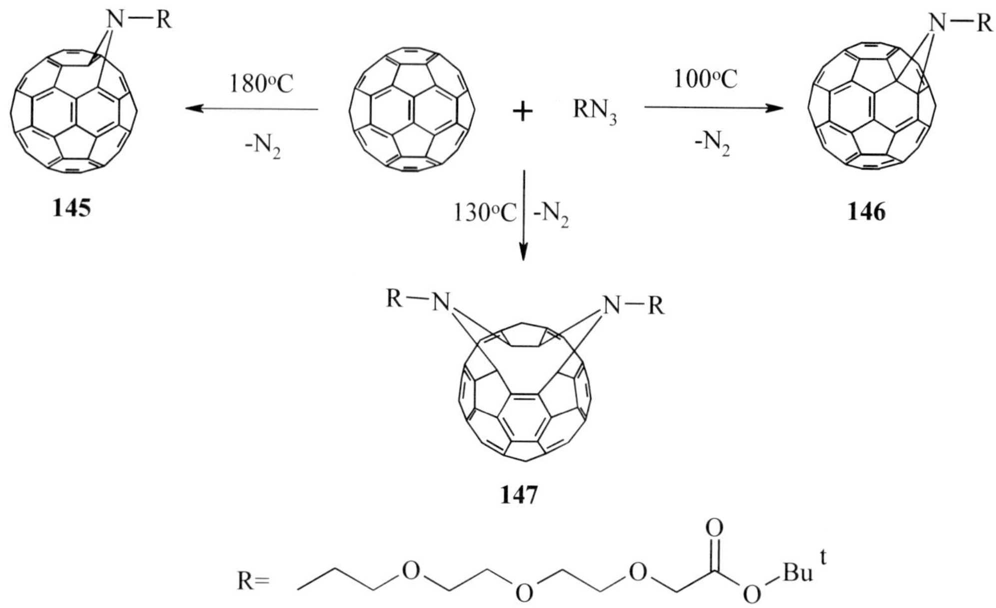

Scheme 50.

2.3. Reactions of Fullerenes with Inorganic and Organometallic Azides

Considering that in chemical reactions fullerenes behave themselves as polyene-like molecules, the authors of this review suggested that the reactions of the carbon clusters with inorganic or organometallic azides may serve to accomplish the synthesis of previously inaccessible azidofullerenes with the azido-group directly linked to the fullerene core.

Thus, the reaction C_{60} with the 100-fold excess of HN_3 generated *in situ* from NaN_3 and H_2SO_4 in a sealed ampoule (40°C, 3 h, chlorobenzene) was found to produce previously undiscribed triazolinofullerene **148** in ~15 % yield. This fullerene, being unstable even at room temperature, gradually decomposed under storage to form, after elimination of molecular nitrogen (N_2), unsaturated fulleroaziridine **42**, the synthesis of which included several stages [23, 24]. With increase in temperature up to 60°C the yield of aziridinofullerene **42** reached 50 %.

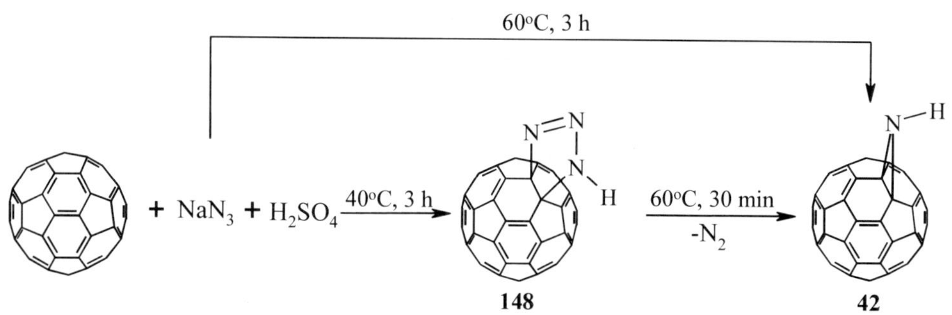

Scheme 51.

Despite the fact that the signal (δ 4.61 ppm) attributable to the aziridine hydrogen in the ^{1}H NMR spectrum of compound **42**, registired in $CDCl_3$, significantly differs from that reported in the literature (5.9 ppm) [23, 24], the authors have unambiguously proved that the first signal belongs to the hydrogen atom of the N-H bond by using the HMBC experiment and also deuterium labeling study of the same reaction with DN_3.

Unsubstituted aziridinofullerene **42** may be used as a precursor for the synthesis of practically important fullerene derivatives (Scheme 52), intended as modern and competitive drugs [23, 24, 109, 110].

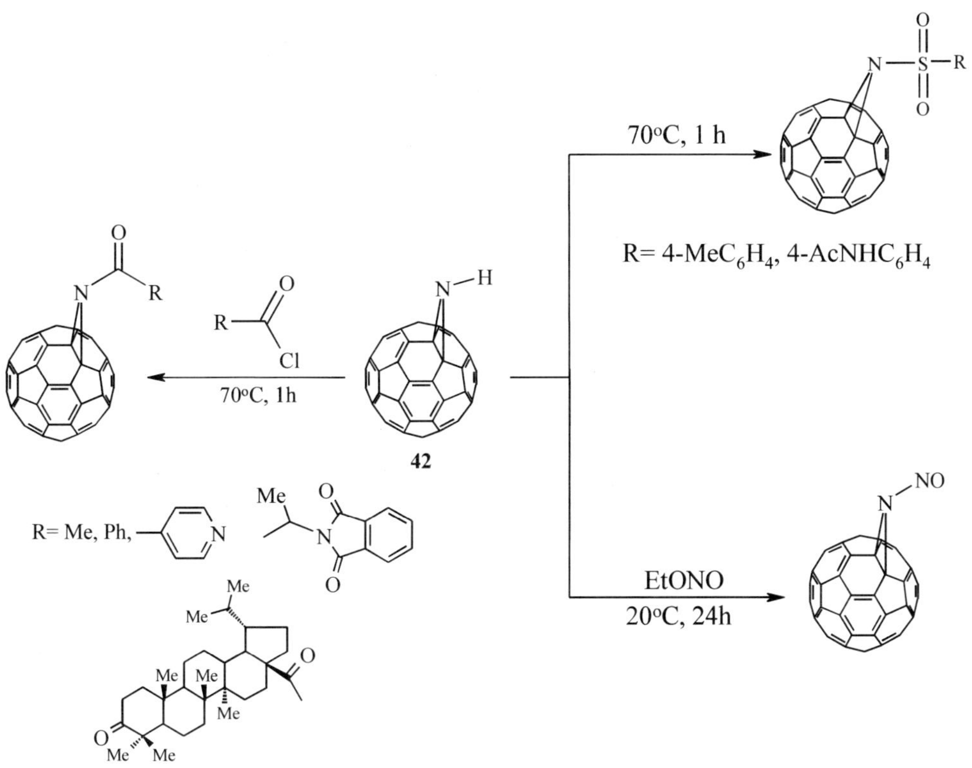

Scheme 52.

In the development of the above cycloaddition between hydrazoic acid and C_{60} fullerene, the reaction of the latter with halogen azides have been studied [111]. Considering that the N–Hal bond is less strong than N–H bond, the authors suggested that the implementation of the reactions between halogen azides (IN_3, BrN_3 and ClN_3) and the carbon clusters may result in previously unreported fullerene azides.

Indeed, the experiments have hown that the reaction of the C_{60} fullerene and the 3-fold excess of freshly prepared IN_3 led to 1-iodo-2-azidofullerene **149** in ~30 % yield (Scheme 53) [111].

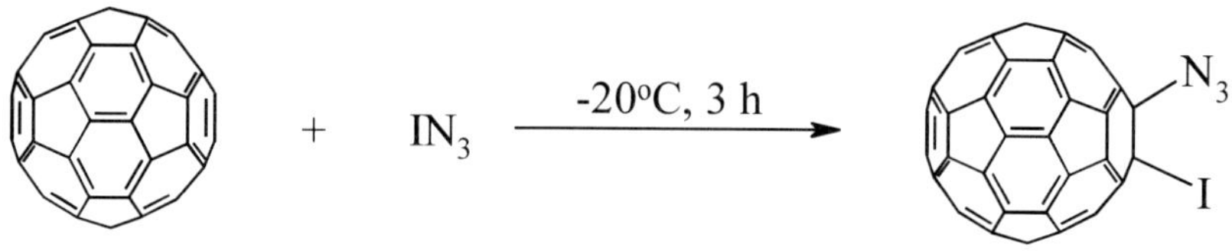

149

Scheme 53.

Analoguously, bromine azide entered into reaction with C_{60} (−40°C, 3 h, chlorobenzene) giving rise to the corresponding adduct **150** in ~30 % yield (Scheme 54).

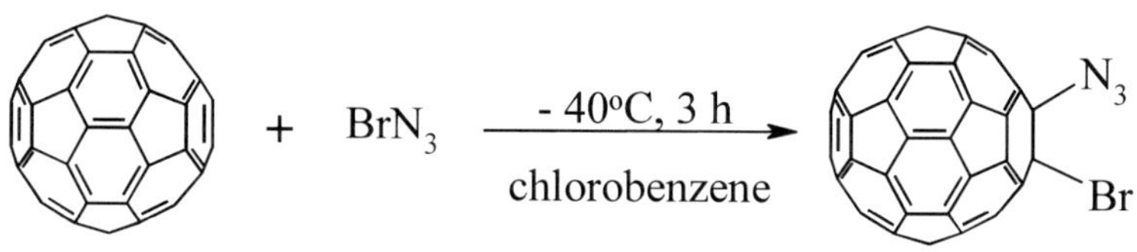

150

Scheme 54.

Taking into account the explosive properties of individual halogen azides, the authors of the work [111] have developed the methodology for the *in situ* generation of $HalN_3$ that implies the interaction of corresponding *N*-halosuccine imides with trimethylsilyl azide ($TMSN_3$) in the presence of equimolar amounts of lanthanide or copper triflates.

Thus, the reaction C_{60} fullerene and the *in situ* generated bromine azide at room temperature in the presence of stechiometric amounts (~100 mol%) of lanthanide (Sm, Yb, La) or copper triflates afforded 1-bromo-2-azido-1,2-dihydrofullerene **150** in 25−30% yield. Without a catalyst the reaction did not occur.

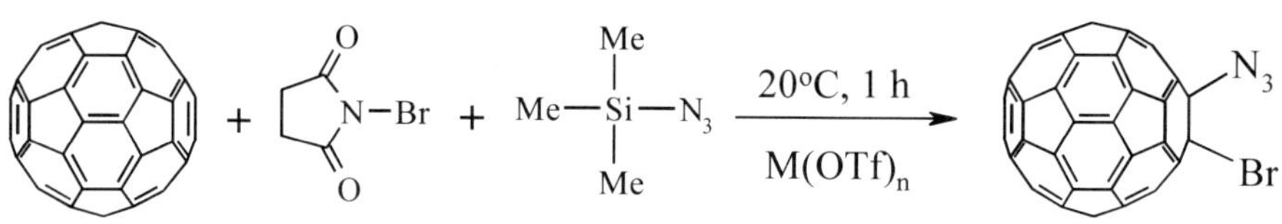

150

M= Sm (n=3), Cu (n=2), Yb (n=3), La (n=3)

Scheme 55.

The attempts to combine C_{60} fullerene with ClN_3 have failed. The starting fullerene remained unchanged in all the experiments.

The first synthesized representatives of azidofullerenes **149** and **150**, being unstable compounds, have low solubility in organic solvents that did not allow unambiguous establishing their structure by NMR and mass spectrometry.

Thus, the authors could not register the *sp*^{3}-hybridized C_{60} carbon atom associated with the azide group and the iodine or bromine atoms because of low solubility of azidofullerenes in organic solvents, and the low signal intensity of the fullerene carbon atoms, to which heteroatoms are attached.

In addition, to prove reliably the structure of 1-iodo-2-azido-1,2-dihydro(C_{60}-I_h)[5,6]-fullerene **149** and 1-bromo-2-azido-1,2-dihydro(C_{60}-I_h)[5,6]-fullerene **150**, the reductive cyclizations of the latters to aziridinofullerene **42** have been carried out (Scheme 56) [111], similar to those described previously for organic halogen azides, which were transformed to appropriate azacyclopropanes [112].

Thus, the reduction of haloazidofullerenes **149** and **150** with 1 M solution of $BH_3{\cdot}THF$ or under treatment with gaseous NH_3 at room temperature for 12 h provided the formation of corresponding aziridinofullerene **42** in quantitative yield. In contrast, when using the reducing agents such as $LiAlH_4$, $NaBH_4$, $N_2H_4{\cdot}H_2O$ or molecular H_2 with of 10% Pd/C, compounds **149** and **150** in the reduction reaction are inert and do not give compound **42**.

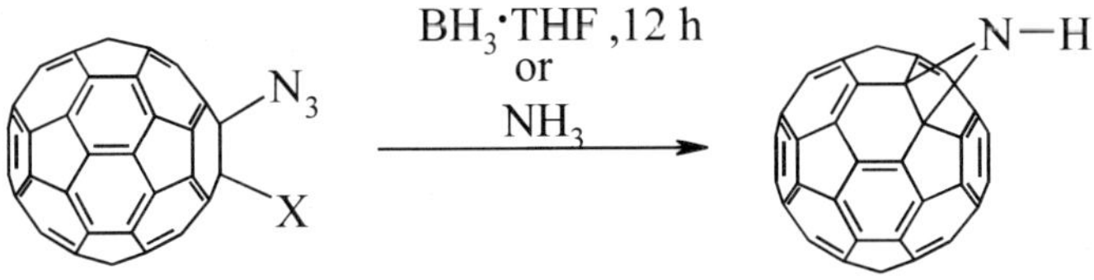

X = I (**149**), Br (**150**) **42**

Scheme 56.

The reaction of haloazidofullerenes **149** and **150** with alkynes should be considered as another promising type of reaction in the chemistry of organic azides, leading to appropriate triazoles ("click reaction"). Lately, this type of reaction has been widely used in the chemistry of fullerenes. In this regard, the authors of this review have studied the ability of the azido group in synthesized azidofullerenes **149** and **150** to form the corresponding triazoles in the reaction with acetylenes.

The experiments showed that 1-halogen-2-azidofullerenes **149** и **150** do not enter into "click" reaction with acetylenes in the presence of $CuSO_4{\cdot}5H_2O$ and sodium ascorbate [113]. That may be due to steric hindrance, which create the bulky fullerene molecule and the halogen atom located near to the azide group.

However, during the search of another catalyst capable of implementing this type of reaction, the authors of [111] have found that the interaction of the synthesized azidofullerenes **149** and **150** with a 5-fold excess of phenylacetylene in the presence of equimolar amounts of Cu $(OTf)_2$ as the catalyst, by analogy with the Sonogashira reaction [114], led to the acetylenic derivative of azidofullerene **151** in quantitative yield.

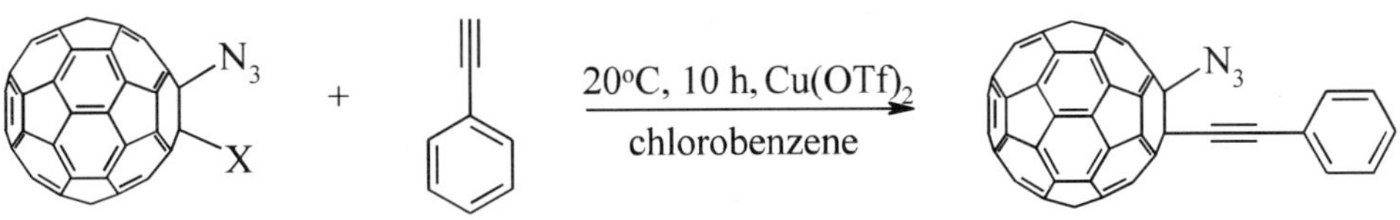

X = I (**149**), Br (**150**) **151**

Scheme 57.

The results obtained in the study of the reaction between the carbon clusters and halogen azides gave the hope that the interaction of C_{60} fullerene with metal azides may lead to previously undescribed diazidofullerenes, by analogy with the reaction of olefins with metallazides [115, 116].

It was established [117] that azides of metals such as Pb, Hg and Sn do not react with C_{60} fullerene with selective formation of azidofullerenes.

Thus, the reaction of C_{60} fullerene with an equimolar amount of lead azide generated *in situ* through the interaction between Me_3SiN_3 and $Pb(OAc)_4$ [116], being extremely rapid and nonselective, resulted in a regioisomeric mixture of adducts of the azide polyaddition to C_{60} **152**, which are difficult to identify due to the low solubility in organic solvents and a high tendency to explode (Scheme 58).

The reaction of C_{60} with mercury and tin azides being synthesized *in situ* via the interaction of $Hg(OAc)_2$ or $SnCl_2$ with Me_3SiN_3 or NaN_3, provided the formation of dihydrofullerene **153** and aziridinofullerene **42** in the yield of 20 and 10 % respectively. Apparently, the treatment of the reaction mixture with $NaBH_4$, which is used for decomposition of mercury azide, caused the formation of the hydride fullerene derivative **42**.

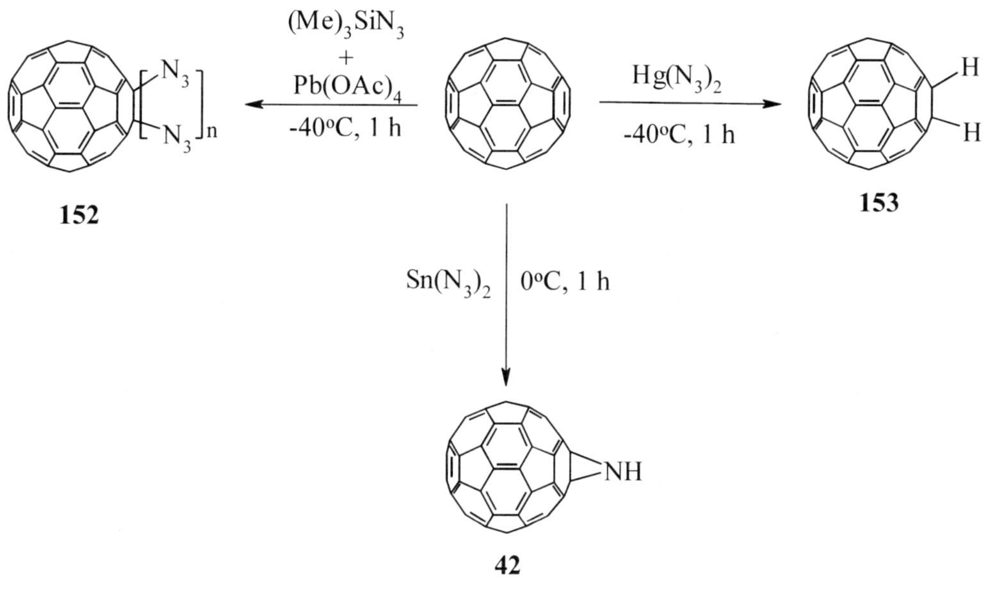

Scheme 58.

In contrast to metal azides, organometallic azides (Et_2AlN_3, $EtAl(N_3)_2$ and Bu_3SnN_3) enter into reaction with C_{60} fullerene giving rise to previously undescided 1-alkyl-2-azido-1,2-dihydrofullerenes **154** and **155** in 20–30 % yield (Scheme 59). Unlike organoaluminum azides, Bu_3SnN_3 is less reactive that makes it necessary to use stoichiometric amounts of $Cu(OTf)_2$ (100 mol. %) [117]. No azidofullerenes were found in the absence of the latter as the catalyst.

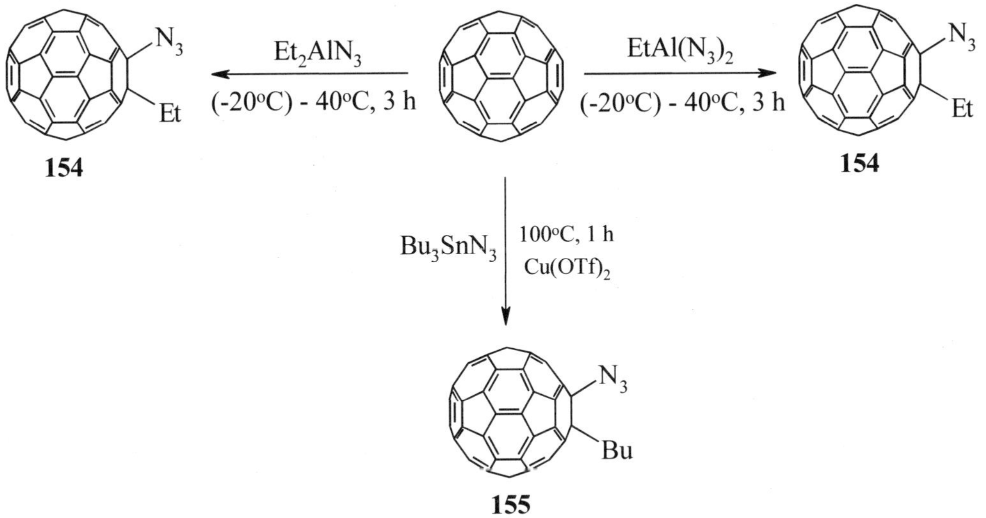

Scheme 59.

To prove more reliably the structure of 1-azido-2-alkylfullerenes **154** and **155** the authors of the work [117] have performed some typical of organic azides chemical transformations.

Thus, the reaction of azidofullerenes **154** and **155** with PPh_3 in dry toluene for 2 h at room temperature led to the formation of the corresponding iminophosphoranes **156** and **157** in quantitative yield (Scheme 60).

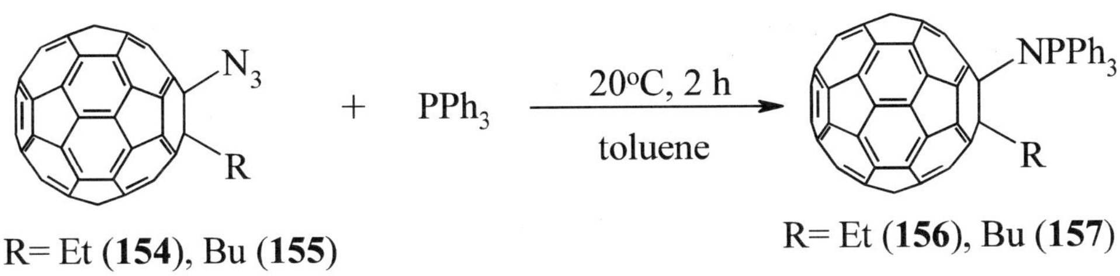

Scheme 60.

When treating the synthesized iminophosphranes **156** and **157** with MeOH in dry toluene at 80°C for 3 h, the corresponding unsaturated aminofullerenes **158** and **159** as a brown powder, slightly soluble in organic solvents, have been obtained in quantitave yield (Scheme 61). It should be noted that the reducing agents such as $LiAlH_4$, $NaBH_4$, $N_2H_4{\cdot}H_2O$, $BH_3{\cdot}THF$ or molecular hydrogen (H_2) with 10% Pd/C are not capable of reducing **154** and **155** to the appropriate aminofullerenes.

In order to increase the solubility of aminofullerene **158** and **159**, they have been involved in the benzoylation reaction by using benzoyl chloride. As a result, the obtained compounds **160** and **161** are readily soluble in organic solvents that allowed reliably proving their structure by NMR spectroscopy [117].

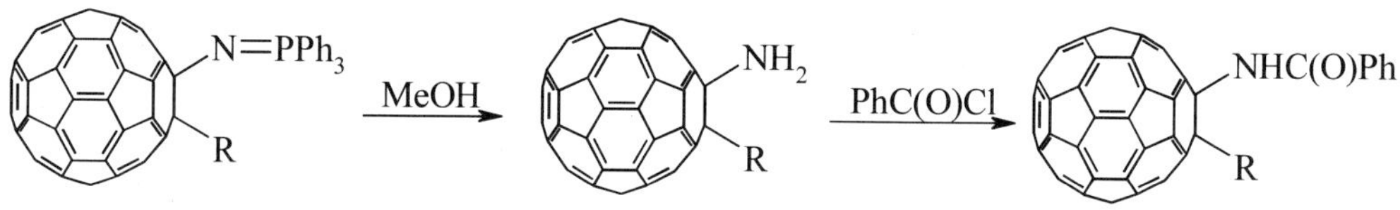

R= Et (**156**), Bu (**157**) R= Et (**158**), Bu (**159**) R= Et (**160**), Bu (**161**)

Scheme 61.

The reaction of 1-azido-2-ethyl-fullerene **154** or 1-azido-2-butyl-fullerene **155** with styrene in the presence of stechiometric amounts of $Cu(OTf)_2$ as the catalyst resulted in the formation of adducts **162** and **163** bearing aziridine fragments in 70 % yield. Without catalyst this reaction failed.

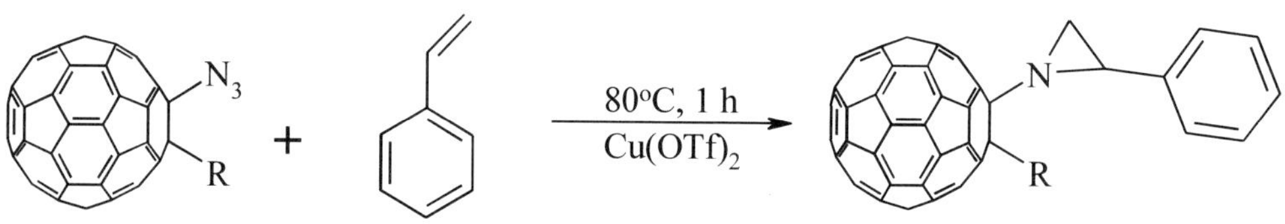

R= **154** Et, **155** Bu R= **162** Et, **163** Bu

Scheme 62.

Unexpected results have been obtained in the reaction between azidoalkylfullerene **154** or **155** and phenyl acetylene. Typically, organic azides react with acetylenes giving rise to triazoles [118]. However, the reaction with azidofullerenes led to acyclic adducts **164** and **165** (Scheme 63).

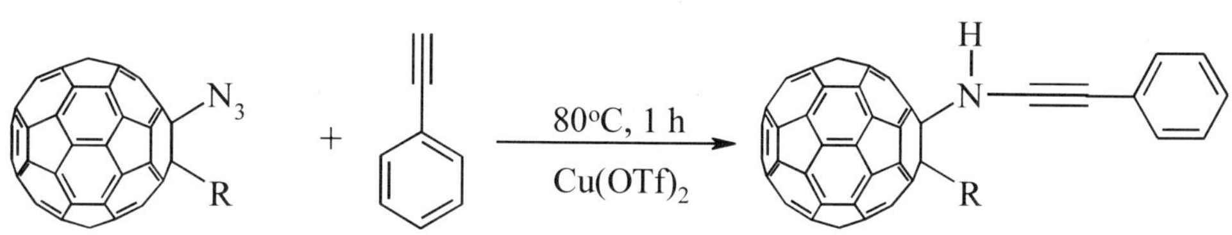

R= **154** Et, **155** Bu R= **164** Et, **165** Bu

Scheme 63.

Thus, the reactions between the carbon clusters and inorganic or organometallic azides, which lead to stable aziridinofullerenes or previously inaccessible azidofullerenes depending on the nature of starting reagents, have been studied for the first time.

Conclusion

As follows from the above survey, cycloaddition of organic azides to the carbon clusters furnishing azahomofullerenes and aziridinofullerenes should be considered as one of the first synthetic transformations of fullerenes since their quantitative production. Nowadays, there exist several variants of this reaction, which differ to each other by the conditions for its performance. This implies the direct interaction between fullerenes and organic azides under thermal or photochemical conditions including microwave irradiation. Using these methodologies, one can obtain both [6,6]-closed and [5,6]-open cycloadducts including stable triazalinofullerenes and bis-adducts with regioselective location of addends.

Most of published research findings in this field are dedicated to the thermal reactions of organic azides with carbon clusters

In recent years, there appeared the works aimed at studying the reactions of fullerenes with inorganic and organometallic azides, which favor the formation of previously undescribed azidofullerenes, being unique precursors in the synthesis of different carbon clusters.

From a practical point of view, the results of studying the reactions of fullerenes with organometallic azides appear to be the most promising, since the use of the latters allows access to very stable azidofullerenes.

Despite the fact that these studies are currently in the initial stage, the chemists have already obtained the data, which are encouraging in terms of creating materials with unique properties. And further development of research into the functionalization of the carbon clusters using organic and inorganic azides can lead to new and unexpected results.

ACKNOWLEDGMENT

The work was financially supported by Grant of Russian Science Foundation (№ 14-13-00296)

REFERENCES

[1] Prato, M.; Maggini, M.; Giacometti, C.; Scorrano, G.; Sandona, G.; Farnia, G. *Tetrahedron* 1996, *52*, 5221-5234.

[2] Bingel, C. *Chem. Ber.* 1993, *126*. 1957.

[3] Camps, X.; Hirsch, A. *J. Chem. Soc., Perkin Trans. 1* 1997, 1595.

[4] Tuktarov, A.R.; Dzhemilev, U.M. *Russ. Chem. Rev.* 2010, *79*, 585-610.

[5] Dzhemilev, U.M.; Tuktarov, A.R. *Metal Complex Catalysis in the Chemistry of Fullerenes* in Handbook on Fullerene: Synthesis, Properties and Applications. Ed. By Verner, R.F.; Benvegnu, C. Nova Science Publishers, 2012, 548.

[6] Sinyashin, O.G.; Romanova, I.P.; Yusupova, G.G.; Kovalenko, V.I.; Yanilkin, V.V.; Azancheev, N.M. *Mendelleev Commun.* 2000, *10*, 96-98.

[7] Romanova, I.P.; Yusupova, G.G.; Fattakhova, S.G.; Nafikova, A.A.; Kovalenko, V.I.; Yanilkin, V.V.; Katkov, V.E.; Azancheev, N.M.; Resnik, V.S.; Sinyashin, O.G. *Russ. Chem. Bull.* 2001, *50*, 445-452.

[8] Cases, M.; Duran, M.; Mesters, J.; Martin, N.; Sola, M. *J. Org. Chem.* 2001, *66*, 433-442.

[9] Prato, M.; Li, Q.C.; Wudl, F.; Lucchini, V. *J. Am. Chem. Soc.* 1993, *115*, 1148-1150.

[10] Grosser, T.; Prato, M.; Lucchini, V.; Hirsch, A.; Wudl, F. *Angew. Chem. Int. Ed. Engl.* 1995, *34*, 1343-1345.

[11] Nuber, B.; Hampel, F.; Hirsch, A. Chem. Commun. 1996, 1799-1800.
[12] Gonzalez, S.; Martin, N.; Swartz, A.; Guldi, D.M. *Org. Lett.* 2003, *5*, 557-560.
[13] Hachiya, H.; Kakuta, T.; Takami, M.; Kabe, Y. *J. Organomet. Chem.* 2009, *694*, 630-636.
[14] Ikuma, N.; Mikie, T.; Doi, Y.; Nakagawa, K.; Kokubo, K.; Oshima, T. *Org. Lett.* 2012, *14*, 6040-6043.
[15] Averdung, J.; Mattay, J. *Tetrahedron* 1996, *52*, 5407-5420.
[16] Ouchi, A.; Hatsuda, R.; Awen, B. Z. S.; Sakuragi, M.; Ogura, R.; Ishii, T.; Araki, Y.; Ito, O. *J. Am. Chem. Soc.* 2002, *124*, 13364-13365.
[17] Ouchi, A.; Awen, B. Z. S.; Hatsuda, R.; Ogura, R.; Ishii, T.; Araki, Y.; Ito, O. *J. Phys. Chem. A* 2004, *108*, 9584-9592.
[18] Ouchi, A.; Awen, B. Z. S.; Luo, H.; Araki, Y.; Ito, O. *Tetrahedron Lett.* 2005, *46*, 6713-6716.
[19] Brabec, C.J.; Cravino, A.; Meissner, D.; Sariciftci, N.S.; Fromherz, T.; Rispens, M.T.; Sanchez, L.; Hummelen, J.C. *Adv. Funct. Mater.* 2001, *11*, 374-380.
[20] Minakata, S.; Tsuruoka, R.; Naqamachi, T.; Komatsu, M. *Chem. Commun.* 2008, 323-325.
[21] Tsuruoka, R.; Naqamachi, T.; Murakami, Y.; Komatsu, M.; Minakata, S. *J. Org. Chem.* 2009, *74*, 1691-1697.
[22] Schick, G.; Grosser, T.; Hirsch, A. *J. Chem. Soc., Chem. Commun.* 1995, 2289-2290.
[23] Schick, G.; Hirsch, A.; Mauser, H.; Clark, T. *Chem. Eur. J.* 1996, *2*, 935-943.
[24] Banks, M. R.; Cadogan, J. I. G.; Gosney, I.; Hodgson, Ph. K. G.; Langrige-Smith, P. R. R.; Millar, J. R. A.; Taylor, A. T. *J. Chem. Soc., Chem. Commun.* 1995, 885-886.
[25] Averdung, J.; Luftmann, H.; Mattay, J.; Claus, K.-U.; Abraham, W. *Tetrahedron Lett.* 1995, *36*, 2957-2958.
[26] Banks, M. R.; Cadogan, J. I. G.; Gosney, I.; Hodgson, P. K. G.; Langridge-Smith, P. R. R.; Rankin, D. W. H. *J. Chem. Soc., Chem. Commun.* 1994, 1365-1366.
[27] Smith, A. B.; Tokuyama, H. *Tetrahedron* 1996, *52*, 5257-5262.
[28] Averdung, J.; Wolff, C.; Mattay, J. *Tetrahedron Lett.* 1996, *37*, 4683-4684.
[29] Thilgen, C.; Diederich, F. *Chem. Rev.* 2006, *106*, 5049-5135.
[30] Guldi, D. M.; Hungerbühler, H.; Carmichael, I. *J. Phys. Chem. A* 2000, *104*, 8601-8608.
[31] Yang, C.; Cho, S.; Heeger, A. J.; Wudl, F. *Angew. Chem. Int. Ed.* 2009, *48*, 1592-1595.
[32] Averdung, J.; Mattay, J. *Tetrahedron* 1996, *52*, 5407-5420.
[33] Benati, L.; Leardini, R.; Minozzi, M.; Nanni, D.; Spagnolo, P.; Strazzari, S.; Zanardi, G. *Org. Lett.* 2002, *4*, 3079-3081.
[34] Deng, L.-L.; Xie, S.-L.; Yuan, C.; Liu, R.-F.; Feng, J.; Sun, L.-C.; Lu, X.; Xie, S.-Y.; Huang, R.-B.; Zheng, L.-S. *Solar Energy Materials and Solar Cells* 2013, *111*, 193-199.
[35] Sinyashin, O.G.; Romanova, I.P.; Yusupova, G.G.; Kovalenko, V.I.; Efremov, Yu.Y.; Vandyukova, I.I.; Badeev, Yu.V.; Azancheev, N.M. *Phosphorus, Sulfur and Silicon* 1999, *147*, 477-477.
[36] Romanova, I.P.; Sinyashin, O.G.; Yusupova, G.G.; Kovalenko, V.I.; Efremov, Yu.Ya.; Badeev, Yu.V.; Vandyukova, I.I.; Arakelyan, I.A. *Russ. Chem. Bull.* 1999, *48*, 2144-2148.

[37] Romanova, I.P. *Elektronoaktseptornye mono- i bis-tsykloaddukty fullerene C_{60}. Sintez, struktura i svoistva*. Dr. A.E. Arbuzov Institute of organic and physical chemistry of RAS. Kazan, Russia 2008.

[38] Romanova, I.P.; Yusupova, G.G.; Nafikova, A.A.; Yakhvarov, D.G.; Larionova, O.A.; Siyashin, O.G. *Russ. Chem. Bull.* 2004, *53*, 144-147.

[39] Ulmer, L.; Mattay, J. *Eur. J. Org. Chem.* 2003, 2933-2940.

[40] Nakahodo, T.; Okada, M.; Morita, H.; Yoshimura, T.; Ishitsuka, M.O.; Tsuchiya, T.; Maeda, Y.; Fujihara, H.; Akasaka, T.; Gao, X.; Nagase, S. *Angew. Chem. Int. Ed.* 2008, *47*, 1298 –1300.

[41] Wu, R.; Lu, X.; Zhang, Y.; Zhang, J.; Xiong, W.; Zhu, S. *Tetrahedron* 2008, *64*, 10694-10698

[42] Hirsch, A. *Sinthesis* 1995, 895-913.

[43] Thilgen, C.; Herrmann, A.; Diederich, F. *Angew. Chem., Int. Ed.* 1997, *36*, 2268-2280.

[44] Thilgen, C.; Diederich, F. *Top. Curr. Chem.* 1999, *199*, 135-171.

[45] Nuretdinov, I.A.; Gubskaya, V.P.; Shishikina, N.I.; Fazleeva, G.M.; Berezhnaya, L.Sh.; Karaseva, I.P.; Sibgatullina, F.G.; Zverev, V.V. *Russ. Chem. Bull.* 2002, *51*, 337-341.

[46] Mc Kenzie, D.R.; Davis, C.A.; Cockayne, D.J.H.; Muller, D.A.; Vassallo, A.M. *Nature* 1992, *355*, 622-624.

[47] Scuseria, G.E. *Chem. Phys. Lett.* 1991, *180*, 451 – 456.

[48] Sabirov, D.Sh.; Khursan, S.L.; Bulgakov, R.G. *J. Mol. Graf. Model.* 2008, *27*, 124 – 130.

[49] Roth, G.; Adelmann, P. *J. Phys. I* (Paris) 1992, *2*, 1541.

[50] Heymfnn, D.; Bachilo, S.M.; Weisman, R.B. *J. Am. Chem. Soc.* 2002, *124*, 6317 – 6323.

[51] Bellavia-Lund, C.; Wudl, F. *J. Am. Chem. Soc.* 1997, *119*, 943-946.

[52] Ishida, T.; Tanaka, K.; Nogami, T. *Chem. Lett.* 1994, 561-562.

[53] Takeshita, M.; Suzuki, T.; Shinkai, S. *J. Chem. Soc., Chem. Commun.* 1994, 2587-2588.

[54] Banks, M.R.; Cadogan, J.I.G.; Gosney, I.; Hodgson, P.K.G.; Langridge-Smith, P.R.R.; Millar, J.R.A.; Taylor, A.T. *Tetrahedron Lett.* 1994, *35*, 9067-9070.

[55] Ramasundaram, S.; Hwang, S. W.; Lee, J.; Jung, J.; Choi, K. J.; Hong, S. W. *Mat. Lett.* 2013, *106*, 213-217.

[56] Fang, H.; Du, C.; Qu, S.; Li, Y.; Song, Y.; Li, H.; Liu, H.; Zhu, D. *Chem. Phys. Lett.* 2002, *364*, 290-296.

[57] Du, C.; Xu, B.; Li, Y.; Wang, C.; Wang, S.; Shi, Z.; Fang, H.; Xiao, S.; Zhu, D. *New J. Chem.* 2001, *25*, 1191-1194.

[58] Liu, F.; Du, W.; Liang, Q.; Wang, Y.; Zhang, J.; Zhao, J.; Zhu, S. *Tetrahedron* 2010, *66*, 5467-5471.

[59] Yassar, A.; Hmyene, M.; Loveday, D.C.; Ferraris, I.P. *Synthetic metals* 1997, *84*, 231-232.

[60] Yashiro, A.; Nishida, Y.; Ohno, M.; Eguchi, S.; Kobayashi, K. *Tetrahedron Lett.* 1998, *39*, 9031-9034.

[61] Kato, H.; Yashiro, A.; Mizuno, A.; Nishida, Y.; Kobayashi, K.; Shinohara, H. *Bioorg. Med. Chem. Lett.* 2001, *11*, 2935-2939.

[62] Yoon, S.; Hwang, S.H.; Ko, W.B. *Curr. Appl. Phys.* 2008, *8*, 774-777.

[63] Okamura, H.; Miyazono, K.; Minoda, M.; Miyamoto, T. *Macromol. Rapid Commun.* 1999, *20*, 41-45.
[64] Xiao, S.; Wang, S.; Fang, H.; Li, Y.; Shi, Z.; Du, C.; Zhu, D. *Macromol. Rapid Commun.* 2001, *22*, 1313-1318.
[65] Wang, S.; Xiao, S.; Li, Y.; Shi, Z.; Du, C.; Fang, H.; Zhu, D. *Polymer* 2002, *43*, 2049-2054.
[66] Hawker, C.J.; Saville, P.M.; White, J.W. *J. Org. Chem.* 1994, *59*, 3503-3505.
[67] Nuber, B.; Hampel, F.; Hirsch, A. *Chem. Commun.* - 1996. - P. 1799-1800.
[68] Lamparth, I.; Nuber, B.; Schick, G.; Skiebe, A.; Grosser, T.; Hirsch, A. *Angew. Chem. Int. Ed. Engl.* 1995, *34*, 2257-2259.
[69] Ikeda, A.; Fukuhara, C.; Shinkai, S. *Tetrahedron Lett.* 1996, *37*, 7091-7094.
[70] Hummelen, I.C.; Prato, M.; Wudl, F. *J. Am. Chem. Soc.* 1995, *117*, 7003-7004.
[71] Hummelen, I.C.; Knight, B.; Pavlovich, J.; Gonzalez, R.; Wudl, F. *Science* 1995, *269*, 1554-1556.
[72] Lund, C.B.; Keshavarz-K, M.; Collins, T.; Wudl, F. *J. Am. Chem. Soc.* 1997, *119*, 8101-8102.
[73] Vougiou Kalakis, G.C.; Prassides, K.; Orfanopoulos, M. *Org. Lett.* 2004, *6*, 1245-1247.
[74] Baran, P.S.; Monaco, R.R.; Khan, A.U.; Schuster, D.I.; Wilson, S.R. *J. Am. Chem. Soc.* 1997, *119*, 8363-8364.
[75] Chen, C.; Zhu, Y.-Z.; Zhao, H.-Y.; Zheng, J.-Y. *Tetrahedron Lett.* 2013, *54*, 1607-1611.
[76] Shi, Z.; Li, Y.; Gong, H.; Liu, M.; Liu, H.; Li, H.; Xiao, S.; Zhu, D. *Org. Lett.* 2002, *4*, 1179-1182.
[77] Xiao, S.; Li, Y.; Fang, H.; Li, H.; Liu, H.; Shi, Z.; Jiang, L.; Zhu, D. *Org. Lett.* 2002, *4*, 3063-3066.
[78] Romanova, I.P.; Kalinin, V.V.; Yakhvarov, D.G.; Nafikova, A.A.; Kovalenko, V.I.; Plekhanov, P.V.; Rusinov, G.L.; Sinyashin, O.G. *Mendeleev Commun.* 2002, *12*, 51-52.
[79] Romanova, I.P.; Yusupova, G.G; Larionova, O.A.; Balandina, A.A.; Latypov, Sh.K.; Zverev, V.V.; Yahvarov, D.G.; Rusinov, G.L.; Sinyashin, O.G. *Russ. Chem. Bull.* 2006, *55*, 502-506.
[80] Romanova, I.P.; Kalinin, V.V.; Nafikova, A.A.; Yahvarov, D.G.; Zverev, V.V.; Kovalenko, V.I.; Rusinov, G.L.; Plekhanov, P.V., Charushin, V.N.; Sinyashin, O.G. *Russ. Chem. Bull.* 2003, *52*, 173-178.
[81] Romanova, I.P.; Yusupova, G.G; Yahvarov, D.G.; Larionova, O.A.; Mogulskaya, N.N.; Sidorova, L.P.; Charushin, V.N.; Zverev. V.V.; Sinyashin, O.G. *Russ. Chem. Bull.* 2003, *52*, 2171-2174.
[82] Hawker, C.J.; Wooley, K.L.; Frechet, J.M.J. *J. Chem. Soc., Chem. Commun.* 1994, 925-926.
[83] Kay, K-Y.; Han, K.-J.; Yu, Y.-J.; Park, Y.D. *Tetrahedron Lett.* 2002, *43*, 5053-5056.
[84] Yuan, D-Q.; Koga, K.; Kourogi, Y.; Fujita, K. *Tetrahedron Lett.* 2001, *42*, 6727-6729.
[85] Chen, Y.; Wang, Y.; Rassat, A.; Sinay, P.; Zhao, Y.; Zhang, Y. *Tetrahedron* 2006, *62*, 2045-2049.
[86] Giacalone, F.; D'Anna, F.; Giacalone, R.; Gruttadauria, M.; Riela, S.; Noto, R. *Tetrahedron Lett.* 2006, *47*, 8105-8108.
[87] Jagerovic, N.; Elguero, J.; Aubagnac, J.-L. *Tetrahedron* 1996, *52*, 6733-6738.

[88] Romanova, I.P.; Yusupova, G.G; Larionova, O.A.; Yahvarov, D.G.; Mogulskaya, N.N.; Sidorova, L.P.; Zverev. V.V.; Charushin, V.N.; Sinyashin, O.G. *Russ. Chem. Bull.* 2005, *54*, 660-665.
[89] Iglesias, M.; Go´mez-Lor, B.; Santos, A. *J. Organomet. Chem.* 2000, *599*, 8-17
[90] Iglesias, M.; Go´mez-Lor, B.; Santos, A. *J. Organomet. Chem.* 2001, *627*, 159-166.
[91] Banks, M.R.; Cadogan, J.I.G.; Gosney, I.; Henderson, A.J.; Hodgson, Ph.K.G.; Kerr, W.G.; Kerth, A.; Langridge-Smith, P.R.R.; Millar, J.R.A.; Mount, A.R.; Parkinson, J.A.; Taylor, A.T.; Thoronburn, P. *Chem. Commun.* 1996, 507-508.
[92] Averdung, J.; Mattay, J.; Jacobi, D.; Abraham, W. *Tetrahedron* 1995, *51*, 2543-2552.
[93] Yan, M.; Cai, S. X.; Keana, J.F.W. *J. Org. Chem.* 1994, *59*, 5951-5954.
[94] Dong, G.-X.; Li, J.-S.; Chan, T.-H. *J. Chem. Soc., Chem. Commun.* 1995, 1725-1726.
[95] Shiu, L.-L.; Chien, K.-M.; Liu, T.-Y.; Lin, T.-I.; Her, G.-R.; Luh, T.-Y. *J. Chem. Soc., Chem. Commun.* 1995. 1159-1160.
[96] Ikeda, A.; Fukuhara, C.; ShinKai, S. *Chem. Lett.* 1997, 407-408.
[97] Shen, C.K.-F.; Chein, K.-M.; Juo, C.-G.; Her, G.-R.; Luh, T.-Y. *J. Org. Chem.* 1996, *61*, 9242-9244.
[98] Shen, C.K.-F.; Yu, H.-H.; Juo, C-G.; Chein, K.-M.; Her, G.-R.; Luh, T.-Y. *Chem. Eur. J.* 1997, 3, 744-748.
[99] Kanakamma, P.P.; Huang, S.-L.; Juo, C.-G.; Her, G.-R.; Luh, T.-Y. *Chem. Eur. J.* 1998, *4*, 2037-2042.
[100] Kawaguchi, M.; Ikeda, A.; Shinkai, S. *J. Chem. Soc., Perkin Trans. V* 1998, 179-184.
[101] Romanova, I.P; Yusupova, G.G.; Larionova, O.A.; Balandina, A.A.; Latypov, Sh.K.; Yakhvarov, D.G., Zverev, V.V.; Sinyashin, O.G. *Russ. Chem. Bull.* 2006, *55*, 697-702.
[102] Romanova, I.P; Yusupova, G.G.; Nafikova, A.A.; Kovalenko, V.I.; Sinyashin, O.G. *Russ. Chem. Bull.* 2002, *51*, 1491-1496.
[103] Prato, M.; Wudl, F.; Grosser, T.; Hirsch, A. *Pat. DE 95-19507502*(to A.-G.Hoechst) 1995.
[104] Imahori, H.; Tamaki, K.; Guldi, D. M.; Luo, C.; Fujitsuka, M.; Ito, O.; Sakata, Y.; Fukuzumi, S. *J. Am. Chem. Soc.* 2001, *123*, 2607-2617.
[105] Kim, B.; Lee, J.; Seo, J. H.; Wudl, F.; Park, S. H.; Yang, C. *J. Mater. Chem.* 2012, *22*, 22958-22963.
[106] Romanova, I.P.; Yusupova, G.G.; Larionova, O.A.; Nafikova, A.A.; Yakhvarov, D.G.; Zverev, V.V.; Efremov, Yu. Ya.; Sinyashin, O.G. *Mendeleev Commun.* 2006, *6*, 309-311.
[107] Romanova, I.P.; Yusupova, G.G.; Balandina, A.A.; Latypov, Sh.K.; Yakhvarov, D.G.; Nifantev, D.V.; Yashunskii, D.V.; Sinyashin, O.G. *Russ. Chem. Bull.* 2007, *56*, 1495-1500.
[108] Akhmetov, A.R.; Tuktarov, A.R.; Dzhemilev, U.M.; Yarullin, I.R.; Gabidullina, L.A. *Russ. Chem. Bull.* 2011, *60*, 1885-1887.
[109] Tuktarov, A.R.; Akhmetov, A.R.; Yarullin, I.R.; Gabidullina, L.A.; Dzhemilev, U.M. *Vserosiiskaya konferenciya "Organicheskii sintez: khimiya i techologiya".* Ekaterinburg, Russia, 2012
[110] Tuktarov, A.R.; Akhmetov, A.R.; Yarullin, I.R.; Gabidullina, L.A.; Dzhemilev, U.M. *unpublished data.*
[111] Dzhemilev, U.M.; Tuktarov, A.R.; Yarullin, I.R.; Akhmetov, A.R. *Mendeleev Commun.* 2013, *23*, 326-328.

[112] Crotti, P.; Di Bussolo, V.; Favero, L.; Pineschi, M. *Tetrahedron* 1997, *53*, 1417-1438.
[113] Pereira de Freitas, R.; Iehl, J.; Delavaux-Nicot, B.; Nierengarten, J.-F. *Tetrahedron* 2008, *64*, 11409-11419.
[114] Sonogashira, K.; Tohda, Y.; Hagihara, N. *Tetrahedron Lett.* 1975, *16*, 4467-4470.
[115] Galle, J.E.; Hassner, A. *J. Am. Chem. Soc.* 1972, *94*, 3930–3933.
[116] Zbiral, E.; Stütz, A. *Tetrahedron* 1971, *27*, 4953-4963.
[117] Akhmetov, A.R.; Yarullin, I.R.; Tuktarov, A.R.; Dzhemilev, U.M. *Tetrahedron Lett.* 2014, in print
[118] Kolb, H.C.; Finn, M. G.; Sharpless, K. B. *Angew. Chem. Int. Ed.* 2001, *40*, 2004-2021.

In: Fullerenes
Editor: Shannon B. Ellis

ISBN: 978-1-63321-385-2

Chapter 6

FULLERENE DERIVATIVES FOR MOLECULAR SWITCH: RECENT ADVANCES AND THEORETICAL INSIGHTS FROM THE POLARIZABILITY

Denis Sh. Sabirov*

Institute of Petrochemistry and Catalysis, Russian Academy of Sciences,
Ufa, Republic of Bashkortostan, Russian Federation

ABSTRACT

Molecular switch is a widespread concept in molecular machinery, which is currently a rapidly developed field of nanotechnology. It consists in the ability of a molecular system to exist in two or more stable states, differing by any physical property. Fullerene derivatives, due to the unique structural and electronic properties, have become very attractive for switchable molecular devices. We have classified the synthesized and tested fullerene-based switches to recognize what novel fullerene derivatives may be useful for this application.

Previously, we have examined polarizability of fullerene derivatives and found that all of them demonstrate remarkably high mean polarizabilities (See: D. Sh. Sabirov, Polarizability of C_{60}/C_{70} Fullerene [2+1]- and [1+1]-Adducts: A DFT-Prognosis//Density Functional Theory: Principles, Applications and Analysis. Editors: J. Morin and J. M. Pelletier. New York: Nova Publishers, 2013. P. 147–170). This property can be a landmark for design of fullerene-based molecular systems with adjustable properties.

In the present paper, we have considered the fullerene-based nanostructures, which can act in two states with different polarizability, i.e., can be potentially useful for application as molecular switch:

1) pairs of isomeric 6.6-closed and 5.6-open fullerene adducts;
2) fullerene dimers with flexible bridge between the cages;
3) reversibly dissociating singly bonded fullerene dimers.

* E-mail: diozno@mail.ru.

Relations of our theoretical findings to the recent experimental works on molecular switch have been analyzed. It makes us hope that our theoretical propositions will be stimulating for the fullerene-based molecular machinery.

Keywords: fullerene C_{60}, fullerene dimers, fullerene adducts, theoretical chemistry, polarizability, split of polarizability, molecular switch, reversibility

1. INTRODUCTION

Since the discovery of the C_{60} fullerene in 1985 [1], the number of studies on its physical and chemical properties grows up as an avalanche. In addition to a fundamental interest in "*a molecular form of carbon*", its diverse applications have been considered, such as gas storage [2], radical scavenging for stabilization of polymers [3], catalysis [4], photovoltaic applications [5], and antiviral pharmaceuticals [6]. The applications above can enter daily life due to the technology [7], which may produce tons of fullerenes if required. However, fullerene derivatives are likely more prospective for various applications than the parent C_{60}. Fullerene by its nature is an electron-deficient alkene [3, 5, 8, 9], so it readily takes part in the reactions of [2+*n*]-cycloaddition and radical addition. These reactions allow attaching diverse chemical groups to the fullerene core and, consequently, regulating physicochemical properties of the resulting compounds (polarity, solubility, redox-properties, propensity for aggregation, sensitivity to light, etc.). In certain cases, adjusting the number of addends in the fullerene adducts is possible as well as the addition pattern [8].

The main difficulty in obtaining pure samples of individual fullerene derivatives deals with non-selectivity of the addition reactions to fullerenes, which is caused by a large number of reactive sites with almost the same reactivity (this also holds true for the next of buckminsterfullerene's kin, e.g., C_{70} fullerene). Therefore, theoretical approaches are developed to preliminarily assess the utility of fullerene derivatives. In this aspect, we should mention computational design of 6.6-closed and 5.6-open $C_{60}CR^1R^2$ adducts with improved electronic properties [10], fullerene-based compounds with desirable static dielectric constants [11], polarizability [12], and hyperpolarizability [13]. Such approaches allow defining structures of the prospective compounds before the synthetic procedures and focusing on the obtaining the desired adducts. Thereby, the cost of scientific endeavor goes down.

Fullerene is a perfect building block for molecular devices because of at least four features. First, it is a rigid bulky moiety; for example, it does not significantly change its volume when charged [14, 15]. Second, it has an outstandingly rich π-electronic system that is not required in additional comments about the usefulness due to its electro- and photoactivity [16]. Third, this is an empty cage that can be filled by guest atoms or molecules; such derivatization of the inside often leads to the change in the outside properties (e.g., encapsulation may change exohedral chemical reactivity [17]). Finally, C_{60} has 30 double bonds, an appropriate number to link it with the desirable amount of the other blocks of a molecular device. These features make fullerene attractive for applications to molecular machinery, in general, and to molecular switch devices, particularly [18, 19].

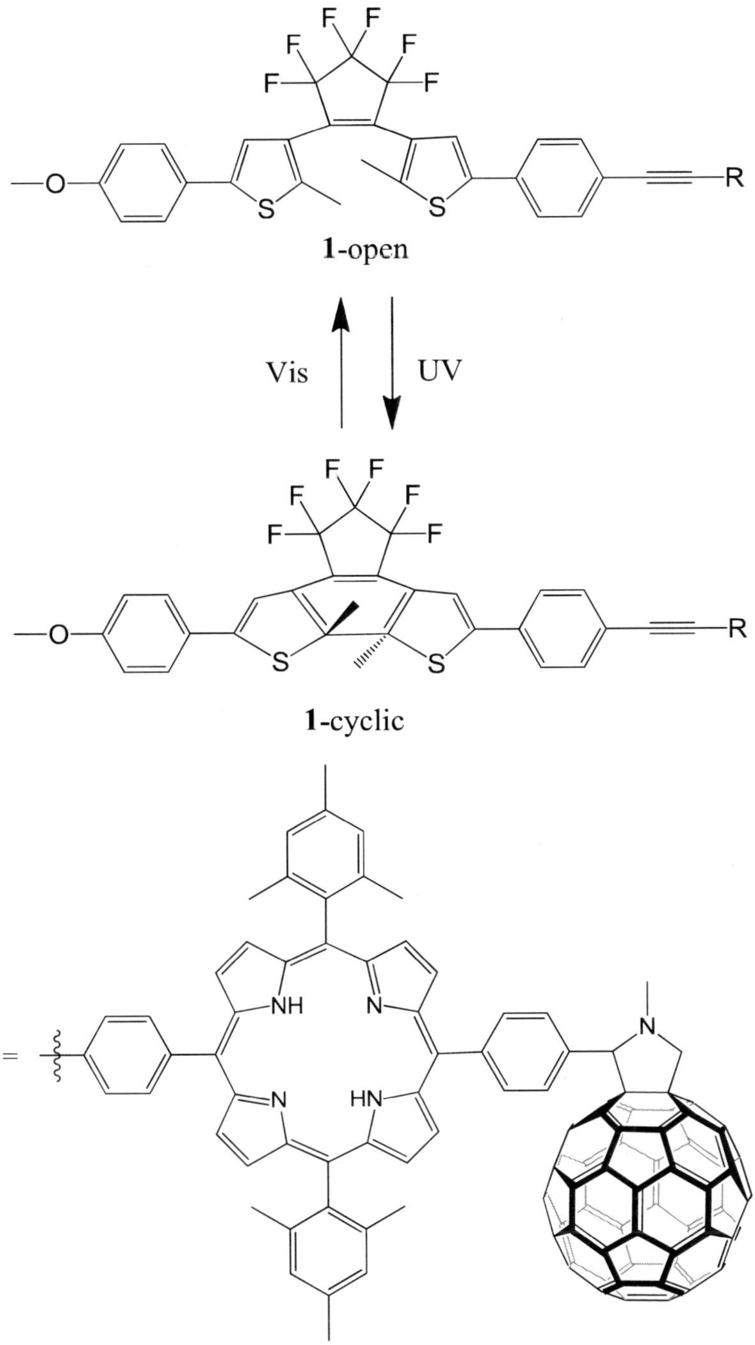

(tested in Ref. 22).

Figure 1. Molecular switch between the open and the cyclic forms of fullerene-containing DTE 1.

In the present prospective paper, we have briefly reviewed the existing molecular switches and considered polarizability as one of the molecular parameters that can facilitate design of novel fullerene-based switchable devices.

2. Fullerene Derivatives for Molecular Switch

In this section, we have reviewed the most interesting fullerene-containing switchable molecular systems. For simplification, we have divided them into several groups, according to the structural types or physicochemical processes that underlie their functioning.

2.1. Fullerene Core Attached to the Well-Known Photochromic Molecular Systems

This approach seems to be the most rational way to design fullerene-containing systems with molecular switch. A classic example of molecular switch is based on the dithienylethylenes (DTE), which can exist in the open and the closed forms (before and after UV treatment, respectively) [20]. These two forms, depending on the substituents, exhibit different photochromic properties [21]. In this context, Liddell et al. have prepared a DTE–porphyrin (P)–fullerene triad with an electron-transfer switch (Figure 1) [22]. When DTE gates are open (1-open), the solution of compound in 2-methyltetrahydrofuran is colorless, and porphyrin excited state donates an electron to C_{60}, giving DTE_{open}–$P^{\bullet+}$–$C_{60}^{\bullet-}$. Irradiation of compound 1 with UV light converts DTE into its closed colored form (1-cyclic). In this form, the porphyrin excited state is quenched by energy transfer to DTE_{closed} in ~2 ps, precluding significant electron transfer to the fullerene core. Visible irradiation returns the system to the colorless form. This photonic switching of photoinduced electron transfer in compound 1 can be cycled many times.

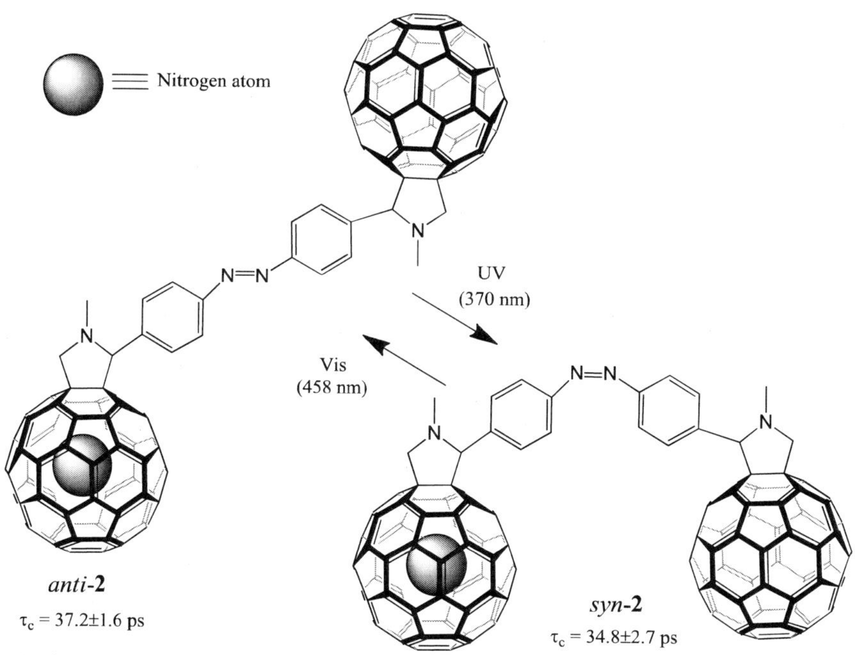

(tested in Ref. 23).

Figure 2. Molecular switch based on fullerene bicage derivative with photoactive azobenzene bridge.

Azobenzene-fullerene adducts provide another opportunity for photochromic switch. Fullerene dimers have the primary interest here. For example, bridging two fullerene cages in molecule 2, the azobenzene moiety retains propensity for photoisomerization (Figure 2) [23]. The main feature of this compound is that one of the cages contains a nitrogen atom inside. Unfortunately, two forms of compound 2 have almost the same absorption spectra. However, they can be distinguished by EPR measurements due to the unpaired electrons of the encapsulated atom. Authors [23] have studied EPR signals of *syn*- and *anti*-isomers and analyzed electron spin lattice and spin-spin relaxation times. Indeed, the molecular rotational correlation time τ_c, deduced from pulse EPR measurements in degassed CS_2, is slightly longer for *anti*-form of compound 2 compared to its *syn*-form.

2.2. Fullerene Core As a Part of Solvent-Switchable Molecular Systems

The bulkiness of C_{60} moiety makes it promising for construction of rotaxanes [18, 19, 24, 25]. Thus, a novel rotaxane has been prepared, in which C_{60} moiety behaves as both a stopper and a photoactive unit [24] (Figure 3). In this molecular system, a cyclic subunit 3 is fixed on the fullerene-containing axis 5. Depending on the solvent, rotaxane (3+5) acts as a molecular shuttle with two states: 1) state A with macrocycle 3, placed in the peptide part of the axis 5 (in dichloromethane); and 2) state B with macrocycle 3, placed in the alkyl part of the axis 5 (in dimethyl sulfoxide, DMSO) (Figure 4).

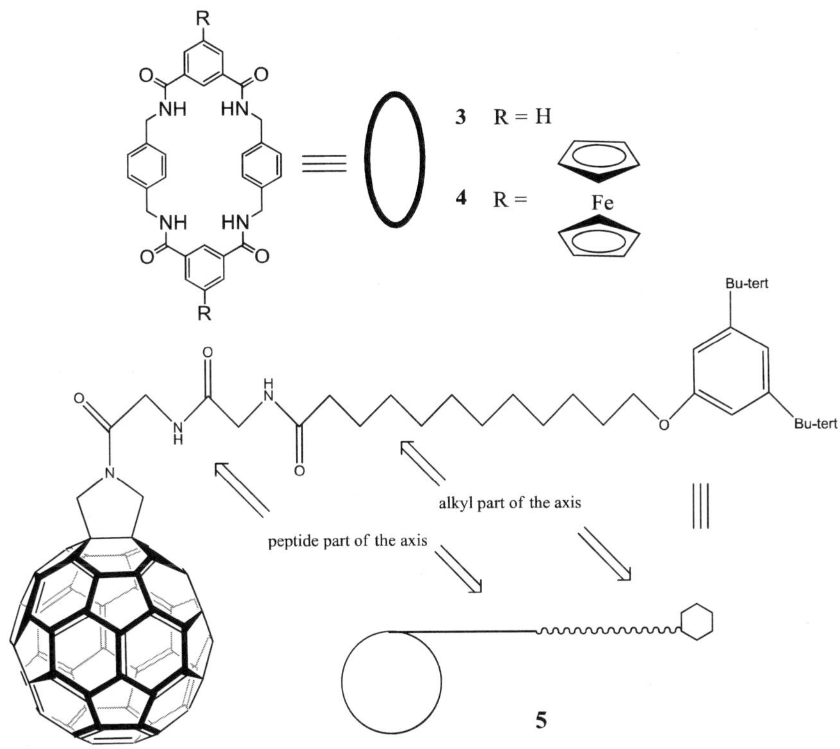

Figure 3. Blocks for molecular shuttling, structural formulae and schematic representations: macrocycles (3 and 4) and thread fullerene derivative 5.

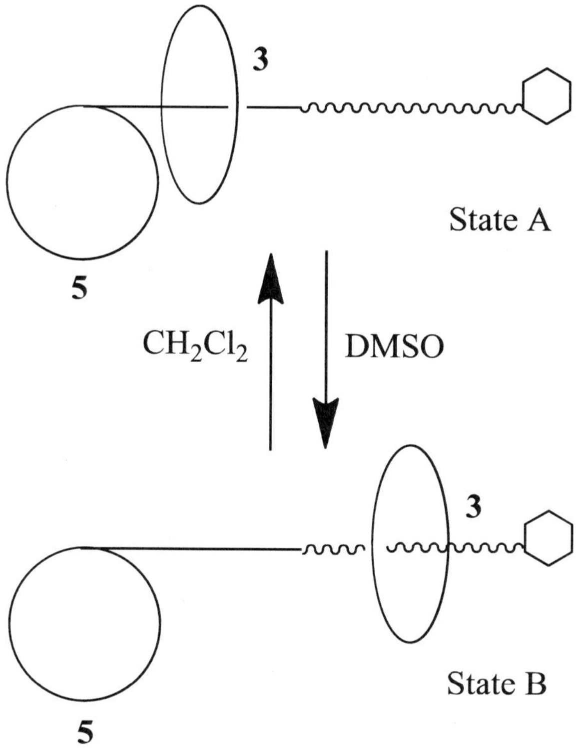

Figure 4. Principal scheme of molecular shuttling, based on compounds 3 and 5.

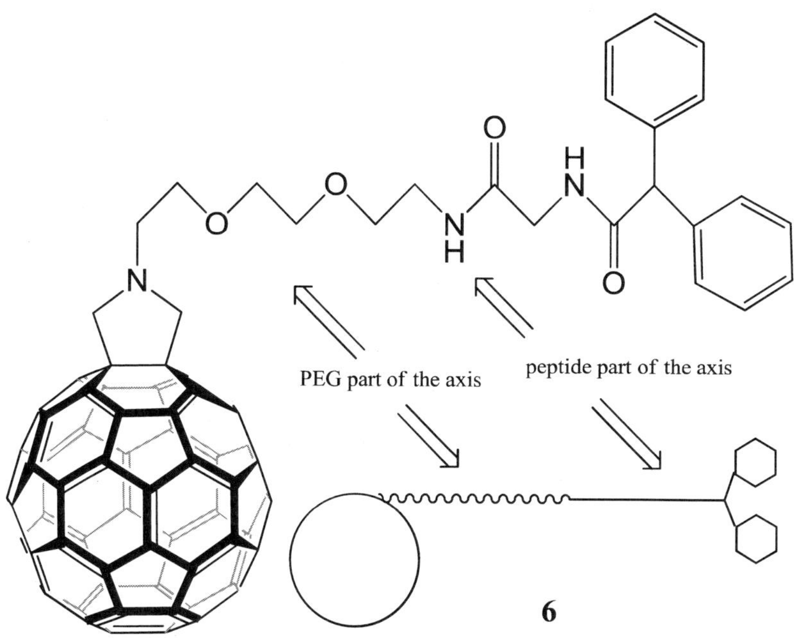

Figure 5. Substituted pyrrolidinofullerene used as a thread compound for molecular switch in Ref. 25.

These states are distinguishable by ^{1}H NMR spectra. Moreover, photophysical properties of the molecular system in two states differ. The proximity of macrocycle 3 to the C_{60} core does not affect fluorescence but has a significant effect on the triplet-triplet spectra. After 337 nm laser excitation of the solutions, triplet lifetimes have been measured. These found to be nearly identical for (3+5) rotaxane in state B and the isolated thread compound 5 (~17 μs). However, the lifetime of the excited state A differs from the values above (~10 μs). It means that closer placing the macrocyclic subunit 3 to the C_{60} core of thread compound 5 leads to the change in photophysical properties. Thereby, A and B states become distinguishable.

Fullerene-containing rotaxanes may exhibit solvent-switchable states differing by the nonlinear optical properties. For example, the rotaxanes, based on macrocycles (3 or 4) and thread fullerene compound 6 (Figure 5), have been prepared and their second-order hyperpolarizabilities (γ) have been studied [25]. It has been found that states A and B of molecular shuttles (6+3) and (6+4) exhibit substantial difference in γ values (Figure 6). The response of (6+4), in which a macrocycle has ferrocene moieties instead of two hydrogen atoms, is enhanced in comparison with rotaxane (6+3) due to electron transfer between the ferrocene residues and the fullerene.

In the mentioned systems, fullerene acts as an electron acceptor because of the high electron affinity [26]. However, very attractive electron donor–acceptor conjugate $Ce_2@I_h$-C_{80}–ZnP has been reported; it is made up by the $Ce_2@I_h$-C_{80} endofullerene and zinc porphyrine (ZnP) [27].

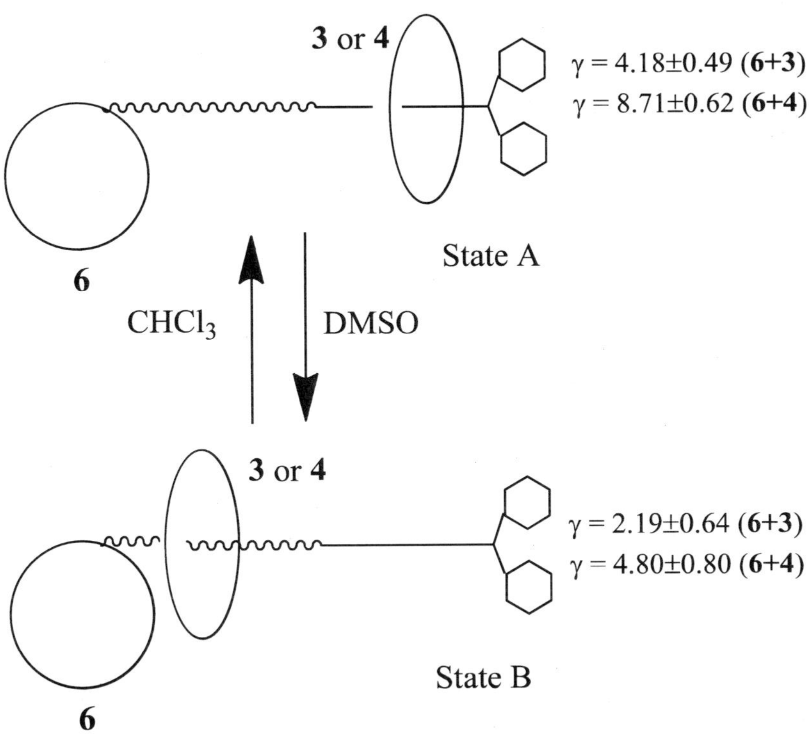

(numerical data taken from Ref. 25).

Figure 6. Principal scheme of molecular shuttling based on rotaxanes (6+3) and (6+4). Second hyperpolarizabilities are in 10^{-32} esu.

This molecular switch can exist in the following states. In nonpolar solvents, the C_{80} fullerene core undergoes reductive charge transfer, so $[Ce_2@I_h\text{-}C_{80}]^{\bullet-}$–$[ZnP]^{\bullet+}$ is formed. Meanwhile, the oxidative charge transfer dominates in polar media that results in $[Ce_2@I_h\text{-}C_{80}]^{\bullet+}$–$[ZnP]^{\bullet-}$.

The system, based on the $Li^+@C_{60}$ endofullerene and tetrathiafulvalene calix[4]pyrroles (TTF-C4P) as complexing agents, have been used for the ion-controlled switch of electron transfer [28]. Binding of chloride anion to a TTF-C4P donor results in electron transfer to the endofullerene to produce the radical pair TTF-C4P$^{\bullet+}$...$Li^+@C_{60}{}^{\bullet-}$. Thus, the electron transfer is switched on by the addition of Cl^-, and it can be switched off by the addition of tetraethylammonia ions, which bind to TTF-C4P more effectively than $Li^+@C_{60}$.

Other examples of switchable fullerene-containing rotaxanes can be found in review [19].

2.3. Switching the States of the Atoms/Molecules Encapsulated by the Fullerene Cage

This opportunity arises from the hollow structure of fullerenes and deals with the control and measurements of the different states of the atoms, putted inside. Their monitoring is possible in the case of endofullerenes with non-zero electric dipole moments [29] that is required for such molecular devices [30]. For example, to control the thermal rotational states of $Tb@C_{82}$ endofullerene, an octanethiol self-assembled monolayer has been introduced between the ehdofullerene and Au (111) surface [31]. Low-temperature scanning tunneling spectroscopy (LT-STM) of the described system have demonstrated hysteresis including negative differential conductance. It has been interpreted in terms of switching the $Tb@C_{82}$ molecular orientation, caused by interaction of its dipole moment and an external electric field.

The conductance switching of single $Sc_3N@I_h\text{-}C_{80}$ molecules in a junction, formed by an atomically ordered Cu(110)-(2×1)-O substrate and LT-STM tip, has been studied [32]. This switch is based on the tunneling electron-driven rotation of a triangular Sc_3N cluster inside $I_h\text{-}C_{80}$ cage among three pairs of enantiomorphic configurations.

Another molecular devices exploiting motion of the encapsulated species inside carbon cages have been studied theoretically, e.g., motion of endofullerene in a bulkier carbon cage in $K^+@C_{60}@C_{640}$ under the external force fields [33], nuclear spin dynamics of $^{15}N@C_{60}$ and $^{31}P@C_{60}$ endofullerenes for the purposes of a nuclear spin quantum computer [34] or electron switch in bicage fluorinated fullerene anions $e^-@C_{20}F_{18}(XH)_2C_{20}F_{18}$ (X = N or B) [35].

2.4. Other Types of Fullerene-Based Switch

Molecular switch is very wide concept, which may include all possible systems demonstrating the reversible transitions between two or more measurable states. Therefore, the fullerene-based molecular switch is not limited by the examples considered above. Thus, the construction of switch devices may be based on the other phenomena, e.g., on the reversible protonation of fullerene azafulleroids [36] or sandwiching the intact fullerene between two carbon nanotubes [37].

3. Polarizability of Fullerene Derivatives and Theoretical Insights into the Fullerene-Based Molecular Switch

3.1. Polarizability Studies in Fullerene Science

Polarizability is a molecular property that describes molecule's ability to acquire induced dipole moment in external electric fields. It is defined as a ratio of the induced dipole moment of a molecule $\boldsymbol{\mu}_{\mathbf{ind}}$ to the electric field $\boldsymbol{E}$ that produces the moment:

$$\boldsymbol{\mu} = \boldsymbol{\alpha} \boldsymbol{E}, \tag{1}$$

where $\boldsymbol{\alpha}$ is polarizability tensor (a 3×3 matrix, symmetric about the main diagonal). Its trace is invariant under coordinate system [38]. Mean polarizability equals to the arithmetic mean of the diagonal elements of $\boldsymbol{\alpha}$ tensor:

$$\alpha = \frac{1}{3}\left(\alpha_{xx} + \alpha_{yy} + \alpha_{zz}\right) \tag{2}$$

Mean polarizability determines many physical properties such as dielectric constant, refractive index, van der Waals constant, ion mobility in gas, *etc.* [38]. It makes information about polarizability irreplaceable for analysis of the structure, physical and chemical properties of molecular and atomic clusters [39]. Polarizability is also a good landmark for design of novel photomodulated systems [40] and electron transfer in molecular functional materials [41].

Fullerenes, due to their rich π-electronic systems, are highly polarizable molecules: the mean polarizabilities of C_{60} and C_{70} equal to ~80 and ~105 Å^3, respectively (see the summarizing tables of theoretical and experimental estimations in Refs 12, 42, 43). Polarizability of fullerene derivatives is studied mainly by theoretical techniques due to many reasons (e.g., their possible dissociation under experiments, low solubility, and low availability).

As is known from the theoretical studies, mean polarizabilities of exohedral fullerene derivatives are higher than the polarizability of the original fullerene if it does not degrade upon chemical functionalization [12, 13, 44–62]. This is true for both mono- and polyadducts, produced by [1+1]-addition [13, 44, 45, 48, 49–52, 54, 55, 57, 60] and [2+*n*]-cycloaddition [13, 46, 47, 51, 53, 56, 58, 61, 62].

High α values of fullerenes and their derivatives have been used for explaining the distinctiveness of physical and chemical processes in the fullerene-containing systems. To demonstrate the diversity, we have accumulated the performed and the possible applications of polarizability to fullerene-containing systems in Table 1. The list is long but not exhausted by the counted processes and systems.

Table 1 shows that polarizability defines numerous processes, which may be important for molecular switch. Thus, knowledge of polarizability is able to provide new ideas to molecular switchable devices.

Table 1. Use of polarizability in the studies of fullerenes and related nanostructures

Field of application	Comments and references
Intermolecular interactions (including collision and energy transfer)	Polarizability is needed for assessing constants of dispersion interaction between C_{60} and molecules (C_6) [63]. It has been used to understand propensity of fullerenes for aggregation in solutions [64] and formation of donor-acceptor complexes [65]. It is also applicable to the collision processes [66] and quenching the electronically excited states owing to energy transfer [67].
Computational material science (including organic solar cells application of fullerene adducts)	Applied to this field, polarizability of fullerenes and related nanostructures is useful in two ways: a) to calculate/explain physicochemical properties of the obtained fullerene/nanotube-containing materials (fullerenes, nanotubes [68, 69] and those with defects [70], carbon nano-onions [71–73], fullerene/nanotube–polymer composites [74], metal [75, 76] and metal-nitride endofullerenes [77], fullerene derivatives [13, 44–62], and graphene nanostructures [78]) or predict their behavior in electric fields; b) to predict novel materials with adjustable properties (e.g., adducts of fullerenes with Si and Ge [79], fullerene dimers [13, 62], lithium [80] and hydrogen [81] endofullerenes, giant fullerenes [43, 82], fullerene@fullerene complexes [83], encapsulated adducts of small fullerenes [84], and "nanotubules" [85]). Particular attention we pay to the suitability of polarizability to organic solar cells, e.g., understanding the mechanisms of functioning of fullerene derivatives [86] and assessing the efficiency of C_{60} derivatives for this application [59].
Computational molecular science	High polarizability of fullerenes is responsible for screening effect when fullerene cage traps atoms and molecules [87]. It can be used to monitor the state of the encapsulated guests [88]. As is known, usual processes (such as conformational transits) may occur in the fullerenes [89, 90] and nanotubes [91] in an unexpected manner. To model such processes, polarizability of fullerenes and related structures can be used (*e.g.*, for calculating their dielectric constants and simulating the influence of the medium in terms of polarizable continuum models [92]).
Chemical properties	Polarizability is considered as an index of reactivity of higher fullerenes in dipolar cycloaddition reactions [93] because it reflects their structures (polarizability ellipsoids of fullerenes replicate the shape of their molecules [94, 95]).
Chromatography	Polarizability of fullerenes is correlated with dipole moments of the respective endometallofullerenes [96]. It may be useful for estimation of their retention times, similar to hydrocarbons [97].
Other or potential applications	Interpretation of Raman spectra [98], estimation of wetting [99, 100] (based on the advances in theory of wetting [101]), characterization of nanomaterials in biological systems [102], and interaction of organic compounds with positrons [103] are performed with polarizability.

3.2. Fullerene Derivatives, Prospective for Molecular Switch, Due to Their Polarizability

Two key points should be taken into account, constructing a molecular switch device. First, the molecular system should exist in two (or more) stable states that can be converted into each other. We illustrate this by the energetic scheme (Figure 7). The proposed reversible system exists in two states A and B, which have nearly the same total energies; *i.e.* difference in their energies $\Delta E = E_A - E_B$ is small. The last statement is a thermodynamic condition for the reversibility. The transits between the states should occur easily. It means that the energetic barriers of both transits A→B (E_1^{act}) and B→A (E_{-1}^{act}) should be sufficiently low. This is a kinetic condition for the reversibility.

Second, states A and B should demonstrate measurable difference in some physicochemical parameters (one or more). These can be spectral characteristics (absorption, ^{1}H NMR or EPR spectra) or specific physicochemical properties (e.g., conductance).

Polarizability is very informative physical quantity and it is sensitive even to small changes in chemical structure of a molecular system [39]. However, it is hard to use it directly to monitor switching in the bistable molecular systems because its measurements usually require a complicated methodology. Meanwhile, polarizability defines other quantities that could be (easily) measured such as refractivity indices, ion mobility, and dielectric constant [38]. In this aspect, it can be useful for screening potential compounds for molecular switch because the difference in polarizabilities of two states of molecular switch may lead to the difference in their polarizability-dependent properties. To find out prospective fullerene derivatives, we have performed a journey through their polarizabilities and selected three types of the fullerene-based molecular systems that may be considered for the application as molecular switches. These are 1) pairs of isomeric 6.6-closed and 5.6-open fullerene adducts; 2) fullerene dimers with flexible bridge between the cages; and 3) singly bonded fullerene dimers.

Pairs of isomeric 6.6-closed and 5.6-open fullerene adducts. It is well-known that [2+1]-addition to C_{60} and C_{70} results in at least two types of adducts, *viz.* 6.6-closed (addition to 6.6 bond) and 5.6-open (addition to 5.6 bond with its simultaneous cleavage) [8, 104].

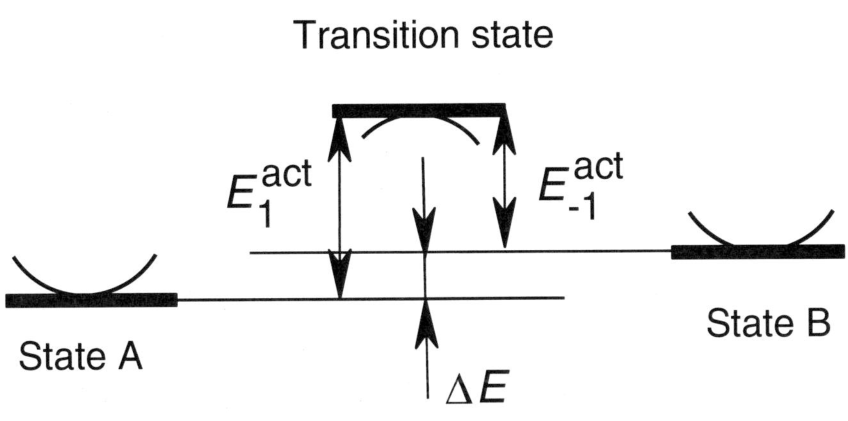

Figure 7. Simplified energetic diagram of molecular switch.

Here, we do not consider 6.6-open fullerene adducts; in general, these unexpected products of addition reactions seem to be unstable [104]. 5.6-Open fullerene adducts (fulleroids) are usually formed in the same reactions in a mixture with their 6.6-closed isomers and then convert to them spontaneously or under thermal treating. In addition, fulleroids arise as intermediates of decomposition of certain fullerene derivatives: e.g., in the case of decomposition of the primary C_{60} and C_{70} ozonides [105, 106] or pyrazolinofullerenes, primary adducts of fullerene and diazomethanes [107]. The preliminary examination of 6.6-closed *vs.* 5.6-open fullerene adducts allows pointing out the minimal difference in their nuclear frameworks. It means that these compounds should be almost isoenergetic though 5.6-open species are less stable and isomerized to their 6.6-closed counterparts [5, 107]. Meanwhile, the electronic structures of the mentioned compounds differ significantly. The π-electronic system of the parent fullerene remains almost intact in 5.6-closed fullerene adducts whereas it loses one π-bond when the 6.6-closed adduct is formed. As we have shown in our previous theoretical works, mean polarizability reflects the last fact [12, 53]. Indeed, 5.6-open adducts (**8**), due to the richer π-electronic systems, have mean polarizabilities, which slightly higher than the respective values of the 6.6-closed isomers (7) (Table 2) [53].

The difference in mean polarizabilities (Δα split) may be increased if several addends are attached to the fullerene core. Such a structure can be useful for molecular switch if it remains propensity for isomerization. Quantum-chemical calculations show that difference in 4–5 $Å^3$ is achieved in the case of $C_{60}X_6$ multiadducts with simple X (Table 2). Unfortunately, these are also small values, so another opportunity should be considered to enhance Δα split, for example, varying X moiety to obtain such a pair of monoadducts that would demonstrate a desirably high split.

Another obstacle to creation of the molecular switch, based on the described possibilities, is caused by the irreversibility the reaction from 7 to 8; i.e., in the most cases, the 6.6-closed fullerene adducts are much more thermodynamically stable than their 5.6-open counterparts. However, this problem seems to be solvable due to the recently demonstrated conversion of the 6.6-closed fullerene adducts into their 5.6-open isomers [108].

Table 2. Mean polarizabilities and their splits for $C_{60}X_n$ (n = 1 and 6) isomers with 5.6-open and 6.6-closed moieties ($Å^3$)

$C_{60}X_n$	Mean polarizability		$\Delta\alpha = \alpha_{5.6\text{-open}} - \alpha_{6.6\text{-closed}}$
	$\alpha_{5.6\text{-open}}$	$\alpha_{6.6\text{-closed}}$	
$C_{60}O$ [a]	83.2	83.9	0.7
$C_{60}NH$ [a]	84.2	84.8	0.5
$C_{60}CH_2$ [a]	85.0	85.5	0.6
$C_{60}O_6$ [b]	89.3	85.2	4.1
$C_{60}(NH)_6$ [b]	95.6	90.7	4.9
$C_{60}(CH_2)_6$ [b]	99.3	94.8	4.5

[a] Taken from Ref. 53. [b] Calculated by the PBE/3ζ method, similar to Refs 53 and 56, for the hexakisadducts with uniform distribution of addends on the fullerene cage.

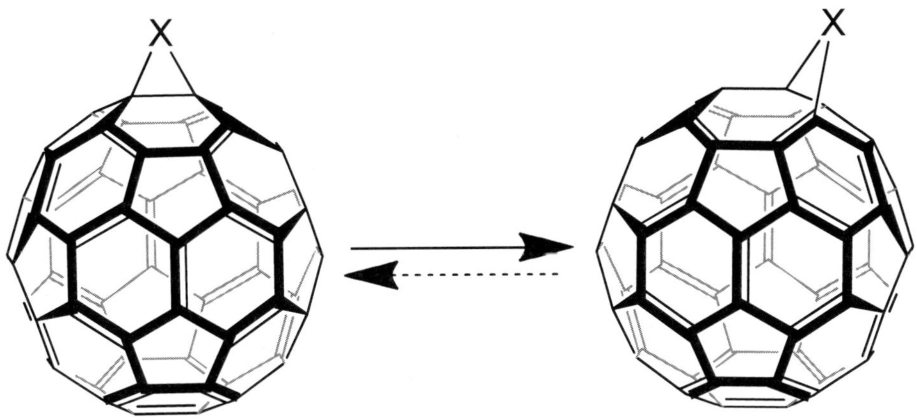

Figure 8. Hypothetical molecular switch, based on 6.6-closed–5.6-open reversible isomerization.

Thus, the systems, shown in Figure 8, can be recommended to test their switching, though with reservations.

Molecular switch based on the reversible transitions in bicage fullerene derivatives. Recently, we have performed theoretical studies of diverse fullerene dimers and oligomers and found that resulting mean polarizability of such multicage structures depends on the maximal remoteness of the cages [58, 62]. For example, the mean polarizability of the linear $(C_{60})_5$ exceeds the respective value of the "+"-shaped isomer by 30.1 $Å^3$ [58]. The centers of the terminal cages are separated by 36.45 Å in the linear $(C_{60})_5$; the same value for the "+"-shaped $(C_{60})_5$ is twice shorter (18.16 Å). Thus, the differences in their structures and polarizabilities are obvious.

Unfortunately, such oligomers are rigid compounds with no transitions between the forms with the closest and the remotest locations of the fullerene cages. However, we have also shown that the described split polarizability may be typical for bicage structures regardless of the type of bridging (for both [2+2]- and [1+1]-dimers [60]).

Therefore, we assume a bicage fullerene system with the cores, connected by a flexible bridge. The initial state of the system (9) is characterized by a short distance between the fullerene cores. If the bridge undergoes a transformation that increases its linear size (*e.g.*, the well-known photo-induced isomerization or conformational change), the system enters state 10 with a greater remoteness of the fullerene fragments (Figure 9). In the second state, the molecular system obtains the higher polarizability and, consequently, the enhanced response to an external electric impact. In contrast to state 9, it can be facilely manipulated by the external electric field.

The dimer, very similar to the described above, has been tested in Ref. 23 (Figure 2, compound 2). In the experiments, one the fullerene cages hosts a nitrogen atom. This was decisive for functioning of dimer 2 as a molecular switch. Authors explain different EPR behavior of the compound 2 in its *syn*- and *anti*-forms with the changing distance from the nitrogen center of the unpaired electrons and the empty cage (it is more distant in *anti*-2, so its influence on EPR characteristics of *anti*-2 isomer is weaker).

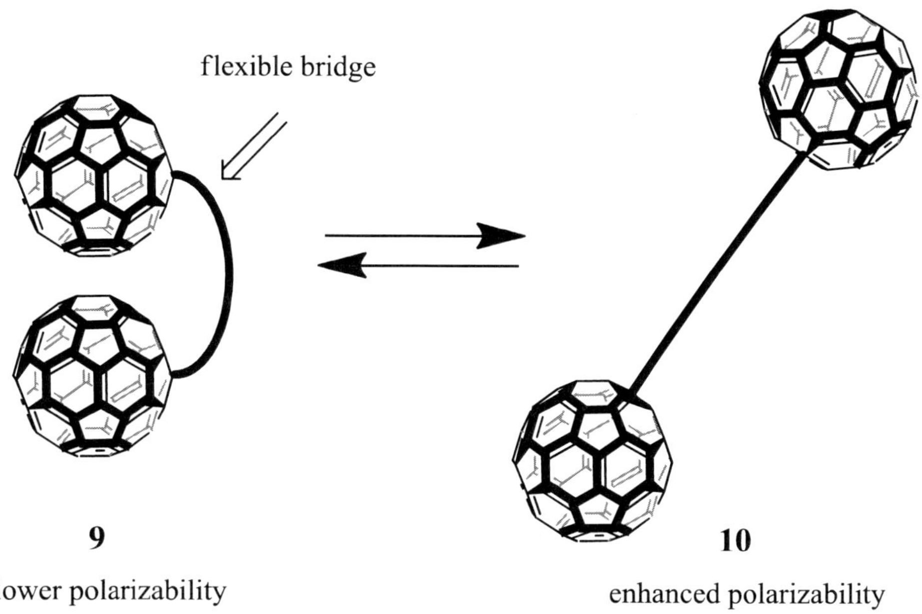

Figure 9. Hypothetical bicage fullerene-containing system existing in two states, characterized with lower (9) and enhanced (10) polarizabilities.

Is there another way to achieve split polarizability? Additionally, we have found that such a split also depends on the number of double bonds of the subunits in the dimer molecules [58]. Thus, the replacement in compound 9 of one of the C_{60} cages by a moiety with a rich π-electronic system should also demonstrate the split. The list of potential species for such enhancement contains compounds with conjugated double bonds, e.g., porphyrins, polycyclic aromatic hydrocarbons, higher fullerenes and carbon nanotubes (Figure 10). Theoretically, if the number of double bonds is larger than 30 (their number in the C_{60} molecule), one should expect the strengthening of $\Delta\alpha$. Additional opportunities may arise when carbon nanotubes are used as a part of fullerene-containing molecular switch because the electron-transfer processes between C_{60} and nanotubes strongly depend on the type of the last one [109].

Molecular switch, based on the reversible dissociation of bicage fullerene derivatives. The reaction of dimerization of fullerenyl radicals $XC_{60}^{\bullet}$ is well-studied [3]. It results in the singly bonded fullerene dimers XC_{60}–$C_{60}X$ (so-called [1+1]-dimers). The interest in such compounds is caused, particularly, by an opportunity of their use for generation of free fullerenyl radicals [110]. In addition, theoretical studies predict that C_{60} singly bonded dimers should demonstrate outstandingly high values of the first [60] and the second [13] polarizabilities. It makes these chemicals prospective for optical applications and nanodevices.

The mentioned reversible reaction can change its direction depending the conditions (Figure 11). EPR spectrometry is the easiest way to monitor in what state, dimeric (11) or radical (12), the system is. Additionally, according to our DFT study [60], the mean polarizabilities of XC_{60}–$C_{60}X$ are more than twice higher compared to those of the respective fullerenyl radicals.

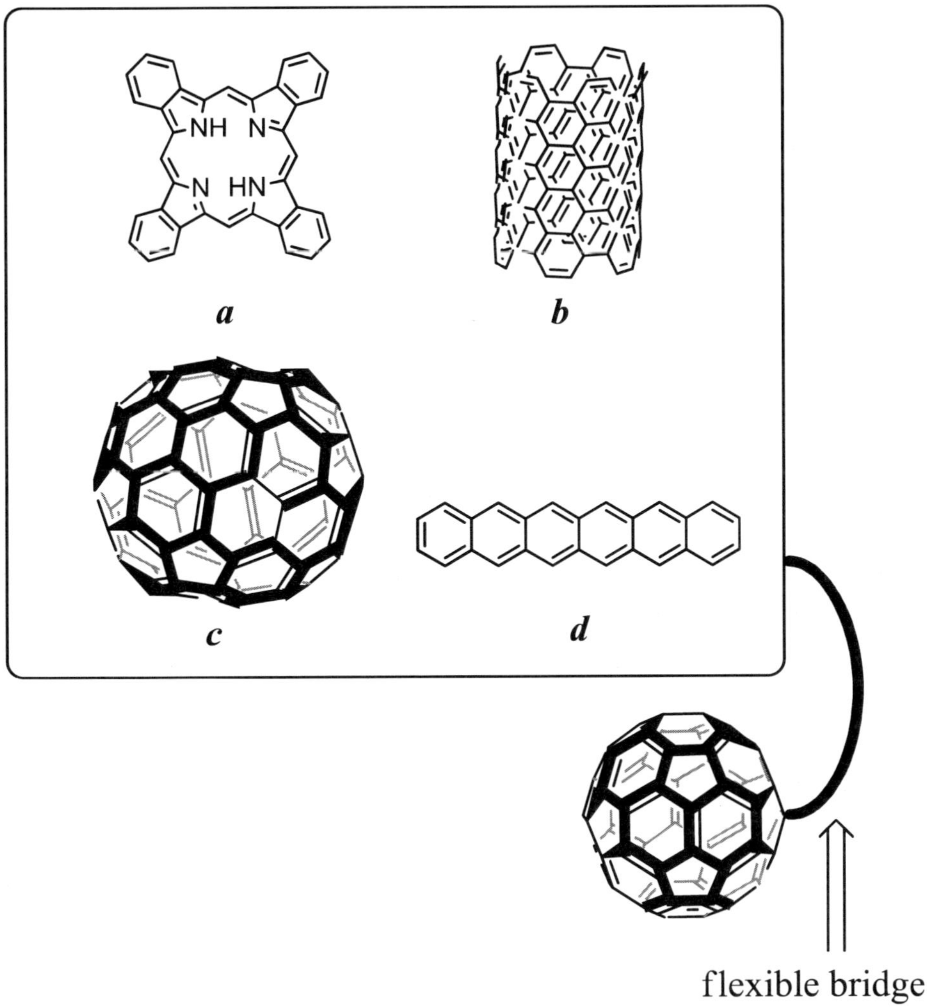

Figure 10. Building blocks for fullerene-containing system with possible split polarizability: porphyrins (*a*), carbon nanotubes (*b*), higher fullerenes (*c*), and polycyclic aromatic hydrocarbons (*d*).

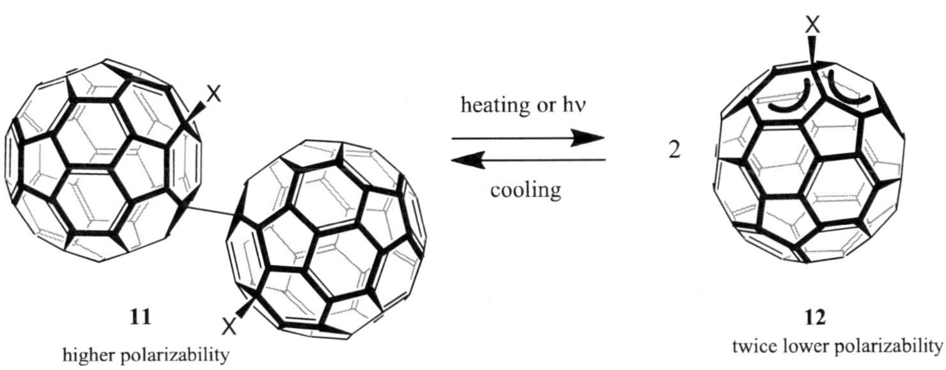

Figure 11. Reversible dissociation of singly bonded fullerene dimers.

It means that the switchable system, shown in Figure 11, should be more facilely manipulated by external electric fields when it is in the dimeric state.

The possible disadvantage of this molecular system is that one of its states (12) has unpaired electron. This may involve $XC_{60}^{\bullet}$ in side processes. However, as shown in experimental and theoretical works, fullerenyl radicals $XC_{60}^{\bullet}$ slowly react with other species, e.g., with molecular oxygen [60, 111, 112].

It is noteworthy that systems, such as described in Figure 11, attract attention in another way. Thus, in a solid-state study [113], it has been shown that thermolysis of the solid fullerene dimers results in both reversible and irreversible generation of radicals $XC_{60}^{\bullet}$ (X = –$CH_2Si(CH_3)_2C_6H_4OCH_2CH(C_2H_5)(C_4H_9)$).

The free radicals, irreversibly generated after the first cooling of the solid, do not recombine because of the emergence of a second state of molecular packing, which produces a solid containing a long-lived radical. This free radical acts as a dopant for the fullerene solid and increases the electron mobility [113].

CONCLUSION

We have examined the main approaches to design the fullerene derivatives with molecular switch: these are attachment of the fullerene cores to the well-known photochromic or solvent-switchable molecular systems, the switch of the states of atoms/molecules, encapsulated by the fullerene cage, and others. All the mentioned systems have been synthesized and successfully tested.

We have demonstrated usefulness of polarizability for studies on physicochemical processes in fullerene-containing systems and offered this quantity as a starting parameter for screening novel compounds, which could act as molecular switches. The split of polarizability of two forms of the potential candidates is the necessary condition for the switch (in addition to the common demands of reversible transitions between the forms). Guiding by this statement, we have selected three prospective types of the fullerene-based systems:

1) pairs of isomeric 6.6-closed and 5.6-open fullerene adducts;
2) fullerene dimers with flexible bridge between the cages;
3) reversibly dissociating singly bonded fullerene dimers.

The systems are listed in order of the increasing split of mean polarizability. Relations of our theoretical findings with the recent experimental works on molecular switch have been analyzed. It makes us hope that our theoretical propositions will be stimulating for fullerene-based molecular machinery.

ACKNOWLEDGMENT

The author is grateful to Igor S. Shepelevich and Dmitry Rudakov (Bashkir State University) for their assistance at all stages of the work.

The work was supported by the Presidium of Russian Academy of Sciences (Program No. 24 'Foundations of Basic Research of Nanotechnologies and Nanomaterials').

REFERENCES

[1] Kroto, H. W.; Heath, J. R.; O'Brien, S. C.; Curl, R. F.; Smalley, R. E. *Nature*, 1985, 318, 162–163.

[2] Popov, A. A.; Yang, S.; Dunsch L. *Chem. Rev.*, 2013, 113, 5989–6113.

[3] Tzirakis, M. D.; Orfanopoulos, M. *Chem. Rev.*, 2013, 113, 5262–5321.

[4] Goldshleger, N. *Fullerene Sci. Technol.*, 2001, 9, 255–280.

[5] Troshin, P. A.; Troshina, O. A.; Lyubovskaya, R. N.; Razumov, V. F. Functional fullerene derivatives, methods of synthesis and perspectives for use in organic electronics and biomedicine, Ivanovo, 2008 (in Russian).

[6] Piotrovsky, L. B.; Kiselev, O. I. Fullerenes in Biology, St. Petersburg, 2006 (in Russian).

[7] Murayama, H.; Tomonoh, S.; Alford, J. M.; Karpuk, M. E. *Fullerenes Nanotubes Carbon Nanostruct.*, 2005, 12, 1–9.

[8] Thilgen, C.; Diederich, F. *Chem. Rev.*, 2006, 106, 5049–5135.

[9] Troshin, P. A.; Lyubovskaya, R. N. *Russ. Chem. Rev.*, 2008, 77, 305–349.

[10] Tokunaga, K. Computational Design of New Organic Materials: Properties and Utility of Methylene-Bridged Fullerenes C_{60}. In: Handbook on Fullerene: Synthesis, Properties and Applications (Editors: Verner, R. F.; Benvegnu, C.), New York: Nova Science Publishers, Inc. 2011, 517–537.

[11] Wang, Y.; Seifert, G.; Hermann H. *Phys. Stat. Sol. A*, 2006, 203, 3868–3872.

[12] Sabirov, D. Sh. Polarizability of C_{60}/C_{70} Fullerene [2+1]- and [1+1]-Adducts: A DFT-Prognosis. In: Density Functional Theory: Principles, Applications and Analysis (Editors: Morin, J.; Pelletier, J. M.). New York: Nova Science Publishers, Inc. 2013, 147–170.

[13] Ma, F.; Li, Z.-R.; Zhou, Z.-J.; Wu, D.; Li, Y.; Wang, Y.-F.; Li, Z.-S. *J. Phys. Chem. C*, 2010, 114, 11242–11247.

[14] Fowler, P. W. *Philos. Mag. Lett.*, 1992, 66, 277–279.

[15] Sabirov, D. Sh.; Zakirova, A. D.; Tukhbatullina, A. A.; Gubaydullin, I. M.; Bulgakov, R. G. *RSC Adv.*, 2013, 3, 1818–1824.

[16] Guldi, D. M.; Prato, M. *Acc. Chem. Res.*, 2000, 33, 695–703.

[17] Osuna, S.; Swart, M.; Solà, M. *Phys. Chem. Chem. Phys.*, 2011, 13, 3585–3603.

[18] Mateo-Alonso, A.; Guldi, D. M.; Paolucci, F.; Prato, M. *Angew. Chem. Int. Ed.*, 2007, 46, 8120–8126.

[19] Sasabe, H.; Takata, T. *J. Porphyrins Phthalocyanines*, 2007, 11, 334–341.

[20] Tian, H.; Yang, S. *Chem. Soc. Rev.*, 2004, 33, 85–97.

[21] Endtner, J. M.; Effenberger, F.; Hartschuh, A.; Port, H. *J. Amer. Chem. Soc.*, 2000, 122, 3037–3046.

[22] Liddell, P. A.; Kodis, G.; Moore, A. L.; Moore, T. A.; Gust, D. *J. Amer. Chem. Soc.*, 2002, 124, 7668–7669.

[23] Zhang, J.; Porfyrakis, K.; Morton, J. J. L.; Sambrook, M. R.; Harmer, J.; Xiao, L.; Ardavan, A.; Briggs, G. A. D. *J. Phys. Chem. C*, 2008, 112, 2802–2804.
[24] Da Ros, T.; Guldi, D. M.; Morales, A. F.; Leigh, D. A.; Prato, M.; Turco, R. *Org. Lett.*, 2003, 5, 689–691.
[25] Mateo-Alonso, A.; Iliopoulos, K.; Couris, S.; Prato, M. *J. Amer. Chem. Soc.*, 2008, 130, 1534–1535.
[26] Reed, C. A.; Bolskar, R. D. *Chem. Rev.*, 2000, 100, 1075–1120.
[27] Guldi, D. M.; Feng, L.; Radhakrishnan, S. G.; Nikawa, H.; Yamada, M.; Mizorogi, N.; Tsuchiya, T.; Akasaka, T.; Nagase, S.; Herranz, M. Á.; Martín, N. *J. Amer. Chem. Soc.*, 2010, 132, 9078–9086.
[28] Fukuzumi, S.; Ohkubo, K.; Kawashima, Y.; Kim, D. S.; Park, J. S.; Jana, A.; Lynch, V. M.; Kim, D.; Sessler, J. L. *J. Amer. Chem. Soc.*, 2011, 133, 15938–15941.
[29] Nuttall, C. J.; Hayashi, Y.; Yamazaki, K.; Mitani, T.; Iwasa, Y. *Adv. Mater.*, 2002, 14, 293–296.
[30] Kornilovitch, P. E.; Bratkovsky, A. M.; Williams, R. S.; *Phys. Rev. B*, 2002, 66, 245413.
[31] Yasutake, Y.; Shi, Z.; Okazaki, T.; Shinohara, H.; Majima, Y. *Nano Lett.*, 2005, 5, 1057–1060.
[32] Huang, T.; Zhao, J.; Feng, M.; Popov, A. A.; Yang, S.; Dunsch, L.; Petek, H. *Nano Lett.*, 2011, 11, 5327–5332.
[33] Kang, J. W.; Hwang, H. J. *Physica E*, 2004, 23, 36–44.
[34] Ju, C.; Suter, D.; Du, J. *Phys. Lett. A*, 2011, 375, 1441–1444.
[35] Wang, X.; Ma, J. *Phys. Chem. Chem. Phys.*, 2011, 13, 16134–16137.
[36] Asselberghs, I.; Zhao, Y.; Clays, K.; Persoons, A.; Comito, A.; Rubin, Y. *Chem. Phys. Lett.*, 2002, 364, 279–283.
[37] Gutierrez, R.; Fages, G.; Cuniberti, G.; Grossmann, F.; Schmidt, R.; Richter, K. *Phys. Rev. B*, 2002, 65, 113410.
[38] Bonin, K. D.; Kresin, V. V. Electric-Dipole Polarizabilities of Atoms, Molecules and Clusters, World Scientific, 1997.
[39] Broyer, M.; Antoine, R.; Benichou, E.; Compagnon, I.; Dugourd, P. *C. R. Physique*, 2002, 3, 301–317.
[40] Bossi, M. L.; Aramendía, P. F. *J. Photochem. Photobiol. C: Photochem. Rev.*, 2011, 12, 154–166.
[41] Painelli, A.; Terenziani, F.; Soos, Z. G. *Theor. Chem. Acc.*, 2007, 117, 915–931.
[42] Alparone, A.; Librando, V.; Minniti, Z. *Chem. Phys. Lett.*, 2008, 460, 151–154.
[43] Calaminici, P.; Carmona-Espindola, J.; Geudtner, G.; M. Köster, A. M. *Int. J. Quantum Chem.*, 2012, 112, 3252–3255.
[44] Hu, Y. H.; Ruckenstein, E. *J. Am. Chem. Soc.*, 2005, 127, 11277–11282.
[45] Yang, Y.; Wang, F.-H.; Zhou, Y.-S. *Phys. Rev. A*, 2005, 71, 013202.
[46] Zhang, C. R.; Liang, W. Z.; Chen, H. S.; Chen, Y. H.; Wei, Z. Q.; Wu, Y. Z. *J. Mol. Struct. THEOCHEM*, 2008, 862, 98–104.
[47] Zhang, C. R.; Chen, H. S.; Chen, Y. H.; Wei, Z. Q.; Pu, Z. S. *Acta Phys. Chim. Sin. (Wuli Huaxue Xuebao)*, 2008, 24, 1353–1358.
[48] Rivelino, R.; Malaspina, T.; Fileti, E. E. *Phys. Rev. A*, 2009, 79, 013201.
[49] Tang, S.-W.; Feng, J.-D.; Qiu, Y.-Q.; Sun, H.; Wang, F.-D.; Chang, Y.-F.; Wang, R.-S. *J. Comput. Chem.*, 2010, 31, 2650–2657.

[50] Tang, C.; Zhu, W.; Deng, K. *Chin. J. Chem.*, 2010, 28, 1355–1358.
[51] Sabirov, D. Sh.; Bulgakov, R. G. *Fullerenes Nanotubes Carbon Nanostruct.*, 2010, 18, 455–457.
[52] Tang, S.-W.; Feng, J.-D.; Qiu, Y.-Q.; Sun, H.; Wang, F.-D.; Su, Z.-M.; Chang, Y.-F.; Wang, R.-S. *J. Comput. Chem.*, 2011, 32, 658–667.
[53] Sabirov, D. Sh.; Bulgakov, R. G. *Chem. Phys. Lett.*, 2011, 506, 52–56.
[54] Tang, C.; Zhu, W.; Zou, H.; Zhang, A.; Gong, J.; Tao, C. *Comput. Theor. Chem.*, 2012, 991, 154–160.
[55] Sabirov, D. Sh.; Garipova, R. R.; Bulgakov. R. G. *Chem. Phys. Lett.*, 2012, 523, 92–97.
[56] Sabirov, D. Sh.; Tukhbatullina A. A.; Bulgakov, R. G. *Comput. Theor. Chem.*, 2012, 993, 113–117.
[57] Sabirov, D. Sh.; Garipova, R. R.; Bulgakov, R. G. *Fullerenes Nanotubes Carbon Nanostruct.*, 2012, 20, 386–390.
[58] Sabirov, D. Sh. *RSC Adv.*, 2013, 3, 19430–19439.
[59] Sabirov, D. Sh. *J. Phys. Chem. C*, 2013, 117, 9148–9153.
[60] Sabirov, D. Sh.; Garipova, R. R.; Bulgakov, R. G. *J. Phys. Chem. A*, 2013, 117, 13176–13183.
[61] Peyghan, A. A.; Soleymanabadi, H.; Moradi, M. J. *Phys. Chem. Solids*, 2013, 74, 1594–1598.
[62] Sabirov, D. Sh.; Terentyev, A. O.; Bulgakov, R. G. *Phys. Chem. Chem. Phys.*, 2014, DOI: 10.1039/C3CP55528C.
[63] Kumar A.; Thakkar, A. *J. Chem. Phys. Lett.*, 2011, 516, 208–211.
[64] Bezmel'nitsyn, V. N.; Eletskii, A. V.; Okun', M. V. *Phys. Usp.*, 1998, 41, 1091–1114.
[65] Konarev, D. V.; Lyubovskaya, R. N. *Russ. Chem. Rev.*, 1999, 68, 19–38.
[66] Narits, A. A. *J. Phys. B: At. Mol. Opt. Phys.*, 2008, 41, 135102.
[67] Bulgakov, R. G.; Galimov, D. I.; Sabirov, D. Sh. *JETP Lett.*, 2007, 85, 632–635.
[68] Zope, R. R. *J. Phys. B: At. Mol. Opt. Phys.*, 2007, 40, 3491.
[69] Jensen, L.; Åstrand, P. O.; Mikkelsen, K. V. *J. Comput. Theor. Nanosci.*, 2009, 6, 270–291.
[70] Sedel'nikova, O. V.; Bulusheva, L. G.; Okotrub A. V. *Phys. Solid State*, 2009, 51, 863–869.
[71] Langlet, R.; Mayer, A.; Geuquet, N.; Amara, H.; Vandescuren, M.; Henrard, L.; Maksimenko, S.; Lambin, Ph. *Diamond Relat. Mater.*, 2007, 16, 2145–2149.
[72] Langlet, R.; Lambin, P.; Mayer, A.; Kuzhir P. P.; Maksimenko S. A. *Nanotechnol.*, 2008, 19, 115706.
[73] Moreau, F.; Langlet, R.; Lambin, P.; Kuzhir, P. P.; Bychanok, D. S.; Maksimenko S. A. *Solid State Sci.*, 2009, 11, 1752–1756.
[74] Latypov, Z. Z.; Pozdnyakov, O. F. *Tech. Phys. Lett.*, 2006, 32, 381–383.
[75] Torrens, F. *Microelectron. Eng.*, 2000, 51–52, 613–626.
[76] Torrens, F. *J. Phys. Org. Chem.*, 2002, 15, 742–749.
[77] He, J.; Wu, K.; Sa, R.; Li, Q.; Wei, Y. *Chem. Phys. Lett.*, 2009, 475, 73–77.
[78] Evangelisti, S.; Bendazzoli, G. L.; Monari, A. *Theor. Chem. Acc.*, 2010, 126, 257–263.
[79] Tang, C.; Zhu, W.; Deng, K. *J. Mol. Struct. THEOCHEM*, 2010, 950, 36–40.
[80] Reis, H., Loboda, O., Avramopoulos, A., Papadopoulos, M. G., Kirtman, B., Luis, J. M.; Zaleśny, R. *J. Comput. Chem.*, 2011, 32, 908–914.
[81] Tang, C.-M.; Cao, Q.-S.; Zhu, W.-H.; Deng, K.-M. *Chin. Phys. B*, 2010, 19, 033603.

[82] Henrard, L.; Senet, P.; Lambin, P.; Lucas A. A. *Synt. Met.*, 1996, 77, 27–30.
[83] Zope, R. R. *J. Phys. B: At. Mol. Opt. Phys.*, 2008, 41, 085101.
[84] Xu, L.; Tang, H.; Li, C.; Li, F.; Li, X.; Tao, S. *Struct. Chem.*, 2013, 24, 463–470.
[85] Pederson, M. R.; Quong, A. A.; Broughton, J. Q.; Feldman J. L. *Comput. Mater. Sci.*, 1994, 2, 536–542.
[86] Wei, H. X.; Li, J.; Cai, Y.; Xu, Z. Q.; Lee, S. T.; Li, Y. Q. ;Tang J. X. *Org. Electron.*, 2011, 12, 1422–1428.
[87] Delaney, P.; Greer, J. C. *Appl. Phys. Lett.*, 2004, 84, 431–433.
[88] Sabirov, D. Sh. *J. Phys. Chem. C*, 2013, 117, 1178–1182.
[89] Kuznetsov, V. V. Russ. *J. Gen. Chem.*, 2013, 83, 1163–1164.
[90] Kuznetsova, M. V.; Kuznetsov, V. V. Russ. *J. Org. Chem.*, 2013, 49, 1845–1847.
[91] Ramachandran, C. N.; De Fazio, D.; Sathyamurthy, N.; Aquilanti, V. *Chem. Phys. Lett.*, 2009, 473, 146–150.
[92] Halls, M. D.; Schlegel, H. B. *J. Phys. Chem. B*, 2002, 106, 1921–1925.
[93] Sabirov, D. Sh.; Bulgakov, R. G.; Khursan, S. L.; Dzhemilev, U. M. *Dokl. Phys. Chem.*, 2009, 425, 54–56.
[94] Sabirov D. Sh. Intermediates of Fullerene Ozonolysis and Fullerene Reactivity in 1,3-Dipolar Cycloaddition, PhD Thesis, Bashkir State University, Ufa, Russia, 2009 (in Russian).
[95] Sabirov, D. Sh.; Bulgakov, R. G.; Khursan, S. L. *ARKIVOC*, 2011, part 8, 200–224.
[96] Fuchs, D.; Rietschel, H.; Michel, R. H.; Fischer, A.; Weis, P.; Kappes M. M. *J. Phys. Chem.*, 1996, 100, 725–729.
[97] Lamparczyk, H.; Wilczyńska, D.; Radecki, A. *Chromatographia*, 1981, 14, 707–708.
[98] Chadli, H.; Rahmani, A.; Sauvajol, J.-L. *J. Phys.: Condens. Matter*, 2010, 22, 145303.
[99] Ugarte, D.; Châtelain, A.; de Heer, W. A. *Science*, 1996, 274, 1897–1899.
[100] Ugarte, D.; Stöckli, T.; Bonard, J. M.; Châtelain, A.; de Heer, W. A. *Appl. Phys.*, 1998, 67, 101–105.
[101] de Gennes, P. G. *Rev. Mod. Phys.*, 1985, 57, 827–863.
[102] Xia, X. R.; Monteiro-Riviere, N. A.; Riviere, J. E. *Nat. Nanotechnol.*, 2010, 5, 671–675.
[103] Danielson, J. R.; Young, J. A.; Surko C. M. *J. Phys. B: At. Mol. Opt. Phys., 2009,* 42, 235203.
[104] Reinov, M. V.; Yurovskaya, M. A. *Russ. Chem. Rev.*, 2007, 76, 715–730.
[105] Heymann, D.; Weisman, R. B. C. R. *Chimie*, 2006, 9, 1107–1116.
[106] Bulgakov, R. G.; Sabirov, D. Sh.; Dzhemilev, U. M. *Russ. Chem. Bull.*, 2013, 62, 304–324.
[107] Tuktarov, A. R.; Korolev, V. V.; Sabirov, D. Sh.; Dzhemilev U. M. *Russ. J. Org. Chem.*, 2011, 47, 41–47.
[108] Minikata, S.; Tsuruoka, R.; Nagamachi, T.; Komatsu, M. *Chem. Commun.*, 2008, 323–325.
[109] Isborn, C. M.; Tang, C.; Martini, A.; Johnson, E. R.; Otero-de-la-Roza, A.; Tung, V. C. *J. Phys. Chem. Lett.*, 2013, 4, 2914–2918.
[110] Lu, S.; Jin, T.; Bao, M.; Yamamoto, Y. *Org. Lett.*, 2012, 14, 3466–3469.
[111] Zhang, Y.-K.; Janzen, E. G.; Kotake, Y. *J. Chem. Soc. Perkin Trans. 2*, 1996, 1191–1196.

[112] Bulgakov, R. G.; Ponomareva, Yu. G.; Sadykov, R. A. *Russ. Chem. Bull.*, 2008, 57, 2028–2029.

[113] Abe, Y.; Tanaka, H.; Guo, Y.; Matsuo, Y.; Nakamura, E. *J. Amer. Chem. Soc.*, 2014, 136, 3366–3369.

In: Fullerenes
Editor: Shannon B. Ellis

ISBN: 978-1-63321-385-2

Chapter 7

CHEMISTRY OF FULLERENE

A. Sáenz, L. I. López, A. Castañeda and L. Farías
Facultad de Ciencias Químicas, Universidad Autonoma de Coahuila,
Saltillo Coahuila, México

ABSTRACT

Fullerenes are organic compounds of high interest, formed by a structure very special of hydrophobic character, presented by a class of spherical molecules of 60 and 70 carbons in general. They present a hybridized sp^2 with a conjugated system, which gives high aromaticity to organized organic structures, and excellent electrical properties. Each fullerene is a closed network of fused pentagons and hexagons. Spherical fullerenes are called buckyballs and the cylindrical ones are called buckytubes, which may have single or multiple walls. All these structural features make them special. They are considered highly stable molecules, ideally involved in highlighting different reactions like: addition, substitution, involving radicals with metal complexes, polymerization, among others, which provide them a covalent surface modification. Such molecules can be obtained naturally or synthetically. They are present in important applications in different areas highlighting electronics, medical, automotive, computer, and biomedical, among others. This chapter aims to present obtaining, reactivity and applications of fullerenes.

INTRODUCTION

Fullerene are very interesting organic molecules. They are composed of C, and were discovered in 1985 [1]. They present important properties, especially biological and electrical properties due to their chemical nature. They have a 60 π-electron system including 12 pentagonal sp^2 carbon arrays and are aromatic. This type of system increases reactivity in organic and inorganic reactions, attracting a great deal of attention in different areas. Recently, the use of fullerenes in special medical areas has increased substantially. Other important applications are electronic and polymers.

Fullerenes are highly symmetric cage molecules. C_{60} o C_{70} is considered a Buckminster fullerene, and is the most important member: a truncated icosahedron with a diameter of 0.7

nm, containing 60 carbon atoms forming 20 hexagons and 12 pentagons. Thus the C_{60} is the fullerene which can perform different modifications [2], Figure 1. They are considered a family of organic cage type molecules, which can be obtained in different ways, natural or synthetic. The synthetic method uses high temperatures, pyrolysis, electrical arc discharge, and others.

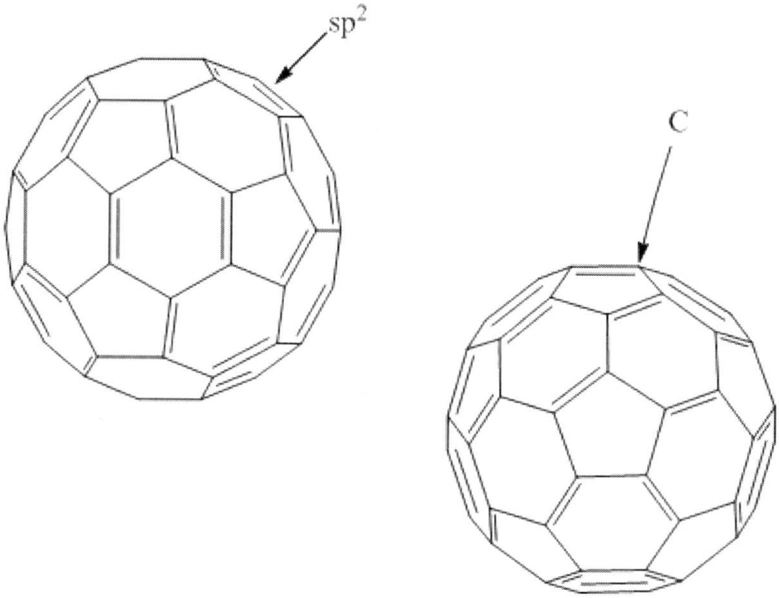

Figure 1. Chemical structure of Fullerene C_{60} and C_{70}.

A lot of information on fullerenes currently exists in the literature, because these types of organic molecules are considered fundamental in nanotechnology. The present chapter provides a short overview of fullerene, including methods of obtaining them, reactions chemistry, and modification surface of the fullerene and applications.

Methods of Obtaining Fullerenes

The versatility of fullerene molecules has led to a large amount of research exploring their properties. Fullerenes are much more abundant than we think, and may be even more abundant than graphite and diamonds.

Since the synthesis, isolation and characterization of fullerenes [3], there has been considerable interest in obtaining this carbon molecule, which occurs naturally on Earth and in the cosmos (e.g., circumstellar shells, medium interstellar, etc.). For example, it has been suggested that the fullerene molecule, C_{60}, might be widely distributed in the universe, particularly in the outflows of carbon stars, due to its thermal stability and photochemical properties [4]. Despite an intense effort carried out by the scientific community to verify this hypothesis, only a few natural occurrences of fullerenes have been reported. Terrestrial discoveries [5] have been limited to the detection of mostly C_{60} and C_{70} whereas extraterrestrial evidence is based on trace findings of C_{60} [6] and a single mass spectrum of a high mass carbon envelope from the Allende, Mexico meteorite [7].

On the Earth, trace amounts of fullerenes have been reported in some unusual rocks such as “shungite,” a highly metamorphosed coal, and “fulgurite,” a glassy rock that forms when lightning hits the ground [8]. Fullerenes have also been detected in deposits associated with two separate events involving the impact of a large bolide (asteroid or comet) with the Earth. Fullerenes have been extracted from shock-produced breccias associated with the 1.85 billion-year-old Sudbury Impact Crater and in clay sediments within the 65 million-year old Cretaceous/Tertiary (KTB) boundary [5, 9]. Possible theories for the presence of fullerenes in the Sudbury and KTB impact deposits are as follows: Fullerenes were synthesized within the impact plume from the carbon contained in the bolide or target rocks: Fullerenes were already present in the bolide and survived the impact event based on their unique noble gas isotopic signature: or fullerenes in the KTB clays formed as a result of global wildfires triggered by the impact event [5]. In the earth, fullerenes can be found in soot and in certain rocks.

Theory of Formation of Fullerenes in Earth

There is no geological or geochemical evidence that the rocks in the Mítov area could be associated with a giant meteoroid impact or global wildfires. It is considered that some other process of high temperature transformation of precursor organic matter was responsible for C_{60} formation. Specifically, it is suggest that fullerene formation at Mítov was related to high temperature pyrometamorphism of precursor bitumen. In order to make an argument Lieber and Chen in 1994 [10] reviewed some possible fullerene forming laboratory processes focusing on pyrolitic transformations and combustion or organic matter.

Laboratory Fullerene Formation by Pyrolysis and Combustion

It was demonstrated that Fullerene could be formed by laser ablation or pyrolysis of suitable carbonaceous precursors. For example, Fullerenes were produced in the gas phase that was formed by laser ablation used at low power densities of the pirolyzed algal product coorongite [11]. Coorongite is a deposit of *Elaeophyton* algae in saltwater bodies near Coorong (Australia). Other authors have described experimental pyrolytic transformation of the aromatic hydrocarbons and the formation of fullerene, such as the generation of C_{60} and C_{70} during pyrolysis of naphthalene in an Ar atmosphere [12]. Cyclopentafused polycyclic aromatic hydrocarbons (CP-PAHs) are important precursors for Fullerene formation such as during flash vacuum thermolysis and other conditions, as well as in industrial furnaces [13-15]. Under high temperature conditions (900-1100°C) many CP-PAH combustion products were reported to have formed in the gas phase for which was proposed various, possible transformation mechanisms.

Additionally, cyclotrimerization of 4,5-dihydrobenzacephenanthrylene $C_{20}H_{14}$ can produce $C_{60}H_{30}$ by heating the monomer with elemental sulfur at lower temperatures (300°C). The trimer could be considered a C_{60} precursor [16]. Thus, PAHs and especially CP-PAHs can be precursors for fullerene during specific laboratory experiments. Fullerenes are high temperature transformation products of petroleum asphaltenes [17]. Fullerene-like particles were found in petroleum soot [18]. Fullerene-like structures and fullerene precursors were also detected by HRTEM in different types of carbon blacks [19]. Unfortunately, independent confirmation of fullerenes in these studies is still missing.

The common traits among these aforementioned studies are the rather specific nature of the precursor and high (900-1200°C) transformation temperatures. Fullerene formation can be very rapid (milliseconds to seconds) slow (minutes to hours) [20] but such times are very short when compared to the duration of typical geological processes such as pyrometamorphism and regional metamorphisms but comparable to, for example, the duration of a lightning strike.

As can be shown, the fullerene scan be formed in nature. However, its artificial production in the laboratory and in industry is much more significant.

Important aspects of these molecules are studied in the field of nanotechnology, hydrogen storage, and as superconducting materials in nanomedicine.

Production Methods of Fullerene

Fullerene production encompasses numerous methods and new routes are continuously being developed. The following sections are devoted to give an overall description of the most established Fullerene production routes.

Laser Ablation

A powerful technique for producing Fullerene involves laser ablation of graphite in a helium atmosphere. This technique was the first used to generate fullerene clusters in the gas phase [19, 20] although initial attempts to prepare C_{60} were unsuccessful with this technique. The early attempts to prepare Fullerenes by laser ablation relied on ablation in an inert gas buffer at room temperature. This technique is not suitable for mass production due to the fact that dense carbon plasma created during laser vaporization cools too quickly. The rapidly cooling plasma does not provide sufficient time for growing carbon fragments to rearrange into the stable fullerene structures. Later, Smalley and coworkers showed that laser ablation of graphite carried out at elevated temperatures could lead to efficient production of C_{60} [21].

The key to this technique is that the graphite is ablated within a high temperature furnace. By carrying out the ablation at an elevated temperature, the plasma cools more slowly and growing carbon clusters have sufficient time to rearrange (anneal) into stable fullerenes. Lieber and Chen [22] have found that fullerene production is most efficient at 1200°C. At lower temperatures, the yield of fullerenes is reduced [4].

There are several features of the laser ablation method that make it an excellent technique for the preparation of Fullerenes. First, it is possible to systematically vary the characteristics of the carbon plasma through variations in laser pulse energy and wavelength [23]. Second, by controlling the buffer gas pressure and the temperature it is also possible to systematically control the growth of clusters from the carbon plasma. Control of the growth and annealing should increase the efficiency and selectivity of the Fullerenes produced. Finally, since the laser light is easily positioned and no external electrical connections to the sample are needed, it is possible to produce Fullerenes from very small amounts of carbon precursors.

Arc Vaporization of Graphite

The Fullerne field was revolutionized by the discovery of simple resistive vaporization of graphite rods, which could produce Fullerenes in substancial yield [24]. This procedure, which is often termed the Krätschmer-Huffman method, is a straightforward and low cost method for generating large quantities of fullerene. Key components are arc vaporization Fullerenes generating high current electrical feedthroughs, at least one movable electrode and a water-cooled surface to collect the vaporized carbon products. These components are housed within a vacuum chamber that is connected to a pumping/gas handling system that can evacuate the chamber and can control the pressure between 10-500 torr. A high current (100-200 A), low voltage AC or DC power supply is needed to drive the evaporation: an arc welder is an inexpensive but effective power supply.

In the experiment, two high purity graphite rods are clamped to the high current feedthroughs, and the chamber is pumped down to $\leq 10^{-3}$ torr and refilled with He gas to a presure of 150-250 torr. The electrodes are positioned so that the carbon rods are just touching, and then vaporization is initiated by passing a high current through the rods. For 6.25 mm diameter rods, currents between 100 and 200 A lead to efficient Fullerene formation. Under these conditions, the rods are consumed at a rate of about 5-10 mm/min. The crude carbon product or soot produced by this vaporization collects on the water-cooled inner surface of the Fullerene apparatus and is readily removed from the walls and collected using a stiff brush. This soot contains a variety of carbon products including C_{60} and larger Fullerenes. The isolation and purification of the fullerene is discussed later [24].

There are several experimental factors that can be varied to maximize the yield of Fullerenes in the soot produced from the vaporization of graphite electrodes. These factors include the vaporization current density and partial helium pressure. The importance of these two experimental variables in determining the yield of fullerenes can be understood by considering the environment in which the clusters form. Specifically, the vaporization of the carbon rods is not driven by heating but rather by an arc discharge between the ends of the graphite rods. These latter conditions produce a plasma, and hence, control of temperature and cooling of the carbon plasma will determine the yield and size distribution of the Fullerene. Increases in the current density will increase the plasma temperature, while increases in the buffer gas pressure will increase the cooling and growth rates of carbon species. An example of the effects of He pressure is found in comparing the carbon products produced from dc discharges in 150 and 500 torr of helium. Using the former conditions Fullerenes are observed in high yield, while at higher He pressures carbon nanotubes are the main product [24].

Several other techniques, including hydrocarbon combustion, low pressure helium sputtering, electron beam evaporation, and inductively coupled RF evaporation of graphite targets have been used to prepare Fullerene containing carbon products. Of these only the combustion technique is sufficiently developed as a general method for the production of large quantities [24].

The Combustion Process

The possibility that Fullerenes could also be formed in sooting flames had been suggested [25] and all-carbon ions having the same mass charge ratios as Fullerenes in carbon vapors had been detected in flames and assumed to have the closed-cage structure of fullerenes [26]. Howard and collaborators [27] reported their observations that both C_{60} and C_{70} Fullerenes were produced in varying amounts from premixed laminar benzene/oxygen/argon flames operated under ranges of conditions including pressures of 12-100 Torr and C/O ratios from 0.717 to 1.072, the critical value for soot formation being 0.760. The proper selection of flame conditions gave Fullerenes in yields up to 0.26% of the fuel carbon burned and allowed control of the C_{70}/C_{60} molar ratio over the range 0.26-5.7. The yield of C_{60} + C_{70} and the C_{70}/C_{60} ratio were found to depend on temperature, pressure, carbon/oxygen ratio, and residence time in the flame. In order to characterize the flame- generated fullerenes, samples of condensable compounds and soot were collected from flames under different conditions and analyzed using high-performance liquid chromatography with diode-array spectrophotometric detection, mass spectrometry, and infrared spectrophotometry.

Frontier Carbon industries started mass production of fullerenes with a capacity of 400 kg per year by the use of such a combustion method [28], and the price of Fullerenes was decreased drastically.

The production capacity was increased. This improved flame-based technology is most suited for mass production of fullerenes, since it is a continuous process and uses inexpensive hydrocarbons as its starting materials, which is similar to that employed in the commercial carbon black production process. Fullerene yield in the resulting soot is around 20%, which demonstrates that the combustion method is a practical technology for fullerene-soot production. The actual commercial usage of Fullerenes has already started.

Isolation and Purification of Empty Fullerenes

It is important to note that the raw material obtained by production techniques such as the Krätschmer-Huffman method, is a soot constituted of a mixture of soluble Fullerenes (Cn, n<100), "giant" Fullerenes [29] (Cn, n>100), nanotubes [30] and amorphous carbon. By sublimation [31] or extraction [24], [29,32] techniques it is then possible to isolate the Fullerenes from the soot. Afterward, repeated chromatography on neutral alumina allows separation of the various Fullerene by-products and isolation of stable solid samples of C_{76}, C_{84}, C_{90}, and C_{94}. Of the Fullerene family, C_{60} and C_{70} are the major isomers obtained with 73 and 23% yield, respectively, a variety of molecules larger than C_{60} and C_{70} representing a total amount of 3 to 4% by weight.

The colors of the Fullerene family vary according to their molecular weight and symmetry: Their colors in solution are magenta (C_{60}), port-wine red (C_{70}), brown (C_{76}, C_{78}) and yellow-green (C_{84}).

CHEMICAL SYNTHESIS OF FULLERENE

Pyrolytic Dehydrogenation or Dehydrohalogenation

As C_{60} consists of six dehydronaphthalene moieties located at the octahedral sites, R. Taylor showed that the pyrolysis of naphthalene produces C_{60} [33], as does corannulene, 7,10-bis(2,2'-dibromovinyl) fluoranthene, and 11,12-benzofluoranthène. In 2001, L. T. Scott [34] and collaborators showed that it is possible to "stitch up" a hydrocarbon incorporating 60 carbon atoms, 13 of the 20 benzene rings, and 75 of the 90 carbon-carbon bonds required to form C_{60}, by the use of a high-energy laser. That method, however, did not permit isolation of the fullerene produced.

One year later, the same authors were able to produce isolable quantities of C_{60}, in 12 steps from commercially available starting materials, by rational chemical methods [35]. A molecular polycyclic aromatic precursor bearing chlorine substituents at key positions forms C_{60} when subjected to flash vacuum pyrolysis at 1100°C.

Currently researchers discovered that asymmetrical fullerenes formed from larger structures settle into stable Fullerenes. The synthesized substance was a particular metallofullerene consisting of 84 carbon atoms with two additional carbon atoms and two yttrium atoms inside the cage [36].

Chemistry of Fullerene

Fullerene are very interesting organic molecules because of their chemical structure, highlighted by the type of hybridization sp^2 and aromaticity.

Fullerene from a unique class of spherical molecule contain a conjugated π system.

These fractures create highly reactive Fullerenes for different reactions; below are listed some of the more important reactions in which Fullerenes play an important role [37].

Sawara et al., reported [38] the synthesis of a hybrid based on ferreocene and Fullerene. It was obtained by the structural arrays of Fullerene. The hybrid was obtained by $Fe(C_{60}Me_5)Cp$, $FeBrCp(CO)_2$ and a thallium complex of pentamethylated C_{60} Fullerene $Ti(C_{60}Me_5)$ using 180°C and 8 and 21h of reactions; this hybrid presents important applications of chemistry of the Fullerene core and nanotechnology.

Fullerenes are important in inorganic reactions. Balch et al., reported details from a 1998 study on the reactions of fullerene and different metal transitions: Lanthanum, Yttrium, Scandium, Hafnium, Zirconium, Titanium, Tantalum, Niobium, Vanadium, Tungsten, Molybdenum, Chromium, Rhenium, Manganese, Osmium, Ruthenium and Iron [39].

Other important reactions that use modified Fullerenes are oxidations, as Balch et al. reported in 1995. The oxidation of Fullerene C_{60} with m-chloroperoxybenzoicacid produces two isomers $C_{60}O$ and $C_{60}O_2$; all of these oxides react with triphenylphosphine PPh_3 [40].

Also redox reactions are developed with Fullerene. Wan et al. reported the production of $[(\eta^5\text{-}C_5Me_5)_2Ni[C_{60}]CS_2$, red dark [41].

Reactions of additions such as synthesis of Aza [60] fullerene $(C_{59}N)_2$ has been reported. In 2010, Vougioukalakis et al. reported the studied details of hetero-Fullerene with potential applications in science and biology [42].

In 2009, efficient photochemical methodology was reported for the direct acylation of Fullerene C_{60}, using appropriate conditions. The radical addition precedes decarbonylation of the intermediate radical even in phenylacetaldehyde. The details of a method developed for the direct acylation of C_{60} includes the novel reactions of C_{60} with carbonyl radicals mediated by tetrabutyl ammonium decatungstate (TBADT) as a photocatalyst. This free-radical acylation method was found to be highly effective when using a wide range of aldehyde, aromatic or alkyl [43].

Another important reaction experienced with Fullerene is the addition of C-centered radicals to Fullerene. The formations of radical (R*) to generate photolytically and thermally obtained RC_{60}* was identified by ^{1}H-RMN and ^{13}C-RMN. The formation of the (R*) in the Fullerene can present in different positions, gendered isomers; this type of reaction presents applications in the area of organic chemistry [44, 45].

Other types of reactions are the photochemical or thermal decompositions in the presence of initiator peroxides like 2,2´-azo-(bisisobutyronitrile) AIBN or dimethyl azo(bisisobutyrate) DMAIB, have also been utilized for the productions of alkyl radical. Ford et al. reported that thermal decompositions of AIBN in a 1,2-dichlorobenzene solution of C_{60} generates 2-cyano-2-propyl radicals, which can be added to C_{60} to afford three regioisomeric fullerene adducts [46-47].

Reactions of the Modifications of the Fullerene

In recent years it has been shown that Fullerene derivatives display a range of biological activities and, specifically, may be used as protective drugs against diseases related with oxidative stress.

In fact, Fullerenes and their organic derivatives are regarded as ‘radical sponges’ as they are able to trap multiple radicals per molecule (Bianco et. al. 2007) [48].

Fullerene-flavonoid conjugated synthesis and antioxidant activities are reported. Chalcone, flavone, and arylideneflavanone derivatives bearing one or two 3,5-di-tert-butyl-4-hydroxyphenyl groups were synthesized from 20,40-dihydroxyacetophenone and 3,5-di-tert-butyl-4-hydroxybenzaldehyde. These flavonoids were converted into the corresponding malonates and then reacted with C60 to yield the Fullerene-flavonoid compounds. These novel C_{60}-flavonoid conjugates are therefore promising leads for broad-spectrum radical scavengers [49].

Dichloromethylene Fullerenes are reported by a nucleophilic addition-substitution route. These compounds are prepared from reactions between Fullerenes with alkali-metal trichloroacetates. The intermediates, $C_{60}(CCl_3)^-$ and $C_{70}(CCl_3)^-$ anions, have been isolated in their protonated forms as ortho-C60(CCl_3)H, as well as three *ortho* and one *para* isomer of $C_{70}(CCl_3)$ (Apenovaet al. 2014), [50].

The synthesis of novel Fullerene C_{60}-fused tetrahydroazepinones and azepinonimines is reported. The reaction involves a possible mechanism of two pathways for the reaction of fullerene C_{60} with N-sulfonylated o-amino-aromatic methyl ketones or O-alkyl oxides by an efficient $Cu(OAc)_2$-mediated N-heteroannulation reaction, [51].

Fullerene-Dendrimer Assemblies

Other Fullerenes modifications are dendritic specimens containing C_{60} as a functional moiety. These structures are related to the capability of dendritic architectures to generate specific properties and functions, resulting from their unique molecular structures. With the ball-shaped structure of C_{60} and the possibility of multiple functionalization, this versatile carbon allotrope has demonstrated its potential not only as a moiety at the center of the dendrimer structure, but also as a functional unit at the surface of such branched structures. When the dendrimer chemistry started at the same time, the first dendritic specimens containing C_{60} as functional moiety are described (Wooley et.al. 1993, Hahn et.al. 2012), [52-53].

The amination of hydrofullerenes has been reported. These derivatives were synthesized by Cu-catalyzed C-H amination of the monofunctionalized hydrofullerenes in good to high yields under mild reaction conditions. The experiments showed that the reaction proceeds through the formation of a fullerene mono radical by an amine followed by coupling with an amine radical generated by a $CuBr_2$ catalyst under an oxygen atmosphere. These types of structures have important optical properties with possibilities in electronic applications [54].

The ferric chloride-catalyzed reaction of C_{60} with various *tert*-butyl-N-substituted carbamates to produce oxazolidino fullerenes has been reported. This transformation proceeds smoothly under mild conditions and exhibits good tolerance to various aliphatic and aromatic carbamates. A possible mechanism for the formation of oxazolidino fullerenes is proposed, [55].

Click Chemistry of Fullerenes

An important reaction to modification of fullerenes is the 1,3-dipolar cyclo additions between an azide and terminal alkyne to give 1,2,3-triazoles [56]. This type of reaction is the most popular in the "Click Chemistry".

Click Chemistry is not a single specific reaction; it is a term that describes reactions that are high yielding, wide in scope, created only by products that can be removed without chromatography, are stereospecific, simple to perform, and can be conducted in easily removable benign solvents. It is conceived to mimic nature, which also generates substances by joining small units.

One of the most popular reactions within the Click Chemistry concept is the azide alkyne Huisgen cycloaddition. The classical azide alkyne Huisgen cycloaddition is a 1,3-dipolar cycloaddition between an azide and a terminal or internal alkyne to give a 1,2,3-triazole [57]. Thus, Huisgen and coworkers studied 1,3-dipolar cycloadditions between azides and terminal alkynes in detail almost 50 years ago [58].

The reaction is thermodynamically favored, but it requires heating, and it does not show regioselectivity, as figure 2 shows that both 1,4-disubstitued and 1,5-disubstituted-1,2,3-triazoles are formed (Meldal et.al. 2008). The catalyzed version of this synthetic methodology was introduced in the early 2000s by Sharpless and Meldal, [60-61], which showed that catalysis by Cu(I) species greatly enhances the reactivity (up to 10^7 rate enhancements), and the reaction can be carried out at room temperature. The catalyzed reaction is extremely regioselective towards the formation of the 1,4-disubstituted triazole.

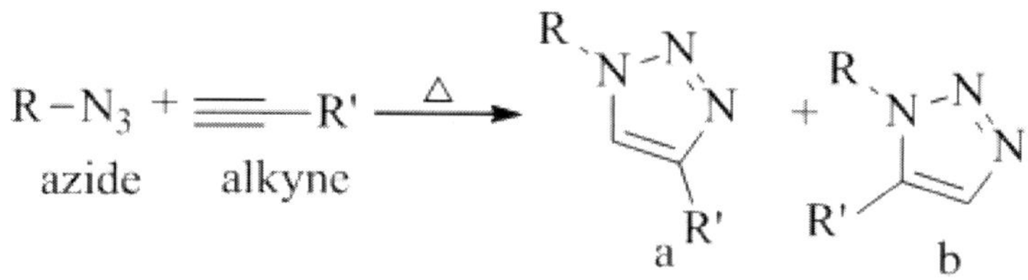

Figure 2. (a) 1,4-disubstituted-1,2,3-triazole and (b) 1,5-disubstituted-1,2,3-triazole obtained by Click Chemistry.

In fact, the Cu(I)-catalyzed azide-alkyne cycloaddition (CuAAC) reaction is used for fullerenes functionalization. Thus, symmetrical C_{60} hexakis-dendritic adducts bearing 12 peripheral azide or alkyne groups functionalized with various ligands, including ferrocene, a porphyrin, and G2 Fréchet-type dendrons were reported [62]. This methodology also enabled preparation of difunctionalized fullerenes via sequential double CuAAC approach starting from a C_{60} hexa-adduct bearing 10 azides and 2 TMS-protected alkyne groups [63].

Biologically relevant ligands such as carbohydrates have been attached to the periphery of C_{60} using this methodology. Some resulting "sugar balls" were shown to be efficient inhibitors of glycosidase activity, and in some cases, a positive multivalent effect was observed [64].

Functionalized fullerenes, especially those based on the C_{60} platform, have been recently investigated as novel materials in nanomedicine. Thus, C_{60} has been used for development of HIV-1 protease inhibitors, drug and gene delivery systems, or novel photodynamic therapy and MRI contrast agents [65].

Polymer-C_{60} hybrids are prepared by the CuAAC reaction. A polystyrene-C_{60} conjugate (PS-C_{60}) was obtained by reacting azido-terminated PS, of different molecular weights, with a monoalkynated fullerene derivative using CuBr, PMDETA and toluene as conditions [66].

Fullerenes and viral nanoparticles (VNP) combinations were explored using classical amide formation conditions and CuAAC reaction. These structures have important biomedical applications, for example in photodynamic tumor therapy. VNP-azide and Alk-PEG-C_{60} were reacted under conditions reactions of $CuSO_4$, sodium ascorbate, THPTA and aminoguanidine, resulting in soluble and stable hybrids with significantly higher loading than amides. In this report the covalent incorporation of C_{60} was checked by Western blotting and STEM and the structural integrity of the VNP PEG-C_{60} hybrid was confirmed by SEC, TEM, STEM, and native gel electrophoresis [67].

Fullerene-capped polystyrene was obtained using the thiol-ene click-chemistry approach.

The methodology was shown to provide mild process conditions with short reaction times for the mono-addition synthesis of such polymer-fullerene conjugates. The first step in the synthesis of C_{60} polystyrene conjugate was to obtain the precursor click component, polystyrene with thiol-ene group (PS-SH).

PS-SH was synthesized by a two-step procedure using bromo-functional polystyrene prepared by atom-transfer radical polymerization. Subsequently, PS-SH was grafted onto the C_{60} nuclei through thiol-ene click reaction in the presence of a known thermo-labile radical source, 2,2-azobisisobutyronitrile (AIBN) [68].

Cycloaddition Reactions with Fullerene

Cycloaddition is one of the most efficient reactions for Fullerene functionalization. Numerous fullerene derivatives have been prepared for studies in material science through various cycloaddition reactions. Modification of C_{60} is already a well-studied process and usually takes place through either the Prato reaction or the Diels-Alder reaction.

Prato Reaction

The Prato reaction was employed for the modification of C_{60} with α,β-unsaturated carboxylic acid-type linker. The Prato reaction is a particular example of the well-known 1,3-dipolar cycloaddition of azomethineylides to olefins. (*E*)-4-(Fullerenopyrrolidin-1-yl)-3-methylbut-2-enoic acid and its corresponding succinimidyl ester, readily obtained through Prato-type modification of C_{60}, were used for the selective N-acylation of polyamines. A very simple, efficient and versatile methodology has been developed for the conjugation of fullerenes with polyamines (PAs). The versatility of the method was demonstrated with preparation of PAs bearing one or two C_{60} units per PA skeleton, as well as selectively monofunctionalized PAs at either the primary or the secondary amino functions [69].

Diels-Alder Reaction

The Diels-Alder reaction is a powerful strategy for their synthesis as well as for their subsequent surface functionalization and is one of the most popular in fullerene organic chemistry. Thus, compounds of type acetalized C_{60} fullerenes were synthesized. This reaction was performed by Diels-Alder reaction of C_{60} with silyletherdiene, on treatment of various aliphatic alcohols under $TiCl_4$ catalyst. The spiro-cyclic acetalized C_{60} fullerenes having five, six, and seven-membered rings were also synthesized by using the corresponding diols under the same condition. The acid-catalyzed hydrolysis easily removed the acetal moiety quantitatively, resulting in a considerable change of solvent solubility of the fullerene C_{70}.

Regioselective Diels-alder reactions directed by carbonyl groups on the rim of open cage fullerene derivatives have been reported. A C_{60} fullerene open-cage derivative with an 18-membered orifice reacts with electron-rich dienes on the rim of the orifice selectively. Both the mono and the bis adduct have been obtained in good yields by controlling the ratio of the reactants and the reaction time. This study shows that single-crystal structures display a head-to-tail and back-to-back dimer for the mono adduct of 1,4-diphenylbutadine and a head-to-tail polymeric chain for the bis adduct of 2,3-dimethylbutadiene [71].

Others Cycloadditions

Cycloaddition is one of the best-studied types of reactions in the organic chemistry of fullerenes. A novel reaction of C_{60} with in situ generated benzyne from 2-amino-4,5-dibutoxybenzoic acid, which affords one monoadduct identified as a [*5,6*]-open annulene adduct is reported. In addition, the theoretical study of the possible two [*5,6*]-closed and [*5,6*]-

open products for the bencene-C_{60} reaction enables one to access the influence of the binding modes in the fullerene framework is presented. This work shows the contrast with *[6,6]*-closed adducts of bencene-$C_{60,}$ the most reported structure [72].

The synthesis of pyrrolidino [C_{60}] fullerenes with complete diastereoselectivities and very high enantioselectivities has been reported. In this study, a range of new dipoles and catalysts are used in 1,3-dipolar cycloadditions of N-metalatedazomethineylides onto C_{60} [73].

The preparation of cyclopentene- and cyclohexene-fused [60] fullerene derivatives is reported by one-step DMAP-catalyzed [3 + 2] and [4 + 2] cycloaddition reactions between C60 and unmodified Morita-Baylis-Hillman (MBH) adducts in the presence of Ac_2O. The MBH adducts bearing aromatic groups only undertake the [3 + 2] process. When the MBH adducts bear an alkyl group, two different reaction pathways could be controlled selectively depending on the conditions [74].

Liu et al. studied the reactions of N-heteroanulations in presence of fullerene, used azo compounds for Cu(OAc), for accelerate of reactions, the chemical functionalizations of fullerene is an important subject in fullerene chemistry for the creations of novel diversified fullerene-based materials with potential applications in nanoscience, electronic devices and biomedical [75].

During the past two decades different functional and structural units have been prepared [76], increasing reports in literature on the reactions with transition-metal/catalyzed-mediated radical reactions for their excellent ability to create novel functional fullerenes and remarkable advantages over the traditional peroxides or light initiation processes, such as high selectivity and efficiency, mild reaction conditions and high compatibility with a wide range of functional groups [45,77].

In 2010 Bogdanov et al. reported the reactions of such cyclic α-diketone as acenaphtenequinone, aceanthrenequinona and N-alkylisatins, with hexaethyltriaminophosphine in the presence of the fullerene [C_{60}], lead to formations of methanofullerene derivatives under mild conditions using compound of phosphor. These types of molecules are used in the pharmaceutical area with antioxidants [78].

Fullerene Applications

Fullerenes have found wide application in different fields of science since their discovery in 1985. Their size, hydrophobicity, three-dimensionality and electronic configurations make them an appealing subject in science, engineering and medicinal fields. Their unique physical, chemical and biological properties have yielded promising contributions to the industries and many industrial applications are now commercialized [79]. Fullerenes have many properties different from either diamond or graphite. There are many applications of practical importance of fullerenes in a wide range of areas and application-oriented patents span a spectrum of potential commercial applications. They include areas such as IT devices, diagnostics, pharmaceuticals, environmental, and energy industries. Few examples include facial creams, moisturizers, lubrications, trace monitors, electronic circuits, electronic devices, sensors, superconductors, catalysts, optical, polymer composites, high-energy fuels, anticancer drug delivery systems using photodynamic therapy, HIV drugs, MRI agents and cosmetics to slow down the aging of human skin.

Medicine and Biology

The therapeutic benefit of photodynamic therapy (PDT) has gained considerable attention in antitumor applications because PDT utilizes external light illumination at the tumor sites for the generation of highly reactive oxygen species (such as singlet oxygen) from photosensitizer drugs systemically delivered to tumors [80, 81]. This method of tumor therapy provides promising modality for the treatment of local tumors illuminated by external light source, associated with ameliorated drug side effects for normal tissues. However, most photosensitizer drugs used in PDT have been outside this expectation and exhibit limited clinical availability due to the lack of multi-functionality for improving drug therapeutic effects [82, 83]. Introduction of a novel photosensitizer drug tailor-made for this purpose can provide a means to overcome the aforementioned issue.

One of important trials in photosensitizer drug design is directed toward the incorporation of stimuli-responsive polymers to photosensitizer drugs. Recently Park et al. [84, 85] reported a photosensitizer drug with intelligent phototoxicities, which was pH dependent.

This photosensitizer drug consists of glycol chitosan grafted with a functional 3-diethylaminopropyl isothiocyanate (DEAP), chlorine 6 (photosensitizer drug), and poly (ethylene glycol), which allowed relatively higher phototoxicity at an extracellular pH ~pH6.8 tumor, while minimizing phototoxicity at physiological pH ~7.4. These pH-dependent properties emphasize the potential of photosensitizer drugs for improved tumor therapy. Fullerene (C_{60}), one of the photosensitizer chemicals, can readily transfer the exited energy to oxygen molecules during the light illumination at 670 nm, resulting in the generation of singlet oxygen in a high yield [82]. Several investigations for photosensitizer drug design have been made by utilizing exohedral derivatization methods using synthetic polymers, peptide/proteins, and antibodies [82, 83].

Recently Kim et al. [86] developed a Glycol chitosan (GC) grafted with 2,3-dimethylmaleic acid (DMA) and Fullerene (C_{60}) conjugates (GC-g-DMA-g-C60) for use in a photosensitizer prodrug. GC-g-DMA-g-C_{60} was prepared via the simple two-step chemical grafting reactions of (i) DMA to free amine groups of GC and (ii) hydroxyl groups of GC-g-DMA to π - π carbon bonds of C_{60}. This conjugate was self-assembled to form polysaccharidicnanogels consisting of a hydrophilic block (GC and DMA) and a hydrophobic block (C_{60}). The GC-g-DMA-g-C60 nanogels also formed multi-nanogel aggregates due to the electrostatic interaction between the pendant carboxylic acid group (due to the DMA) and the residual free amine group of GC at pH 7.4. Interestingly, the nanogel aggregates can be disintegrated at pH 5.0 due to the reduction of electrostatic interaction resulting from the cleavage of the DMA blocks at pH 5.0. Upon 670 nm light illumination, photo-responsive properties of the nanogel aggregates allowed different singlet oxygen generation according to the pH condition: the reduced singlet oxygen generation (due to increased photo-interference effect between C_{60} molecules close-packed in nanogel aggregates) at pH 7.4, but the elevated singlet oxygen generation (due to the disintegration of nanogel aggregates) at pH 5.0. GC-g-DMA-g-C60 nanogel aggregates respond to pH 5.0 (~endosomal pH) and can be a good candidate for endosomal pH targeting and in vivo photodynamic therapy in various malignant tumor cells, but the authors don´t make *in vivo* taste.

Asada et al. [87] studied the PDT anti-cancer effect of a water-soluble complex of fullerene C_{60}: polyethylene glycol (PEG) (C_{60} – PEG), on fibro sarcomas. Their results demonstrated that PEG–Fullerene C_{60} induced significant tumor cell-preferential cytocidal effects by visible light irradiation over the normal cells derived from the same tissue type (human connective tissue). It also showed that PEG–Fullerene C_{60} was satisfactory under the basic conditions of a photosensitizer with characteristics of no cytotoxicity by itself and no production of diverse reactive oxygen species and as effective cytotoxic agents only through irradiation. They suggest that PEG–Fullerene C_{60} is a promising photosensitizer candidate for PDT [88]. On the other hand, Shi et al. [89] modified or "decorated" iron oxide nanoparticles (IONPs) onto the surface of Fullerene C_{60}, and then PEGylation was performed to improve the solubility and biocompatibility of C60-IONP, obtaining a multi-functional C_{60}-IONP-PEG nanocomposite with strong super-paramagnetism and powerful PDT capacity. After that, they formed a drug delivery system by conjugating Hematoporphyrinmonomethyl ether (HMME) (anti-cancer drug) with C_{60}-IONP-PEG. The new C_{60}-IONP-PEG/HMME demonstrated an excellent magnetic targeting ability in cancer therapy. This group also tested their compound *in vitro* and *in vivo*. They found that C_{60}-IONP-PEG/HMME has an enhanced cancer cell killing effect compared with free HMME. Also, it suggests that C_{60}-IONP-PEG can be used as a T_2-contrast agent for *in vivo* magnetic resonance imaging.

In addition, a Fullerene composite can be used as a Prodrug as presented by Fan et al [81]. This study provides useful insights into designing and improving the applicability of fullerene for other targeted cancer prodrug systems; which demonstrates active targeting, pH-responsive chemotherapy, and PDT properties. The system was based on water-dispersible fullerene C_{60} aggregates. Incorporating an anticancer drug, doxorubicin (DOX), and a targeting ligand (folic acid) onto fullerene with a total size of approximately 135 nm that produces a more specific anticancer prodrug.

Fullerenes and its derivatives are well known as a new class of antioxidants and they have attracted considerable attention in biologic applications due to their high reactivity toward radicals [90], especially reactive oxygen species (ROS) such as superoxide, hydroxyl radical, peroxyl radicals, and nitric oxide [91]. The antioxidant property is based on the fact that fullerenes possess large amounts of conjugated double bonds and low lying lowest unoccupied molecular orbital (LUMO) which can easily take up an electron, making an attack of radical species highly possible. In other words the Fullerene can react with many super oxides without being consumed considered as the most efficient radical scavenger or radical sponges.

These harmful radicals attack lipids, proteins, DNA, and other biologic tissues and organs. It has been found that water-soluble Fullerenes can be used as potential antioxidants and neuroprotective drugs against degenerative diseases related to oxidative stress. Thus, water-soluble Fullerenes, including host–guest inclusion complexes, are promising candidates for practical use as antioxidants [92,93].

It has been reported that Fullerenes are capable of inhibiting MC and PBB allergic responses [94]. *In vitro* and anaphylaxis *in vivo* suggests that these molecules have previously unrecognized anti-allergic properties. Follow-up studies demonstrated that the tyrosine phosphorylation of Syk was dramatically inhibited in Ag-challenged cells first incubated with Fullerenes. In addition, fullerene preincubation significantly inhibited IgE-induced elevation in cytoplasmic reactive oxygen species levels. Furthermore, fullerenes prevented the in vivo release of histamine and drop in core body temperature in vivo using a MC-dependent model of anaphylaxis. Fullerenes may represent a new way to control diseases largely influenced by these cells including type I hypersensitivity.

Studies [95,96] of malonic acid C_{60} derivatives (carboxyfullerenes), indicated that they are capable of eliminating both superoxide anion and H_2O_2, and were effective inhibitors of lipid peroxidation, as well. Carboxyfullerenes demonstrated robust neuroprotection against excitotoxic, apoptotic and metabolic insults in cortical cell cultures. They were also capable of rescuing mesencephalic dopaminergic neurons from both MPP1 and 6-hydroxydopamine-induced degeneration. Although there is limited in vivo data on these compounds to date, it has also been reported that systemic administration of the C3 carboxyfullerene isomer delayed motor deterioration and death in a mouse model of familial amyotrophic lateral sclerosis (FALS), which suggests that these novel antioxidants are potential neuroprotective agents for other neurodegenerative disorders, including Parkinson's disease.

Ryan et al. [94] showed that fullerenes could prevent allergic responses in vitro and in vivo, probably through the inhibition of ROS. In addition, oxidative stress through the production of ROS is thought to be a significant underlying factor in aging and has been shown to be present in human hair. The continuous melanin synthesis in the growing (anagen) hair follicle generates high oxidative stress, which leads to their selective premature aging and apoptosis.

In addition, a direct correlation between hair loss and oxidative stress has been implied in studies of patients with alopecia. The same group performed studies to determine if fullerene-based compounds could affect hair growth and hair follicle numbers, in mouse and human skin. They found that in shaved mice, fullerenes increase the rate of hair growth as compared with mice receiving vehicle only. In SKH-1 hairless mice Fullerene derivatives given topically or subdermally markedly increased hair growth. This was paralleled by a significant increase in the number of hair follicles in fullerene-treated mice as compared with those mice treated with vehicle only. The fullerenes also increased hair growth in human skin sections maintained in culture. These results will be translated into new ways to induce hair growth in the approximately 20% to 50% of individuals in whom minoxidil and finasteride have no effect [97].

In 2011, Inui et al. [98] reported the use of Fullerene gel against acne vulgaris. Oxidative stress plays a major role in acne formation, suggesting that oxygen radical scavengers are potential therapeutic agents. They studied the effectiveness of Fullerene gel in treating acne vulgaris, performing an open trial using a Fullerene gel twice a day; at 4 and 8 weeks, the mean number of inflammatory lesions (erythematous papules and pustules) significantly ($P < 0.05$) decreased from 16.09 ± 9.08 to 12.36 ± 7.03 (reduction rate 23.2%) and 10.0 ± 5.62 (reduction rate 37.8%), respectively. The number of pustules, consisting of accumulation of neutrophils, was significantly ($P < 0.05$) decreased from 1.45 ± 1.13 to 0.18 ± 0.60 (reduction rate 87.6%), and further in vitro assays of sebum production in hamster sebocytes revealed that 75 μM polyvinylpyrrolidone-Fullerene inhibits sebum production, suggesting that fullerene suppresses acne through decreasing neutrophil infiltration and sebum production. After treatment for 8 weeks, the water content of the skin significantly ($P < 0.05$) increased from 51.7 ± 7.9 to 60.4 ± 10.3 instrumental units. Therefore, the fullerene gel may help in controlling acne vulgaris with skin care benefit.

If the carbon sphere is perfectly centered, the distance between its surface and that of the enzyme allows the formation of van der Waals interactions. The same research group [99] performed the first synthesis of a fullerene inhibitor of HIVP. Other research has been reported [100-111]. The compound showed to fit smartly in the active site of the viral protease with strong van der Waals interactions on the hollow enzyme surface. In addition, fullerenes and its derivatives have antimicrobial activity against *Candida albicans*, *Bacillus subtilis*, *Escherichia coli*, *Mycobacterium avium*, *Mycobacterium tuberculosis*, *Staphylococcus spp.*, *Streptococcus spp.*, *Enterococcus faecalis* (Gram positive); *Klebsiellapneumoniae*, *Pseudomonas aeruginosa*, *Salmonella typhi* (Gram negative)[83, 112-120].

Material Applications, Lubricants, Photovoltaic and More

Fullerene C_{60} nano-oil is proposed as a promising lubricant to enhance the performance of domestic refrigerator compressors [121]. Recently, the stability of fullerene C_{60} nanoparticles dispersed in a mineral oil and the lubrication properties of the nano-oil were investigated experimentally. It was confirmed that the nanoparticles steadily suspend in the mineral oil at stationary conditions for a long period. The friction coefficients of the nano-oil significantly decrease with increasing the concentration of nanoparticles in the mineral oil, especially at the lower applied loads. The friction coefficients of the nano oil with the concentration of 1 and 3 g/L are 12.9 and 19.6% lower than that of pure mineral oil. The applications of the nano-oil with the specific concentration of 3 g/L to two domestic refrigerator compressors were examined by compressor calorimeter experiments. The results show the COPs of two compressors were improved by 5.6% and 5.3%, respectively, when the nano-oil was used instead of pure mineral oil. Another study about fullerenes as additives in raw oil showed an improved lubrication performance on the friction surfaces by reducing metal surface contacts. In addition, when the viscosity of base oil was low, improving the AW and EP characteristics was greater with the addition of Fullerene nanoparticles. Also, it leads to reduce the friction coefficient between the friction surfaces. However, the addition of fullerene nanoparticles did not have a great influence on improving the lubrication characteristics when the viscosity was high [122]. Another novel application for the graphene – C_{60} hybrid is as a high performance lubricating film in MEMS [123].

Fullerene C_{60} thin films can be prepared by the thermal evaporation technique. The optical Band gap at room temperature is found to be 2.30 eV, which gradually decreases with lowering the temperature and reaches to 2.27 at 77 K, also the conductivity of the film increases with lowering the temperature [124]. A multi-functionality to the composite was made with alumina and reinforced with Fullerene soot [125]. In addition, 2 wt% Ni serves as an efficient catalyst for in situ transformation of soot into diamond and nanoparticles of fullerite and graphitic carbon during thermomechanical processing. The carbon nanostructures act as an effective reinforcement in the composite that result in an enhancement of the fracture toughness to 16.5M $Pa{\cdot}m^{1/2}$. Electrical resistivity decreases by 15 orders of magnitude.

The improvement of composite properties is attributed to the formation of a homogeneous carbon coating on the alumina particles. A small amount of ruby is present as an impurity in the alumina, and serves as a convenient pressure sensor through its stress-dependent luminescence and gives additional multi-functionality to the composite [125].

Titanium dioxide (TiO_2)–Fullerene C_{60} composite is prepared from TiO_2 and b-cylcodextrin (CD) encapsulated C_{60}. The absorption of the composite extends to the visible light region due to the charge transfer from CD and C_{60} to TiO_2. The composite shows a reduced charge recombination compared to that of the bare TiO_2 and TiO_2/CD. Photocatalytic studies in the presence of scavengers reveal that the composites produce a higher amount of reactive oxygen species (ROS). The enhanced photocatalytic activity of the composites is attributed to the visible light responsiveness, reduced charge recombination and increased formation of ROS. The photodegradation of 4-CP is significantly faster in the presence of the composite with 1.5 wt. % C_{60} and is attributed to the synergistic effect of higher adsorption and increased ROS formation. The ROS formation by C_{60} is possible because of the nonaggregated state of C_{60} molecules in the composite and is assigned to the method, which employs CD molecules to disperse C_{60} in the composite. [126].

A few years ago, Zuev et al. [127] synthesized by in situ polymerization composites with PA6 and fulleroid nanofillers. The mechanical properties of composites were improved by incorporation of the 0.001–0.1 wt.% Fullerenes or Fullerene soot. The addition of fullerenes into PA6 matrix leads to its selective crystallization only in the α-form crystals that can explain the reinforcement effect of fullerenes. For nanocomposites with fulleroid fillers the friction coefficient is practically two times lower than that of the neat PA6. Electrical volume resistivity of the composites decreases at loading of fulleroid fillers. The minimum electrical volume resistance at 0.1 wt% loading of fullerene soot is about $10^7\Omega$ cm. In addition, mechanical testing of Fullerene C_{60}/epoxy polymer nanocomposites with different C_{60} loading (0.0–0.08 wt. %), shows that compared with neat epoxy, mechanical and toughening properties of the composites are greatly improved.

The presence of Fullerene hampers molecular motions in the epoxy composites leads to hardening of polymer chains and impedes secondary relaxation processes, which explains the increase of Young's modulus of the nanocomposites, the activation energy of β-relaxation process decreases for nanocomposites. This finding allows confirmation of the energy-dissipating mechanism that leads to enhanced toughness of nanocomposites [50].

Thermal and thermo-oxidative degradation of Fullerene C_{60}/high density polyethylene (HDPE) composites were studied by Zaho et al. under different atmospheric conditions [51-53]. In N_2 the addition of C_{60} increases the onset decomposition temperature by about 10 C with more heavy compounds (more than 34 carbon). Also, the thermal stability of HDPE in air is remarkably improved with the addition of C_{60}. When the content of C_{60} is 2.5 wt.% the onset decomposition temperature increases by about 91°C. The results of viscoelastic behavior and gel content reveal that C_{60} can trap the alkyl radicals and alkyl peroxide radicals to inhibit hydrogen abstraction to suppress the chain scission and preserve the long chain structure.

However, in the absence of C_{60} or with low C_{60} concentration, hydrogen abstraction occurs, resulting in the formation of a series of alkyl radicals and alkyl peroxide radicals, which accelerates the chain scission and plays a leading role in the thermal oxidative degradation. In addition, use of Fullerene C_{60} as fillers improve the flame retardancy of HDPE [54].

Photo induced electron transfer (PET) between nanocarbon architecture and photosensitizers has received considerable attention due to its potential applications for optoelectronics, photonics, sensors, and other functional nanoscale molecular devices [55-58]. The composite of polymethyl methacrylate (PMMA) with fullerene is a pressing problem because they represent promising materials for applications in which the properties of an electron subsystem are important. The photo luminescence properties of the composite show that the interaction between PMMA and fullerene, realized through the carbonyl group of

MMA, can be initiated by the molecular oxygen present in the composite in the form of an impurity giving rise to the formation of oxygen-containing complexes with Fullerene [136].

Organic solar cells provide considerable advantages over their inorganic counterparts such as possible cost-effective and environmentally compatible manufacturing, lightweight applications, and flexibility. Tremendous scientific efforts were made in this field over the last few years, leading to certified power conversion efficiencies currently up to 8.3%. The most commonly used material system for organic solar cells consist of conjugated polymers as the electron donor and fullerene derivate as the electron acceptor materials, blended in a bulk hetero-conjunction to overcome the strong excitonic behavior of photo generated charge carriers [137-138]. In addition, fullerenes act as photo-sensitizing electron acceptors with respect to a wide variety of electron donors; besides, the fullerenes act as good ground state electron acceptors in the presence of light-absorbing electron donors such as porphyrins and phthalocyanines [132], ferrocene, thiophene, and carbon nanotubes, and additional light harvesting antenna molecules. In these systems, Fullerene acts as an efficient electron acceptor because of its intriguing characteristics that allow for the acceleration of photo induced charge separation and a charge recombination deceleration in the electron transfer process [135]. Numerous molecular donor - acceptor arrays containing accessory pigments, from dyads to hexads, have been synthesized to address the detailed kinetics of energy and electron transfer; in a few cases, these have been successfully applied to solar energy conversion [139-140]. Among them, a covalently linked fullerene – porphyrin - ferrocene triad system was prepared explosively to obtain a long-lived charge separation state. A lifetime extension can be achieved by the addition of a secondary intermolecular electron transfer to a porphyrincation radical on the ferrocene, which acts as an additional electron donor in this system. Device performance continues to increase, and the field continues to attract investment from large and small companies alike. Where cell efficiencies were 5% in 2005, they have since increased to 12% in 2013 [79].

Organic Photo-detectors (OPD) are light-sensing devices that can be used in applications that range from motion control to medical imaging. Like OPV, they can be thin, lightweight, flexible, printable, and low-cost. Conventional detectors are typically made from silicon-based semi-conductors, and therefore use vacuum deposition processes typically at elevated temperatures. In contrast, OPD's manufacturing processes are based on low-temperature, atmospheric and solution-based printing on glass or plastic substrates [141-143].

Since Fullerenes were discovered at the end of last century, many new findings and important aspects on these carbon molecules have been accumulated to form a new exciting scientific field. Recent developments suggest that many proposed fullerene applications are practical technologies in a wide range of areas such as IT devices, diagnostics, pharmaceuticals, environmental and energy industries. The Fullerene field is going to contribute to industries. In the past, the high production cost of fullerenes has been the main obstacle in the development of the fullerene market. We are now seeing a very rapid decline in price, which will open the door to a host of other applications. In addition, many industrial applications of fullerenes are now being commercialized.

CONCLUSION

In conclusion the Fullerenes are very interesting organic molecules because of their physical and chemical properties, and their preparation may be natural or synthetic. This type of organic molecule presents a very stable chemical structure, highly attractive to different areas: chemical organic and inorganic, presenting an aromatic system π-conjugates sp^2 hybridization, which helps to present great stability. Their chemical stability makes them attractive and useful molecules important in many different areas such as medical, biomedical, pharmaceutical, nanotechnology, electrical, and optoelectronic, among others. These possibilities offer great expectations in investigation and development of these organic molecules.

ACKNOWLEDGMENTS

We thank the Universidad Autonoma of Coahuila, Faculty of Chemical Sciences and CONACyT for their support for the realization of this chapter, especially basic science projects CB-SEP-CONACYT: CB-2011-01-167903 and CB-2011-01-0167668.

REFERENCES

[1] Kroto, H.W.; Heath, J. R.; O´Brien, S. C.; Curl, R.F.; Smalley, R. E. C_{60}: Buckminsterfullerene. *Nature*, 1985, 318, 162-163.

[2] Tzirakis, M. D.; Orfanopoulos, M. Radical reactions of fullerenes: from synthetic organic chemistry to materials science and biology. *Chemical Reviews*, 2013, 113, 5262-5321.

[3] Saunders M, Jimenez-Vazquez H A, Cross R J, Poreda R J., Stable compounds of helium and neon: he@c60 and ne@c60. *Science*, 1993, 259, 1428–1430.

[4] Kroto H., Fullerenes: An extraterrestrial carbon carrier phase for noble gases. *Science*, 1988, 242, 1139–1143.

[5] Heymann D., Chibante L. P. F., Brooks R. R., Wolbach W. S., Smalley R. E. Fullerenes in the cretaceous-tertiary boundary layer. *Science*, 1994, 256, 645–647.

[6] Becker L., Bada J. L., Winans R. E., Bunch T. E. Fullerenes in Allende meteorite. *Nature,* 1994, 372, 507.

[7] Becker L, Bunch T E, Allamandolla L. J. Higher fullerenes in the Allende meteorite. *Nature*, 1999, 400, 227–228.

[8] Buseck P. R., Tsipursky S. J., Hettich S. Fullerenes from the geological environment. *Science*, 1992, 257, 215–217.

[9] Becker L, Poreda R J, Bada J. L. Extraterrestrial helium trapped in fullerenes in the Sudbury impact structure. *Science*, 1996, 272, 249–252.

[10] Lieber Charles M, Chen Chia-Chun. Preparation of fullerenes and fullerenes based *Materials. Solid State Physics*, 1994, 48, 109-148.

[11] Rose H. R., Dance I.G., Fisher K.J., Smith D.R. Willet G.D., Wilson M.A. Calcium inside C_{60} and C_{70} from coorongite, a Precursor of Torbanite. *Chemical Communications*, 1993, 11, 941-942.

[12] Taylor R., Langley G.J., Kroto H.W. and Walton D.R.M. Formation of C_{60} by Pyrolysis of Naphthalene. *Nature*, 1993, 366, 728-730.

[13] Crowley C., Taylor R., Kroto H.W., Walton D.R. M., Cheng P.C. and Scott L.T. Pyrolytic Production of fullerenes. *Synthetic Materials*, 1996, 77, 17-22.

[14] Ledesma E.B., Kalish M.A., Wornat M.J., Nelson P.F., Mackie J.C. Observation of cyclopenta-fused and ethynyl-substituted PAH during the fuel rich combustion of primary tar from a bituminous coal. *Energy Fuels*, 1999, 13, 1167-1172.

[15] Wornat M.J., Vriesendorp F.J. J., Lafleur A.L., Plummer E. F., Necula A., Scott L.T. The identification of new ethynyl-substituted and cyclopenta-fused polycyclic Aromatic Hydrocarbons in the products of antracene pyrolysis. *Polycyclis Aromatic Compound*, 1999, 13, 221-240.

[16] Sarobe M., Fokkens R.H., Cleij T.J., Jenneskens L.W., Nibbering N. M. M., Stas W. and Versluis C. S-8-mediated cyclotrimerization of 4,5-dihydro-benz [1] acephen-anthrylene: trinaphtho-decacyclene $C_{60}H_{30}$ isomers and their propensity towards cyclodehydrogenation. *Chemistry Physics Letters*, 1999, 313, 31-39.

[17] Camacho-Bragado G.A., Santiago P., Martin-Almazo M., Espinosa M., Romero E. T., Murgich J., Lugo V. R., Lozada-Cassou M., Jose Yacaman M. Fullerenic Structures derived from oil asphaltenes. *Carbon*. 2002, 40, 2761-2766.

[18] Lee T.H., Yao N., Chen T.J. and Hsu W. K. Fullerene-like carbon particles in petrol soot. *Carbon,* 2002, 40, 2275-2279.

[19] Cataldo F., Poniter-Johnson M.A., Recent discoveries in carbon black formation and morphology and their implications on the structure of interstellar carbon dust. *Fullerenes, Nanotubes and Carbon Nanostructures*, 2002, 10, 1-14.

[20] Kroto H. W., Heath J.R., O'Brien S. C., Curl R.F., Smalley R.E., C_{60} Buckminsterfullerene. *Nature*, 1985, 318, 162.

[21] O'Brien S.C., Heath J.R., Curl R.F., Smalley R.E. Photophysics of buckminsterfullerene and other carbon cluster ions. *Journal of Chemical Physics*, 1988, 88, 220.

[22] Lieber Charles M., Chen Chia-Chun, Preparation of fullerenes and fullerenes based Materials. *Solid State Physics,* 1994, 48, 109-148.

[23] CheungJ. T., Sankur H., Growth of thin-films by laser-induced evaporation, CRC critical reviews in solid state and materials. *Science*, 1988, 15, 63-109.

[24] Kratschmer W., Lamb L.D., Fostiropoulos K., Huffman D.R., Solid C_{60}: a new form of carbon. *Nature*, 1990, 347, 354.

[25] Zhang Q. L., O'Brien S. C., Heath J. R., Liu Y., Curl R. F., Kroto H. W. Smalley R. E., Reactivity of large carbon clusters: spheroidal carbon shells and their possible relevance to the formation and morphology of soot. *Journal of Physical Chemistry*, 1986, 90, 525-528.

[26] Gerhardt P., Löffler S., Homann K. H., Polyhedral carbon ions in hydrocarbon flames. *Chemical Physics. Letters*, 1987, 137, 306-310.

[27] Howard J. B., McKinnon J. T., Makarovsky Y., Lafleur Y., Johnson M. E., Fullerenes C_{60} and C_{70} in flames. *Nature*, 1991, 352, 139.

[28] Howard J. B., McKinnon J. T., Johnson M. E., Makarovsky Y. Lafleur Y., Fullerenes C_{60} and C_{70} in flames. *Journal of Physical Chemistry*, 1992, 96, 6657-6662.

[29] Parker D. H., Chatterjee K., Wurz P., Lykke K. R., Pellin M. J., Stock L. M. Hemminger J. Fullerene and giant fullerenes: synthesis, separation, and mass spectrometric characterization. *Carbon*, 1992, 30, 1167-1182.

[30] Lijima S., Helical microtubules of graphitic carbon. *Nature,* 1991, 354, 56.

[31] C. Yeretzian, J. B. Wiley, K. Holczer, T. Su, S. Nguyen, R. B. Kaner and R. L. Whetten, Partial separation of fullerenes by gradient sublimation. *Journal of Physical Chemistry*, 1993, 97, 10097-10101.

[32] Wetten R. L., Alvarez M. M., Anz S. J., Schriver K. E., Beck R. D., Diederich F., Rubin Y., Ettl R., Foote C. S., Darmanyan A. P., Arbogast J. W., *Materials Research. Society Symposium Proceedings,* 1990, 206, 639.

[33] Taylor R., Langley G. J., Kroto H. W. Formation of C_{60} by pyrolysis of naphthalene. *Nature*, 1993, 366, 728.

[34] Boorum M. M., Vasil'ev Y. V., Drewello T., Scott L. T., Groundwork for a rational synthesis of C60: cyclodehydrogenation of a $C_{60}H_{30}$polyarene. *Science*, 2001, 294, 828-831.

[35] Scott L. T., Boorum M. M., McMahon B. J., Hagen S., Mack J., Blank J., Wegner H. Meijere A. De, A rational chemical synthesis of C_{60}. *Science,* 2002, 295, 1500-1503.

[36] Zhang, J., Bowles, F. L., Bearden, D. W., Ray, W. K., Fuhrer, T., Ye, Y., Dixon, C., Harich, K., Helm, R. F., Olmstead, M. M., Balch, A. L., Dorn, H. C. A missing link in the transformation from asymmetric to symmetric metallofullerene cages implies a top-down fullerene formation mechanism. *Nature Chemistry*, 2013, 5(10), 880–885.

[37] Buhl, M. Hirsch, A.; Spherical Aromaticity of fullerene. *Chemical Reviews*, 2001, 101, 1153-1183.

[38] Sawamura, M.; Kuninobu, Y.; Toganoh, M.; Matsuo, Y.; Yamanaka, M.; Nakamura. E. Hybrid of ferrecene and fullerene. *Journal American Chemical Society*, 2002, 124, 9354-9355

[39] Balch, A. L.; Olmstead, M. M. Reactions of transition metal complexes with fullerenes (C_{60}, C_{70}, etc.) and related materials. *Chemical Reviews*, 1998, 98, 2123-2165.

[40] Balch, A. L; Costa, D. A.; Noll, B. C.; Olmstead, M. M. Oxidation of buckminsterfullerene with *m*-chloroperoxybenzoic acid. characterization of a Cs isomer of the diepoxide $C_{60}O_2$. *Journal American Chemical Society*, 1995, 117, 8926-8932.

[41] Wan, W. C.; Liu, X.; Sweeney, G. M.; Broderick, W. E. Structural evidence for the expected Jahn-Teller distortion in monoanionic C_{60}: synthesis and X-ray crystal structure of decamethyl nickelocenium buckminsterfulleride. *Journal American Chemical Society*, 1995, 117, 9580-9581.

[42] Vougioukalakis, G.; Roubelakis, M. M.; Orfanopoulos, M. Radical reactivity of Azo [60] fullerene: preparation of monoadducts and limitations. *Journal Organic Chemical,* 2010, 75, 4124-4130.

[43] Orfanopoulos, M.; Tzirakis, M.D. Acyl radical reactions in fullerene chemistry: direct acylation of [60] fullerene through an efficient decatungstate-photomediated approach. *Journal American Chemical Society*, 2009, 131, 4063-4069.

[44] Morton, J. R.; Preston, K. F.; Krusic, P. J.; Hill, S. A. Wasserman, E. ESR studies of the reaction of alkyl radical with fullerene (C60), *Journal Physical Chemical,* 1992, 96, 3576-3578.

[45] Tzirakis, M. D.; Orfanopoulos. Radical Reactions of fullerenes: from synthetic organic chemistry to materials science and biology, *Chemical Reviews*, 2013, 113, 5262-5321.
[46] Ford, W.T.; Nishioka, T.; Qiu, F.; D´Souza, F. Choi, J. Dimethyl azo(bisisobutyrate) and C60 produce 1,4 and 1,16-di(2-carbomethoxy-2-propyl)-1-x-dihydro[60]fullerenes. *Journal Organic Chemical*, 2000, 65, 5780-5784.
[47] Ford, W.T.; Nishioka, T.; Qiu, F.; D´Souza, F. Choi, J. Structure determination and electrochemistry of products from the radical reaction of C_{60} with azo(bisisobutyronitrile) *Journal Organic Chemical*, 1999, 64, 6257-6262.
[48] Bianco, A., Da Ros, T. Biological applications of fullerenes. In Fullerenes. Principles and applications; Langa F., Nierengarten J.-F., Eds.; RSC: Cambridge, UK, 2007.
[49] Enes R.F., Farinha A.S.F., Tome A.C., Cavaleiro J.A.S., Amorati R., Petrucci S., Pedulli G.F. Synthesis and antioxidant activity of [60] fullerene-flavonoid conjugates. *Tetrahedron*. 2009, 65, 253-262.
[50] Apenova M.G., Akhmetov V.A., Belov N.M., Goryunkov A.A., Ioffe I.N., Lukonina N.S., Markov V. Yu, Sidorov L.N. Alkali-metal trichloroacetates for dichloromethylenation of fullerenes: nucleophilic addition-substitution route. *Journal Chemical Asian*. 2014, 9, 915-923.
[51] Liu T.-X., Zhang Z., Liu Q., Zhang P., Jia P., Zhang Z., Zhang G. Synthesis of [60]Fullerene-fused tetrahydroazepinones and azepinonimines via $Cu(OAc)_2$-promoted N-heteroannulation reaction. *Organic Letters*. 2014, 16, 1020-1023.
[52] Wooley K.L., Hawker C.J., Frechet J.M.J., Wudl F., Srdanov G., Shi S., Li C., Kao M. Fullerene-bound dendrimers-soluble, isolated carbon clusters. *Journal American Chemical Society*, 1993, 115, 9836-9837.
[53] Hahn U., Vögtle F., Nierengarten J.-F. Synthetic Strategies towards Fullerene-Rich Dendrimer Assemblies. *Polymers*. 2012, 4, 501-538.
[54] Si W., Lu S., Bao M., Asao N., Yamamoto Y., Jin T. Cu-Catalyzed C-H Amination of Hydrofullerenes Leading to1,4‑Difunctionalized Fullerenes. *Organic Letters*. 2014, 16, 620-623.
[55] You X., Wang G.-W. Ferric Chloride-Catalyzed Reaction of [60] Fullerene with tert-Butyl N‑Substituted Carbamates: Synthesis of Oxazolidino [4,5:1,2] [60] fullerenes. *Journal Organic Chemistry*, 2014, 79, 117-121.
[56] Lallana E., Sousa-Herves A., Fernandez-Trillo F., Riguera R., Fernandez-Megia E. Click Chemistry for drug delivery nanosystems. *Pharm Res*. 2012, 29, 1-34
[57] Kolb H.C., Sharpless, B.K. The growing impact of click chemistry on drug discovery. 2003, 8, 1128-1137.
[58] Huisgen R., Szeimies G., Möbius L. 1.3-Dipolare cycloaddition, XXXII. Kinetik der Addition en organischerazide and CC-mehrfachbindungen. *Chem. Ber*. 1967, 100, 2494-2507.
[59] Meldal, M., Tornøe, C.W. Cu-catalyzed azide-alkyne cycloaddition. *Chemical Reviews*, 2008, 108, 2952-3015.
[60] Tornøe C.W., Christensen C., Meldal, M. Peptidotriazoles on solid phase: [1,2,3]-triazoles by regiospecific copper(I)-catalyzed 1,3-dipolar cycloadditions of terminal alkynes to azides. *Journal Organic Chemical Society*. 2002, 67, 3057-3064.

[61] Rostovtsev V.V., Green L.G., Fokin V.V., Sharpless K.B. A stepwise huisgen cycloaddition process: copper(I)-catalyzed regioselective "ligation" of azides and terminal alkynes. *Angew. Chem. Int. Ed.* 2002, 41, 2596-2599.

[62] Iehl J., Pereira de Freitas R., Delavaux-Nicot B., Nierengarten J-F. Click chemistry for the efficient preparation of functionalized [60] fullerene hexakis-adducts. *Chem Commun.* 2008, 2450-2452.

[63] Iehl J., Nierengarten J-F. A click-click approach for the preparation of functionalized [5:1]-hexaadducts of C60. *Chem Eur Journal.* 2009, 15, 7306-7309.

[64] Compain P., Decroocq C., Iehl J., Holler M., Hazelard D., Mena Barragán T. Glycosidase inhibition with fullerene iminosugar balls: a dramatic multivalent effect. *Angew Chem, Int* Ed. 2010, 49, 5753-5756.

[65] Partha R., Conyers J.L. Biomedical applications of functionalized fullerene-based nanomaterials. *International Journal of Nanomedicine.* 2009, 4, 261-275.

[66] Zhang W.-B., Tu Y., Ranjan R., van Horn R.M., Leng S., Wang J. "Clicking" fullerene with polymers: synthesis of [60] fullerene end-capped polystyrene. *Macromolecules.* 2008, 41, 515-517.

[67] Steinmetz N.F., Hong V., Spoerke E.D., Lu P., Breitenkamp K., FinnM.G.Buckyballs meet viral nanoparticles: candidates for biomedicine. *Journal American Chemical Society*, 2009, 131, 17093-17095.

[68] Iskin B., Yilmaz G., Yagci Y. Mono-addition synthesis of polystyrene-fullerene (C60) conjugates by thiol-ene Chemistry. Chem. E *Chem Eur Journal.* J. 2012, 18, 10254-10257

[69] Magoulas G.E., Garnelis T., Athanassopoulos C.M., Papaioannou D., Mattheolabakis G., Avgoustakis K., Hadjipavlou-Litina D. Synthesis and antioxidative/anti-inflammatory activity of novel fullerene-polyamine conjugates. *Tetrahedron*, 2012, 68, 7041-7049.

[70] Kokubo K., Masuda H., Ikuma N., Mikie T., Oshima T. Synthesis and characterization of new acetalized [60] fullerenes. *Tetrahedron Letters.* 2013, 54, 3510-3513.

[71] Xu L., Zhang Q., Zhang G., Liang S., Yu Y., Gan L. Regioselective Diels-Alder Reactions Directed by Carbonyl Groups on the Rim of Open-Cage Fullerene Derivatives. *European Journal Organic Chemistry* 2013, 7272-7276.

[72] Kim G., Lee K.C., Kim J., Lee J., Lee S.M., Lee J.C., Seo J.H., Choi W.-Y., Yang C. An unprecedented [5,6]-open adduct via a direct benzyne-C60 cycloaddition. *Tetrahedron*, 2013, 69, 7354-7359.

[73] Maroto E.E., Filippone S., Suárez M., Martínez-Álvarez R., de Cózar A., Cossío F.P, Martín N. Stereodivergent synthesis of chiral fullerenes by [3 + 2] cycloadditions to C_{60}. *Journal American Chemical Society*, 2014, 136, 705-712.

[74] Yang H.-T., Ren W.-L., Miao C.-B., Dong C.-P., Yang Y., Xi H.-T., Meng Q., Jiang Y., Sun X.-Q. DMAP-catalyzed [3 + 2] and [4 + 2] cycloaddition reactions between [60] Fullerene and unmodified Morita-Baylis-Hillman adducts in the presence of Ac_2O. *Journal Organic Chemistry*, 2013, 78, 1163-1170.

[75] Liu T.-X., Zhang Z., Liu Q., Zhang P., Jia P., Zhang Z., Zhang G. Synthesis of [60]Fullerene-fused tetrahydroazepinones and azepinonimines via $Cu(OAc)_2$-promoted N-heteroannulation reaction. *Organic Letters.* 2014, 16, 1020-1023.

[76] Hirsch, A. Addition Reactions of Buckminsterfullerene (C_{60}). *Synthesis*, 1995, 895-913.

[77] Lu, S.; Jin, T.; Bao, M. Yamamoto, Y. J. *Journal American Chemical Society*, 2011, 133, 12842-128.

[78] Romanova, I. P.; Bogdanov, A. V.; Mironov, V.F.; Gulnara, R.; Shaikhutdinova, Larionova, O. A.; Latypov, S. K.; Balandina, A.A.; Yakhvarov, D. G.; Gubaidullin, A. T.; Saifina, A. F.; Sinyashin, O. G. Deoxygenation of someα-dicarbonyl compounds by tris(diethylamino)phosphine in the presence of fullerene C_{60}. *Journal Organic Chemistry*, 2010, 76, 2548-2557.

[79] Nano-C, I. *Nano-C: Fullerenes: Applications*. 2001-2008 Nano-C, Inc.; Available from: http://nano-c.com/fullereneapp.html. March 2014.

[80] Zhu, J., et al., Tumor-Inhibitory Effect and Immunomodulatory Activity of Fullerol C60(OH)x.*Small*, 2008, 1168-1175.

[81] Fan, J., et al., Water-Dispersible Fullerene Aggregates as a Targeted Anticancer Prodrug with both Chemo- and Photodynamic Therapeutic Actions. *Small*, 2013. 9, 613-621.

[82] Chen, Z., et al., Applications of Functionalized Fullerenes in Tumor Theranostics. *Theranostics*, 2012. 2, 38 - 250.

[83] Bosi, S., et al., Fullerene derivatives: an attractive tool for biological applications. *European Journal of Medicinal Chemistry*, 2003. 38, 913-923.

[84] Park, J.H., et al., Self-assembled nanoparticles based on glycol chitosan bearing hydrophobic moieties as carriers for doxorubicin: In vivo biodistribution and anti-tumor activity. *Biomaterials*, 2006. 27, 119-126.

[85] Park, S.Y., et al., An artificial photosensitizer drug network for mitochondria-selective photodynamic therapy. *Chemical Communications*, 2012. 48, 2522-2524.

[86] Kim, S., et al., Acid pH-activated glycol chitosan/fullerene nanogels for efficient tumor therapy. *Carbohydrate Polymers*, 2014. 101, 692 - 698.

[87] Asada, R., et al., Photodynamic anti-cancer effects of fullerene [C60]–PEG complex on fibrosarcomas preferentially over normal fibroblasts in terms of fullerene uptake and cytotoxicity. *Mol Cell Biochem*, 2014. 390 175–184.

[88] Liao, F., Y. Saitoh, and N. Miwa, Anticancer Effects of Fullerene C-60 Included in Polyethylene Glycol Combined With Visible Light Irradiation Through ROS Generation and DNA Fragmentation on Fibrosarcoma Cells With Scarce Cytotoxicity to Normal Fibroblasts. *Oncology Research*, 2011. 19, 203-216.

[89] Shi, J., et al., PEGylated fullerene/iron oxide nanocomposites for photodynamic therapy, targeted drug delivery and MR imaging. *Biomaterials*, 2013. 34, 9666-9677.

[90] KRUSIC, P.J., et al., Radical Reactions of C60. *Science*, 1991. 254, 1183-1185.

[91] Saeidnia, S. and M. Abdollahi, Toxicological and pharmacological concerns on oxidative stress and related diseases. *Toxicology and Applied Pharmacology*, 2013. 273, 442-455.

[92] Matsubayashi, K., et al., Effects of pin-up oxygen on 60 Fullerene for enhanced antioxidant activity. *Nanoscale Research Letters*, 2008. 3, 237-241.

[93] Kokubo, K., et al., Radical-scavenging Ability of Hydrophilic Carbon Nanoparticles: From Fullerene to Its Soot. *Fullerenes Nanotubes and Carbon Nanostructures*, 2014. 22, 250-261.

[94] Ryan, J.J., et al., Fullerene nanomaterials inhibit the allergic response. *Journal of Immunology*, 2007. 179, 665-672.

[95] Dugan, L.L., et al., Fullerene-based antioxidants and neurodegenerative disorders. *Parkinsonism & Related Disorders*, 2001. 7, 243-246.
[96] Dugan, L.L., et al., Buckminsterfullerenol free radical scavengers reduce excitotoxic and apoptotic death of cultured cortical neurons. *Neurobiology of Disease*, 1996. 3, 129-135.
[97] Zhou, Z., et al., Fullerene nanomaterials potentiate hair growth. *Nanomedicine-Nanotechnology Biology and Medicine*, 2009. 5, 202-207.
[98] Inui, S., et al., Improvement of acne vulgaris by topical fullerene application: unique impact on skin care. *Nanomedicine-Nanotechnology Biology and Medicine*, 2011. 7, 238-241.
[99] Sijbesma, R., Srdanov, F. Wudl, J. A. Castoro, Charles Wilkins, Simon H. Friedman, Dianne L. De Camp, George L. Kenyon., Synthesis of a fullerene derivative for the inhibition of HIV enzymes. *Journal of the American Chemical Society*, 1993. 115, 6510-6512.
[100] Cheng, Y., et al., Structure-based design of carbon nanotubes as HIV-1 protease inhibitors: Atomistic and coarse-grained simulations. *Journal of Molecular Graphics & Modelling,* 2010. 29, 171-177.
[101] Dellinger, A., et al., Application of fullerenes in nanomedicine: an update. *Nanomedicine,* 2013. 8, 1191-1208.
[102] Fedorova, N.E., et al., Carboxylic fullerene C-60 derivatives: efficient microbicides against herpes simplex virus and cytomegalovirus infections *in vitro. Mendeleev Communications,* 2012. 22, 254-256.
[103] Leonis, G., et al., Computational Studies of Darunavir into HIV-1 Protease and DMPC Bilayer: Necessary Conditions for Effective Binding and the Role of the Flaps. *Journal of Chemical Information and Modeling*, 2012. 52, 1542-1558.
[104] Lin, C.-M. and T.-Y. Lu, C-60 Fullerene Derivatized Nanoparticles and their Application to Therapeutics. *Recent Patents on Nanotechnology*, 2012. 6, 105-113.
[105] Luo, Z., et al.,Development of Calixarenes, Cyclodextrins and Fullerenes as New Platforms for Anti-HIV Drug Design: An Overview. *Mini-Reviews in Medicinal Chemistry,* 2013. 13, 1160-1165.
[106] Ma, H. and X.-J. Liang, Fullerenes as unique nanopharmaceuticals for disease treatment. *Science China-Chemistry,* 2010. 53, 2233-2240.
[107] Martin, D. and M. Karelson, The Quantitative Structure Activity Relationships for Predicting HIV Protease Inhibition by Substituted Fullerenes. *Letters in Drug Design & Discovery*, 2010. 7, 587-595.
[108] Nakamura, S. and T. Mashino, Water-soluble Fullerene Derivatives for Drug Discovery. *Journal of Nippon Medical School*, 2012. 79, 248-254.
[109] Tanimoto, S., et al.,Target-Selective Photodegradation of HIV-1 Protease and Inhibition of HIV-1 Replication in Living Cells by Designed Fullerene-Sugar Hybrids. *Chemistry-an Asian Journal*, 2012. 7, 911-914.
[110] Tuktarov, A.R., L.L. Khuzina, and U.M. Dzhemilev, Covalent binding of fullerene C-60 to pharmacologically important compounds. *Russian Chemical Bulletin*, 2011. 60, 662-666.
[111] Tzoupis, H., et al., Binding of novel fullerene inhibitors to HIV-1 protease: insight through molecular dynamics and molecular mechanics Poisson-Boltzmann surface area calculations. *Journal of Computer-Aided Molecular Design*, 2011. 25, 959-976.

[112] Marcorin, G.L.; Da Ros, T.; Castellano, S.; Stefancich, G.; Bonin, I.; Miertus, S.; Prato, M. Design and Synthesis of Novel [60] Fullerene Derivatives as Potential HIV Aspartic Protease Inhibitors. *Organic Letters*, 2000. 2, 3955-3958.

[113] Aoshima, H. and K. Kokubo, Antimicrobial agent against methicillin resistant or sensitive Staphylococcus aureus, Staphylococcus epidermidis, acne microbe, Candida and Malassezia, contains fullerene hydroxide as active ingredient, *Univ Osaka; Vitamin C60 Bioresearch Kk.*

[114] Aoshima, H., et al., Antimicrobial Activity of Fullerenes and Their Hydroxylated Derivatives. *Biocontrol Science*, 2009. 14, 69-72.

[115] Brisebois, P.P., et al., Comparative study of the interaction of fullerenol nanoparticles with eukaryotic and bacterial model membranes using solid-state NMR and FTIR spectroscopy. *European Biophysics Journal with Biophysics Letters,* 2012. 41, 535-544.

[116] Dinesh, R., et al., Engineered nanoparticles in the soil and their potential implications to microbial activity. *Geoderma,* 2012. 173, 19-27.

[117] Hadduck, A.N., et al., Does Aqueous Fullerene Inhibit the Growth of Saccharomyces cerevisiae or Escherichia coli? *Applied and Environmental Microbiology*, 2010. 76, 8239-8242.

[118] Kumar, A. and S.K. Menon, Fullerene derivatized s-triazine analogues as antimicrobial agents. *European Journal of Medicinal Chemistry*, 2009. 44, 2178-2183.

[119] Lee, J., Mackeyev, Y.; Cho, M.; Li, D.; Kim, J-H.;Wilson, L. J.; Alvarez, P. Photochemical and Antimicrobial Properties of Novel C-60 Derivatives in Aqueous Systems. *Environmental Science & Technology*, 2009. 43, 6604-6610.

[120] Patel, M.B., et al., Novel Cationic Quinazolin-4(3H)-one Conjugated Fullerene Nanoparticles as Antimycobacterial and Antimicrobial Agents. *Archiv Der Pharmazie*, 2013. 346, 210-220.

[121] Xing, M., R. Wang, and J. Yu, Application of fullerene C60 nano-oil for performance enhancement of domestic refrigerator compressors. *International Journal of Refrigeration,* 2014. 40, 398 - 403.

[122] Ku, B.-C., et al., Tribological Effects of Fullerene (C-60) Nanoparticles Added in Mineral Lubricants According to its Viscosity. *International Journal of Precision Engineering and Manufacturing,* 2010. 11, 607-611.

[123] Pu, J., et al., Fabrication of novel graphene-fullerene hybrid lubricating films based on self-assembly for MEMS applications. *Chemical Communications*, 2014. 50, 469-471.

[124] Qaiser, D., et al., Comparative study of optical parameters of fullerene C-60 film at different temperatures. *Optics Communications*, 2010. 283, 3437-3440.

[125] Fals, A.E., V.G. Hadjiev, and F.C.R. Hernandez, Multi-functional fullerene soot/alumina composites with improved toughness and electrical conductivity. *Materials Science and Engineering a-Structural Materials Properties Microstructure and Processing,* 2012. 558, 13-20.

[126] Ali, M.M. and K.Y. Sandhya, Visible light responsive titanium dioxide-cyclodextrin-fullerene composite with reduced charge recombination and enhanced photocatalytic activity. *Carbon,* 2014. 70, 249-257.

[127] Pikhurov, D.V. and V.V. Zuev, The effect of fullerene C60 on the dielectric behaviour of epoxy resin at low nanofiller loading. *Chemical Physics Letters*, 2014. 601, 13 - 15.

[128] Zhao, L., et al., Thermal and thermo-oxidative degradation of high density polyethylene/fullerene composites. *Polymer Degradation and Stability*, 2013. 98, 1953-1962.

[129] Zhao, L., et al., Influence of fullerene on the kinetics of thermal and thermo-oxidative degradation of high-density polyethylene by capturing free radicals. *Journal of Thermal Analysis and Calorimetry*, 2013. 114, 1287-1294.

[130] Zhao, L., et al., Thermal Stability and Rheological Behaviors of High-Density Polyethylene/Fullerene Nanocomposites. *Journal of Nanomaterials*, 2012.

[131] Liu, H., et al., Thermal degradation and flammability properties of HDPE/EVA/C-60 nanocomposites. *Thermochimica Acta*, 2010. 506, 98-101.

[132] Wasielewski, M.R., Photo induced electron transfer in supramolecular systems for artificial photosynthesis. *Chemical Reviews,* 1992. 92, 435-461.

[133] D'Souza, F. and O. Ito, Photo sensitized electron transfer processes of nanocarbons applicable to solar cells. *Chemical Society Reviews*, 2012. 41, 86-96.

[134] Ahmed, R.M. and S.M. El-Bashir, Structure and Physical Properties of Polymer Composite Films Doped with Fullerene Nanoparticles. *International Journal of Photoenergy,* 2011.

[135] Lee, M.H., J.W. Kim, and C.Y. Lee, [60] Fullerene eporphyrine ferrocene triad self-assembled monolayers (SAMs) for photovoltaic applications. *Journal of Organometallic Chemistry,* 2014. 761, 20 – 27.

[136] Pushkarchuk, A.L., et al., Thermal destruction and defect formation in polymeric fullerene-containing nanocomposites. *Journal of Engineering Physics and Thermophysics*, 2012. 85, 798-801.

[137] Wilken, S., et al., ITO-free inverted polymer/fullerene solar cells: Interface effects and comparison of different semi-transparent front contacts. *Solar Energy Materials and Solar Cells*, 2012. 96, 141-147.

[138] He, Y., et al., High-performance polymer photovoltaics based on rationally designed fullerene acceptors. *Solar Energy Materials and Solar Cells*, 2013, 118, 71-178.

[139] Luo, C.,Guldi, D.M.; Imahori, H.; Tamaki, K., Sakata Y. Sequential Energy and Electron Transfer in an Artificial Reaction Center: Formation of a Long-Lived Charge-Separated State. *Journal of the American Chemical Society*, 2000. 122, 6535-6551.

[140] Lee, C.Y., et al., Remarkably Efficient Photocurrent Generation Based on a [60] Fullerene–Triosmium Cluster/Zn–Porphyrin/Boron–Dipyrrin Triad SAM. *Chemistry – A European Journal*, 2010. 16, 5586-5599.

[141] Li, H., J. Choi, and T. Nakanishi, Optoelectronic Functional Materials Based on Alkylated-pi Molecules: Self-Assembled Architectures and Non-assembled Liquids. *Langmuir*, 2013. 29, 5394-5406.

[142] Baeg, K.-J., et al., Organic Light Detectors: Photodiodes and Phototransistors. *Advanced Materials,* 2013. 25, 4267-4295.

[143] Zielinska, A., et al., Controlled self-assembly of alkylated-π compounds for soft materials — Towards optical and optoelectronic applications. *Current Opinion in Colloid & Interface Science,* 2014.

In: Fullerenes
Editor: Shannon B. Ellis

ISBN: 978-1-63321-385-2

Chapter 8

STARS, THE SOURCE OF FULLERENES

Citlalli Rios and Roberto Salcedo
Instituto de Investigaciones en Materiales, Universidad Nacional Autónoma de México, Circuito Exterior s/n, Ciudad Universitaria, Coyoacán, México D. F.

Philosophy [nature] is written in that great book which ever is before our eyes -- I mean the universe -- but we cannot understand it if we do not first learn the language and grasp the symbols in which it is written. The book is written in mathematical language, and the symbols are triangles, circles and other geometrical figures, without whose help it is impossible to comprehend a single word of it; without which one wanders in vain through a dark labyrinth."
Galileo Galilei

ABSTRACT

The first natural source of fullerenes appears to be giant systems, where extreme thermodynamic properties may provide an appropriate medium for the birth of these interesting molecules. It may be difficult to find these conditions on the surface of a planet; however, the stars are very interesting bodies with particular characteristics that make them a suitable laboratory for this purpose. The aura of a star comprises a very large region (millions of kilometers), where there are clouds of gases and dust that provide appropriate conditions for these reactions. These conditions can lead to the formation of some light elements (such as atoms) and more complex compounds such as fullerenes and other organic molecules, depending on the proximity of a specific region of the aura to the bulk, as well as the quality of the irradiation of the star. Recent studies have demonstrated that the structures of the compounds found in stars are similar to certain organic substances that occur on Earth. This document aims to show how the stars may constitute the principal natural source for fullerenes and related molecules.

1. Introduction

The stars comprise very complicated stellar systems, which have been the object of study since the birth of humanity. It is known that many ancient civilizations dedicated considerable resources to the study of the sky and particularly the stars. Ancient buildings such as the Kukulkan temple at Chichen Itza, Mexico or the pyramids in Egypt have been identified as observatories and indeed agriculture procedures were developed on the basis of astronomical studies. At the present time, astronomy relies on the participation of several disciplines, permitting more precise observations than those made previously. These disciplines include chemistry, physics, spectroscopy, etc. The application of these sciences reveals many aspects that were unknown in the past.

Matter is drastically transformed along the trajectory from the center of the star towards space; the study of these changes involves firstly the application of Physics which classifies all the different elemental particles and their characteristics up to the formation of stable atoms, subsequently Chemistry takes second place, involving the study of the formation of molecules with all relevant consequences, finally molecules reach the planets and Biology studies the final stage, which is life.

In our day, astronomy amasses all this information and processes the results, in order to describe a particular body. We are thus able to derive the composition of a far planet or the electromagnetic properties of interstellar bodies, making observations directly from our planet or from specialized ships or satellites. At the present time, the morphology, structure and thermodynamic properties of the stars have been precisely determined and the diffusion of this information is now included in mass media communication, so that the common man can read about these aspects during his daily life.

Ideas have been put forward concerning the possible benefits of exploring the environment of the stars, taking into account that we are very close to a star; the Sun.

Likewise, important studies have involved investigations concerning the composition of the stars, resulting in different zones within the stars being identified that manifest very different temperatures and varied composition, as well as regions where only elemental particles exist, or others which are colder and where elements or molecules can be produced.

Studies focusing on the composition of stellar dust constitute a huge area of research; thanks to this work scientists have identified many types of substances known in our planet, including essential compounds such as amino acids, porphyrines, or precursors of nucleic acids, which have been detected in outer space [1]. New types of compound studied in this context include nanostructures of carbon; particularly fullerenes and nanotubes [2]. Some scientists have developed synthetic methods to obtain these types of molecules; however, it seems they exist freely in the vicinity of the stars. Therefore, it is important to establish whether the only natural source of these materials is the stars and this is the purpose of this report.

2. FULLERENES AND THE STRUCTURE OF RELATED MOLECULES

The story of the first nanocompound is very interesting, as it was related by the author himself. C_{60} was discovered in a fortuitous event [3]; professor Kroto was visiting Rice University and he and his colleagues had planned an experiment to obtain lineal carbon chains, consisting of between 5 and 9 members. The experiment was carried out and the result, as was later proved, was fullerene C_{60}. A small bonus accompanied this procedure because C_{70} was also discovered during the same experiment, representing a point of justice for the author because C_{60} takes the form of a soccer ball, whereas C_{70} takes the form of a rugby ball, which are the main sports in England.

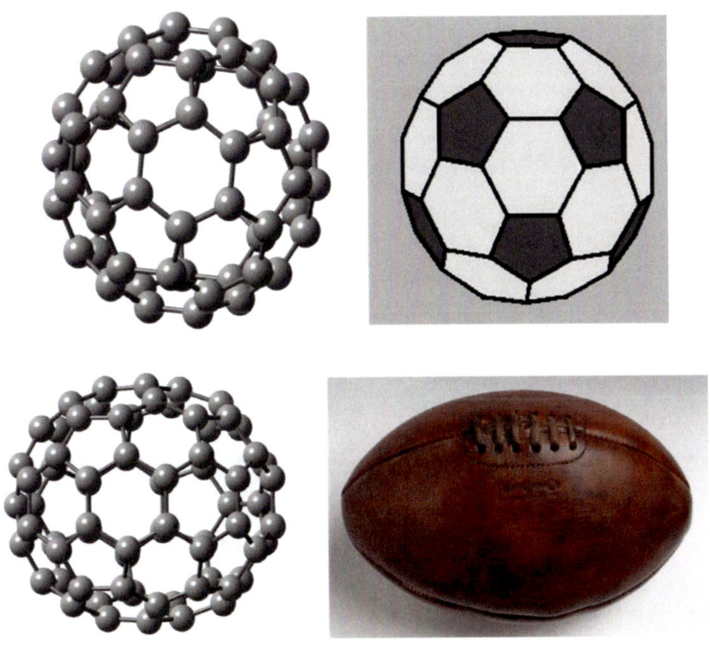

Figure 1. Comparison among Fullerenes and sport balls.

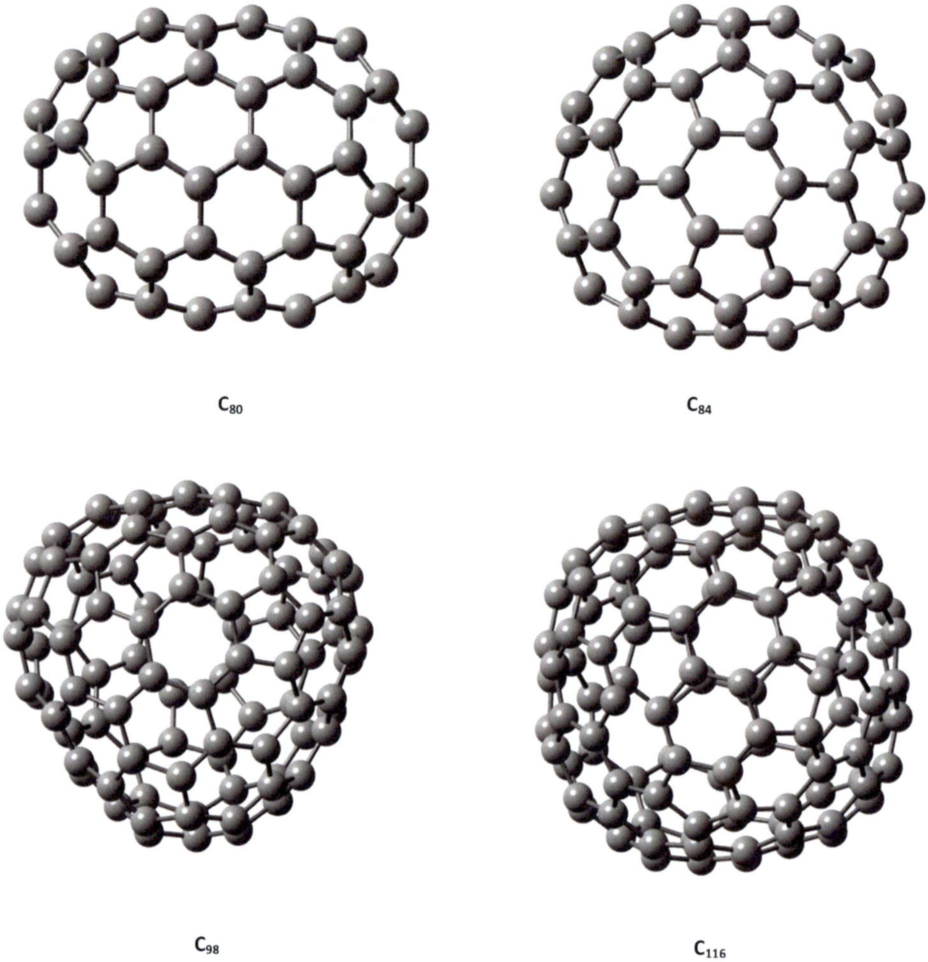

Figure 2. Large Fullerenes.

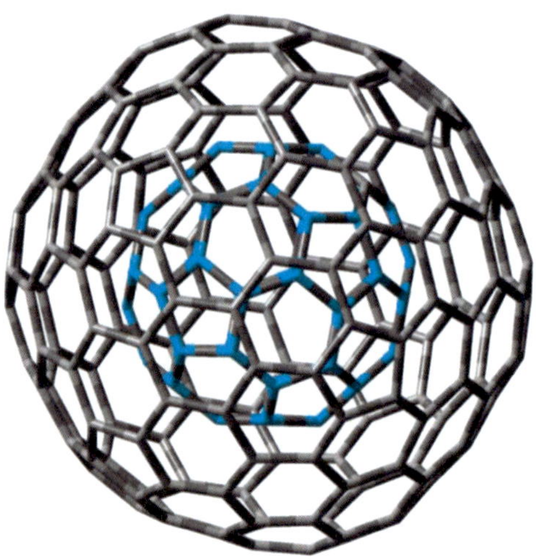

Figure 3. Onion.

A short time later, Krätschmer [4] and his co-workers developed a method for preparing a synthesis of both compounds, therefore fullerenes became commercial items and from that day it has been possible to attain these kinds of substances for laboratory use.

The next task was to investigate whether these two molecules were unique or if it was possible to prepare cages of different sizes. With respect to little cages there are no stable derivatives; therefore, C_{60} is the smallest known carbon cage that exhibits ample stability. However, with respect to large stable molecules, it is possible to find cages containing from 70 to 100 carbon atoms and likewise all of these with the exception of C_{70} are able to present collections of isomers [5-8] and prepared fullerenes have even been circumscribed into other large fullerenes, forming a type of derivative known as an "onion" (see figure 3) [9].

The stability of all these cages is closely related to an empirical rule named "the isolated pentagon rule" [10, 11], which establishes that any stable case with regard to size must be formed by an unlimited number of hexagonal faces and exactly twelve pentagonal faces, surrounded only by hexagons.

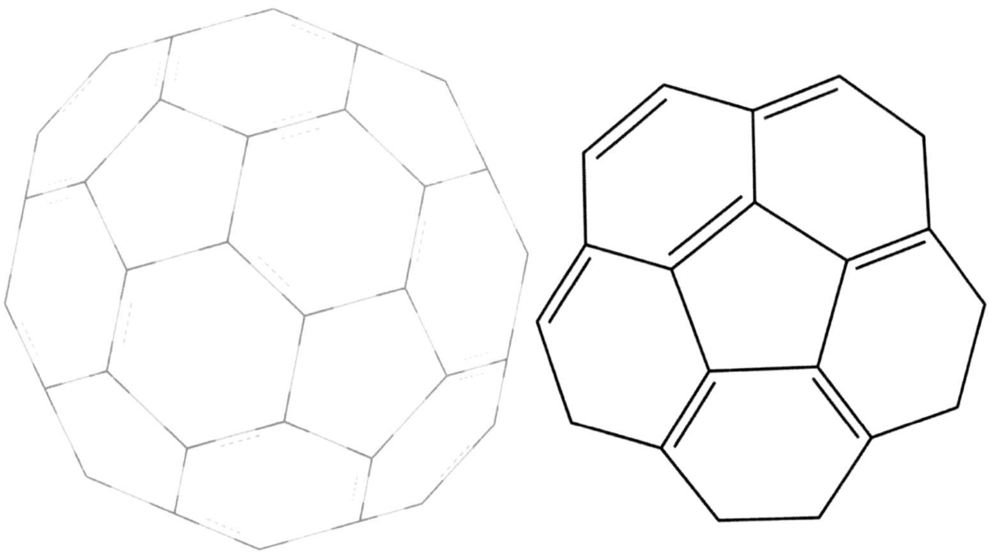

Figure 4. Isolated pentagon rule.

The main chemical concept for establishing the stability of fullerenes is aromaticity. A reader might suppose that the word "aromaticity" implied a particular odor; however so far the scent of fullerenes has never been registered. When we say "aromaticity" the correct interpretation refers to a very important property of certain organic molecules that is defined in the next phrase: the ability of a compound to sustain an induced electronic ring current. This definition is not completely adequate however, as this phenomenon presents significant problems for interpretation; moreover an indisputable and very important aspect is that this characteristic provides considerable stability to any compound where it is found; from molecules as simple as benzene, comprising only six carbon atoms and six hydrogen atoms, up to very complicated molecular systems found in some proteins, nucleic acids or polyaromatic hydrocarbons (PAH´s). These last examples are very important in this discussion because they comprise large organic systems that arise from the fusion of several aromatic rings and their analysis can lead to very similar results to those corresponding to fullerenes.

There are many chemical substances that manifest the property of aromaticity; however PAH´s, graphene, as well as the principal substance dealt with in this work, i.e. fullerenes consist of aromatic molecules and here we outline the importance of this topic.

Figure 5. Aromaticity.

It has been suggested that all hexagonal faces in a fullerene cage are aromatic, whereas all pentagonal faces tend to be slightly anti-aromatic; likewise, it has also been suggested that these molecules can have a sort of "Spherical aromaticity" [12]. However, the important point is that hexagonal benzene fragments must strongly contribute to stabilizing the cage. From this a question arises; is it possible to prepare a material that contains only these stable hexagonal faces?

The answer is yes, but a curious phenomenon arises from this combination; the joined hexagonal faces present curvature at the surface, having the tendency to form cylindrical species, new structures which are known as "nanotubes" [13] and the chemistry of these species is closely related to that of fullerene.

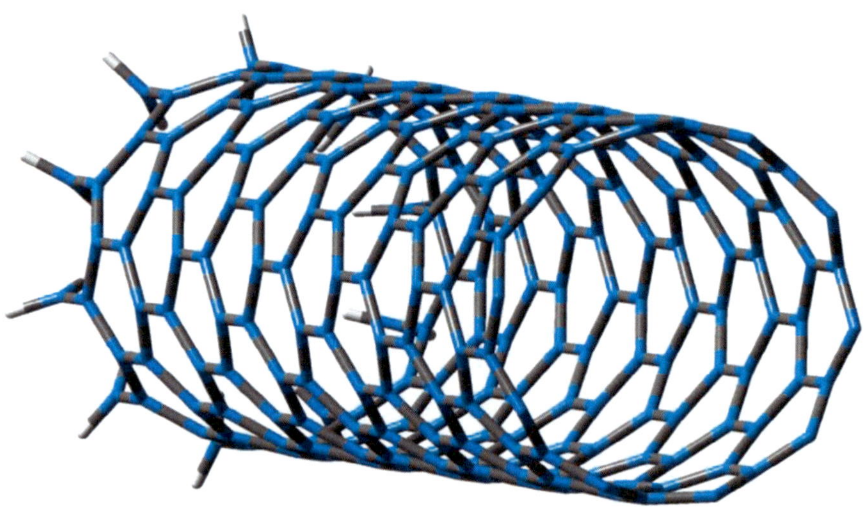

Figure 6. Nanotube.

Based on Mülliken´s theory [14], all these molecules have specific regions where electrons are located; these regions are named molecular orbitals. Remembering that electrons are the building blocks for the formation of bonds between atoms, the most important molecular orbitals are those in which the electrons culminate their occupation i.e. the last orbital with electrons is named the HOMO (Highest Occupied Molecular Orbital), whereas the first unoccupied molecular orbital is the LUMO (Lowest Unoccupied Molecular Orbital), a set known as the frontier molecular orbitals. As a result of these definitions, Fukui´s theory arises [15], which establishes that any chemical reaction should take into account one of these two regions; one for reagents and the other one for its counterpart in the reaction.

Similarly, there is a very important branch of Mathematics dedicated to the interpretation of chemical reactions, comprising the group theory that can classify molecules on the basis of their symmetry. Every specific symmetry (and this can include molecules) has a character table, the factors included in these tables correspond to the sets of molecular orbitals in each case and these are known as irreducible representations. Correspondingly, the combination of Fukui´s theory and the use of character tables provide a straight forward interpretation of chemical reactivity. Fullerenes are very symmetric molecules and all these tools are very useful for understanding the chemical nature of these molecules

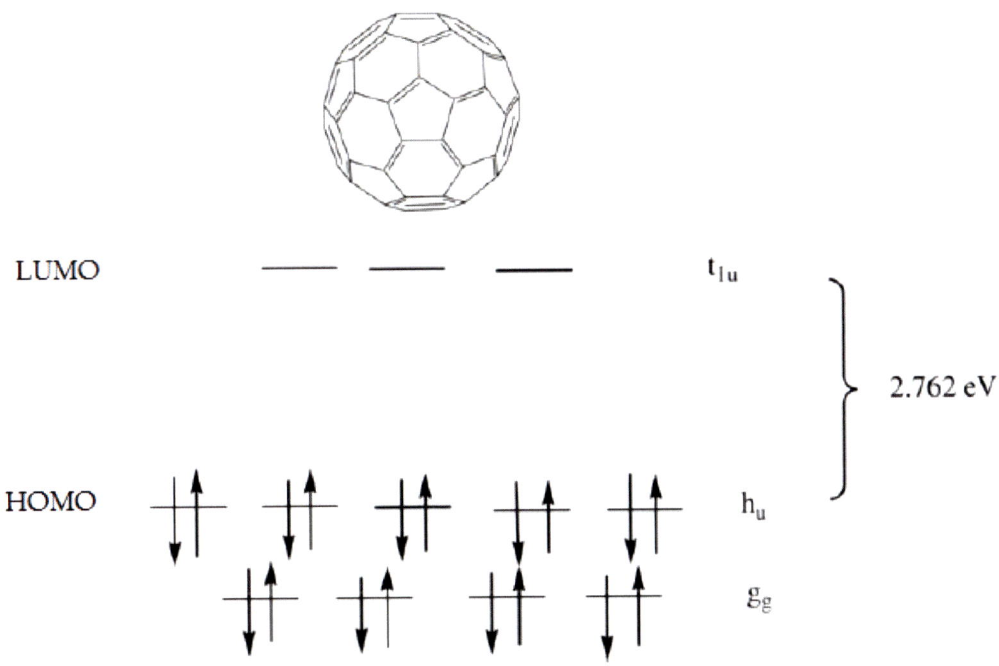

Figure 7. Molecular orbitals interaction diagram.

The first important characteristic of fullerene C_{60} is it´s large symmetry; these molecules belong to the I_h point group, which is the largest group found in the character tables, due to this fact molecules can have very large degenerated sets in their molecular orbital order, in particular the HOMO is a h_u irreducible representation and the consequence is that five molecular orbitals with the same energy make up this set (Also the HOMO-1 set is very near and is a fourfold degenerated set belonging to the irreducible representation g_g) (see the diagram in Figure 7). Subsequently, the large accommodation of electrons in this orbital (and the next) causes an attraction which places the next orbital i.e. the LUMO at a very near location with respect to the HOMO [16].

The consequences of this electronic arrangement are as follows; firstly, the fact that the energy gap between HOMO and LUMO is short ensures that Fullerene acts as a good electronic semi-conductor species [17]. Secondly, this intrinsic electronic characteristic makes these molecules useful for participating in the electronic transport mixtures that lead to the design of photovoltaic cells and related devices [18]. Thirdly, the LUMO which is a triple degenerated set is very available for being occupied by electrons, therefore, the molecule manifests the behavior of a strong oxidant agent and this feature can make an important contribution to synthesis and catalytic procedures.

Besides the features outlined above there are a large quantity of organic, inorganic, organometallic and bio derivatives with their own particular properties and different applications. Particularly, it is important to note the group of the endohedral derivatives, in which the Fullerene species act as the host for little fragments that can go from a single atom (which can be hydrogen, some alkaline metal or more commonly, lanthanides) to complicated systems such as nitride, sulfide or carbide derivatives (Figure 8) [19-22], these types of compounds have been used in different applications such as contrast mediums in medical tests or some types of photovoltaic devices [23, 24].

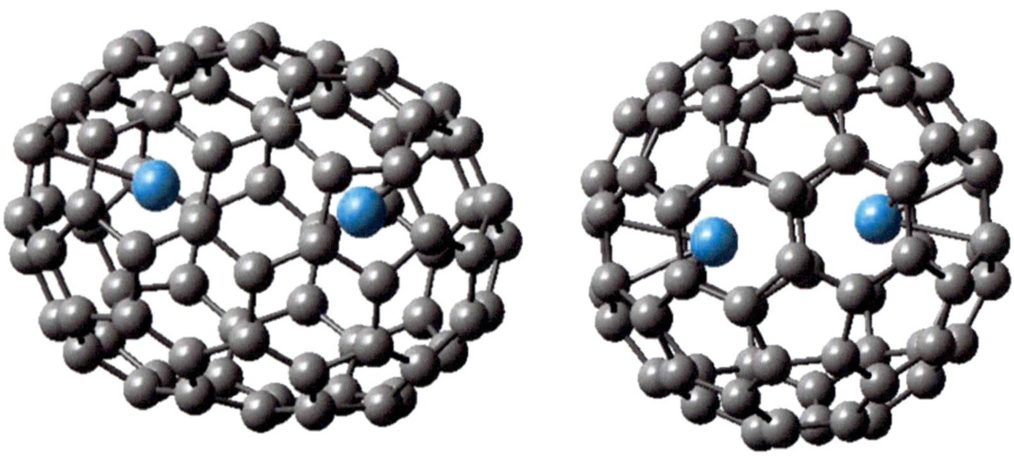

Figure 8. Examples of endohedral derivatives $C_{80}@La_2$, left in this figure D_{5d}, right I_h isomers.

These molecules are the object of several research projects and studies worldwide, with a large variety of applications that can be addressed in very different fields such as pharmaceutics, electronics, building, catalysis, develop of support surfaces or even fuel storage, the particular interest of some group can open a certain field of study, therefore it is expected that the manufacturing of these substances will give place to a particular industry in the near future.

Fullerenes represent a great novelty in our world, however, it seems that they have been around for a long time in the vicinity of the stars; indeed the environment that needs to be developed for the synthesis of these derivatives can only be created in outer space, specifically near the stars, therefore if somebody asks about a possible natural source of fullerene and nano substances the only credible answer to give to this person is the stars, and in the following we describe all aspects of this idea.

3. Radiation Emitted by the Stars. DIB's

3.1. Stars; Emitters of Light

A star manifests a considerable gradient of temperatures from its center to its outer aura. The highest temperature can be $1x10^5$ K, whereas the coldest can be $5x10^3$ K. Within this large range of temperatures, many different interactions between elemental particles can take place, therefore, the entire electromagnetic spectrum must be represented in the emission of light in such a way that a gamma ray can be produced from the same source as a microwave.

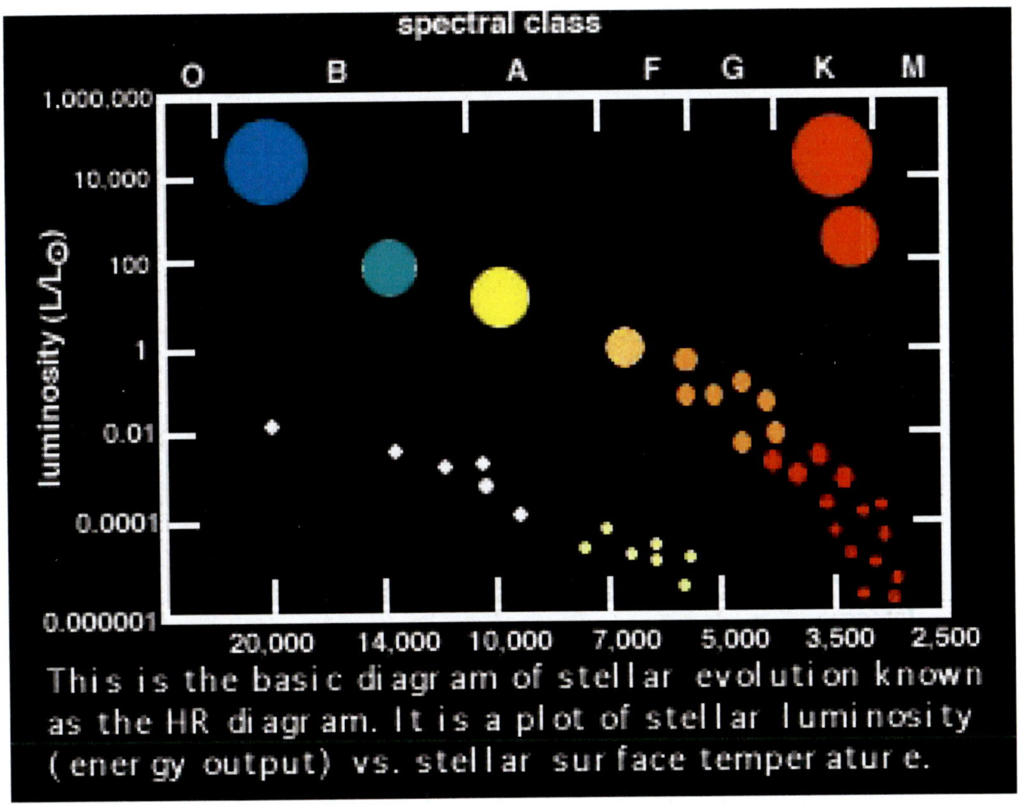

Figure 9. Plot of luminosity vs. temperature in which the evolution of the color is appreciated.

Energy interactions within the bulk of the star have such an elevated value that it is impossible to identify a particular kind of matter as the source of these radiations, indeed only elemental particles or light atoms are expected to reside in these regions of the star, however, the star expels matter to space and this matter finds zones that are cold enough to produce different kinds of molecules. These molecules interact with the star and can emit light at very different frequencies.

If we imagine a person who looks at the sun, he can only see visible light, because the light of high frequency is deflected by shields (the atmosphere, solar wind, van Allen belts, etc.), moreover, this emitted light can be detected with special spectroscopic devices that provide results, indicating to scientists the kind of matter from which this light emanates.

High frequency emissions are classified into several types of radiation, depending on their intrinsic wave length. The largest frequency corresponds to gamma rays (υ=1x10^{19} – 1x10^{20} Hz) and UV rays (υ=1x10^{15}-1x10^{18} Hz). Moderated frequencies correspond to the visible region (υ=1x10^{14} Hz), the IR (υ=1x10^{12}- 1x10^{14} Hz), microwaves (υ=1x10^{8}-1x10^{12} Hz) and radio waves (υ=1x10^{8}-1x10^{2} Hz).

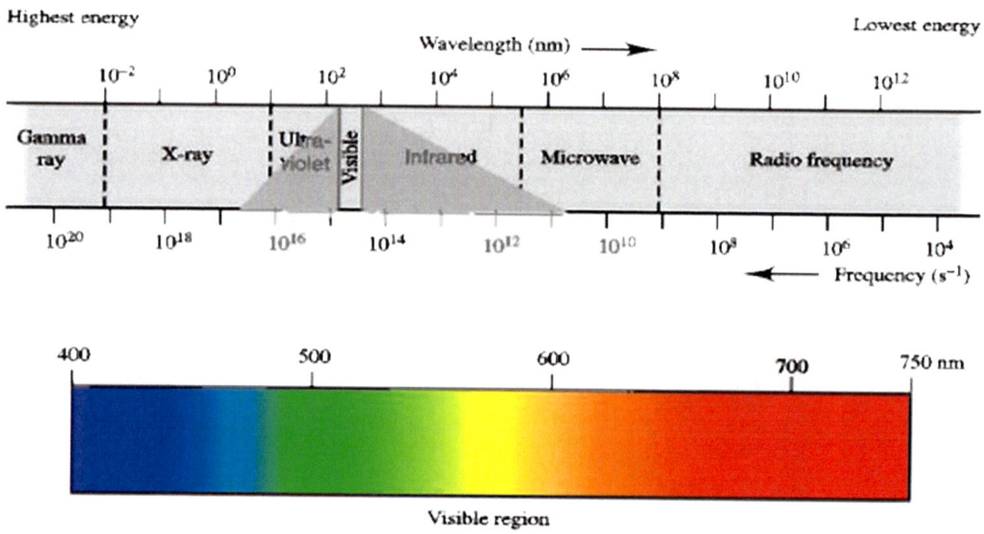

Figure 10. Electromagnetic Spectra.

Spectroscopic techniques exist for analyzing all these types of radiation. There are scientists who are able to combine a telescope with a spectroscopic analyzer and then provide testimony concerning the nature of matter in outer space. This attribute can be applied to the matter of the planets, comets, or stars and differences are marked accordingly.

3.2. Physical and Chemical Reactions in the Stars [25]

Elemental particles are involved in very complicated interactions in the main part of a star; these high temperatures are expected to give rise to moderately free protons, neutrons and electrons (as well as many other particles). Hydrogen is presented only as a free proton, which can react with another hydrogen atom in a fusion process that produces a deuterium atom with a proton and a neutron. Meanwhile, two of these species of deuterium that are obtained succeed to a new fusion process that yields a helium nucleus (the name of this element comes from the name of the Sun God in Greek mythology because this element was first detected in the sun, before it was discovered on Earth).

Due to high temperatures, helium atoms experiment a transmutation process, yielding heavier elements that are expelled to the vicinity of the star, these heavier elements are Oxygen, Nitrogen, Carbon, Silicon, Sulfur and Phosphorus. In the final era of its life, reactions in the stars result in the transmutation to Iron and the production of metallic oxides.

The relative abundance of these elements can have significant influence on the color of a star.

There are several factors that contribute to the change in the color of a star, firstly temperature, a newborn star can develop around $1x10^6$ K at its center, whereas a star near to its death only develops $3x10^3$K, therefore the transmutations and the light emissions of elemental particles are very different in this wide range of thermodynamic conditions, secondly the age of the star determines the type of atoms that are produced in its environment and likewise, the energy promotions of the particles contribute to the change in the color of the star. Correspondingly, it is to be expected that blue stars will only generate elemental particles; hydrogen and helium, whereas deep red stars can even produce metal atoms.

Figure 11. Color and temperatures of the stars.

Traveling from the center to the crown of the star, the temperature decreases until reaching certain diffuse zones known as clouds of dust; in these regions chemical reactions are thought to be able to occur, with elements expelled from the bulk being combined to form simple molecules, moreover, these molecules must have strong resistance to new reactions if they are to have a relatively long life. These molecules can be detected by spectroscopic techniques and are the subject of our discussion

3.3. Specific Techniques of Analysis

The interaction of light with matter is totally dependent on the frequency, as well as on the energy of a specific wave (complying with Planck´s law) and it does not matter which type of material is present in a particular case. This interaction can be destructive when the energy and the frequency are very high, or act solely as a perturbation when wave lengths are longer. A very useful technique is enacted when the frequency has the ability of making perturbations on the valence electrons of substances; these perturbations correspond to a very large region that comprises the visible and near UV zones.

The age of stars can be measured on the basis of their colors, as described previously; the youngest stars are violet or blue and the color alters with time, going from yellow, to orange and finally in the oldest stars, to red. It is possible to relate these different colors to the nature of the emissions of photons, as well as the kind of matter that gives place to specific lines in the spectra. Thus color was the attractive feature that made astronomers focus their attention on these topics and this was the reason for spectroscopic studies concerning the stars. However, high temperatures at the bulk of the star imply that matter cannot form chemical bonds, whereas far from the center of stars, the environment may be suitable for the formation of simple molecules, which can also be analyzed using the same spectroscopic methods.

3.4. DIB´s

Diffuse Interstellar Bands are derived from the interaction between matter and light in the vicinity of the star; they are absorption features in the near UV down to the visible and then there are also UIRs (unidentified IR emissions). Both these phenomena are the subject of various research projects [1].

In 1985, reports claimed [26, 27] that these emissions are produced by polycyclic aromatic hydrocarbons and not by graphite or chains of carbon. The spectroscopic features [1] are the infrared emission bands (UIRs) between 3 and 15 nm, the diffuse interstellar absorption bands (DIBs) observed in the optical range between 400 and 1000 nm, with a strong absorption feature at 220 nm.

The IR spectra of Fullerenes have shown important signals around 1100-1300cm^{-1} [28]; in this region it is also very common for polyaromatic hydrocarbon signals to occur. This is not a fortuitous feature because this signal can be associated with a vibration coming from the electronic resonance found in aromatic systems and both kinds of molecules coincide in terms of this characteristic, therefore it is impossible to define the real source of these signals, only by applying an IR spectroscopic analysis.

The visible and UV regions manifest very similar characteristics, because at this point there are specific electronic excitations for each particular chemical species. However, there are several different organic compounds in the clouds of dust and therefore is very difficult to separate the specific bands that are overlapped, so that results are not conclusive for a specific substance with respect to others.

Polyaromatic Hydrocarbons can easily be formed in environments of graphite and hydrogen at high temperatures and with important light emissions, a situation which can be compared with those that exist in a deep oil well, where high pressure and elevated temperatures lead to the formation of asphalthenes [29]. The reason for this organic "selectivity" is that PAHs are very stable species and are very resistant to reactions or to becoming solvent, preventing them from being destroyed in outer space. These characteristics are fundamental to the assumption that stardust is mainly formed from these types of substances, although for many years prior to the discovery of Fullerenes, the predominant idea among astronomers was that this family of compounds was totally responsible for DIB´s.

New discoveries in astronomical spectroscopic methods have made it possible to identify carbon structures which do not contain hydrogen atoms in a Planetary Nebula [30], furthermore the IR presents two specific bands that can only be assigned to Fullerenes and therefore this is a strong indicator that these species are formed in the same region as PAHs.

A theoretical study has demonstrated that Fullerene molecules are thermodynamically superior to several Polyaromatic Hydrocarbons, with values up to ten times greater in terms of energy; this study indicates that these thermodynamic conditions can promote reactions which result in cages rather than planar systems [31]. They also enable Fullerenes as well as

PAHs to be formed. Spectroscopic analysis made in laboratory with pure substances can be compared to those derived from astronomical observation, bearing in mind that the spectra of clouds of dust near the stars is not the result of pure substances. Even so, the comparison had demonstrated that it is an undisputed fact that both these types of molecules can be produced in the vicinity of the stars.

A simple comparison of IR and electronic spectra of known molecules, some PAHs, C_{60} and C_{70} with two known examples of DIB´s is presented in figures 12 and 13. Evidently there is an appreciable difference when comparing pure substance spectra and the resulting mixture of substances present in dust clouds. However, many of the important peaks that correspond to pure substances are also produced by this mixture.

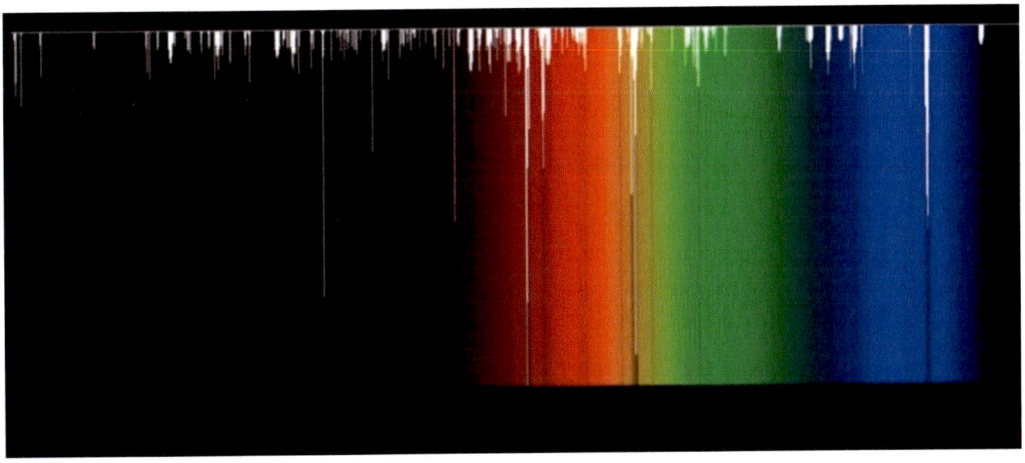

Figure 12. DIB´s.

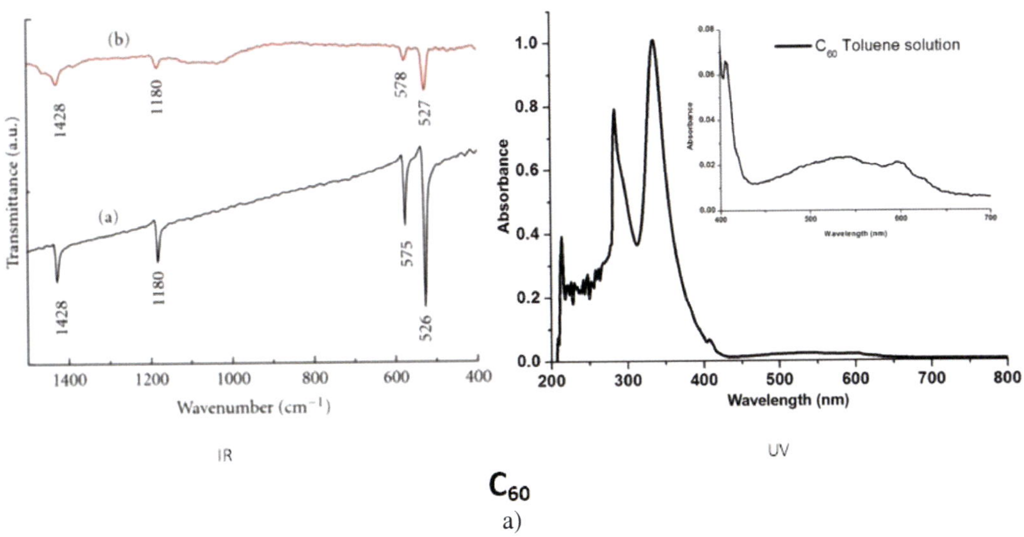

Figure 13. (Continued).

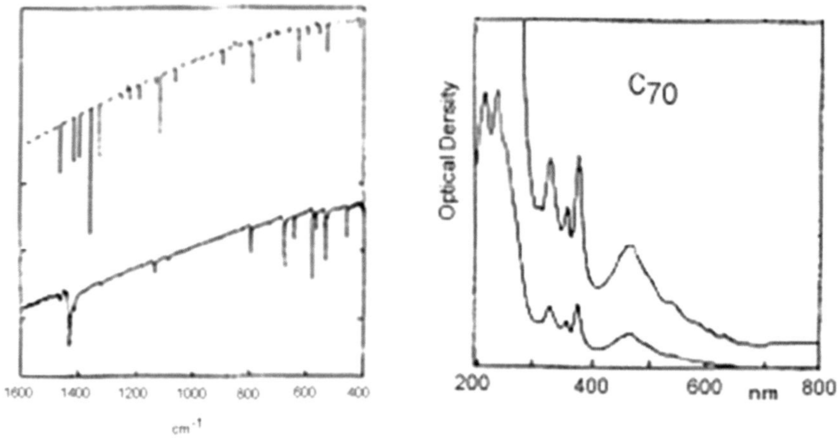

IR UV

C_{70}

b)

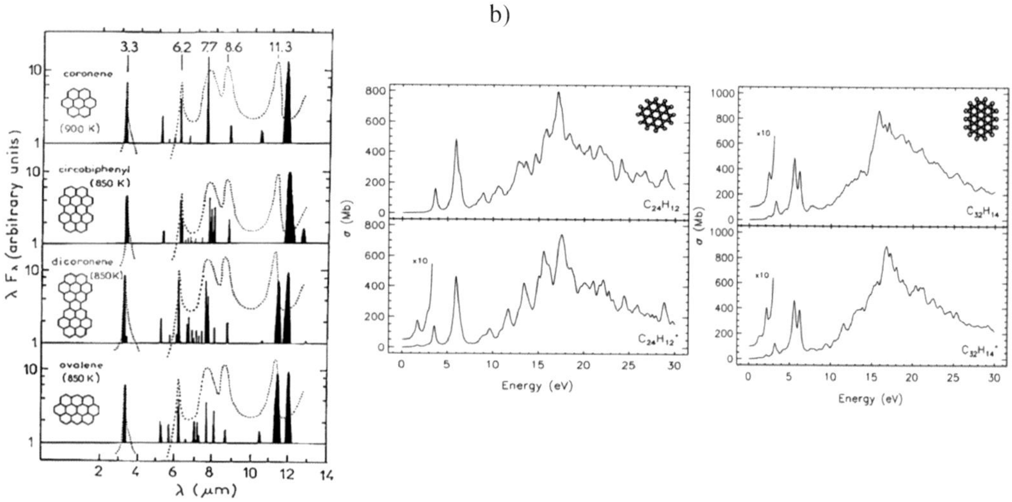

IR UV

PAH´s

c)

Figure 13. IR and UV spectra of a). C_{60} b). C_{70} c). PAH´s.

4. The Birth of Fullerene

4.1. The Interstellar Cycle of Carbon [32]

Inside some types of stars, several elementary reactions take place to produce different types of elements, depending on the intrinsic thermodynamic conditions of a particular star.

In the first second after the Big Bang, the universe cooled to 10^{10}K and when the temperature decreased to 10^{9}K, nucleosynthesis made possible the formation of helium. This general procedure (that occurs in all stars) constitutes the fusion of hydrogen into helium, which is an exothermic reaction.

$$4^1H \rightarrow^4 He + 2e^+ + 2\upsilon_e$$

Where e^+ represents a positron and υ_e a neutrino

This is the most important process that takes place in any star because it provides helium atoms, which are the source of energy that they need for survival and also results in their brilliance.

In some cases, helium can be fused in order to yield a beryllium atom which during a second stage, declines to form a new carbon atom.

$$2^4He + (99 \pm 6keV) \rightarrow^8 Be$$

$$^8Be +^4He \rightarrow^{12} C + \gamma$$

This is the main source of carbon in the interstellar medium.

The formation of certain types of elements depends on two main factors, the mass and the temperature, therefore relatively small stars evolve to give a dense white dwarf, with a temperature of 20,000K and which produces only hydrogen and helium. Contrastingly, massive stars (one magnitude order the size of our sun) can expand to form red giants, in which helium can form carbon or other elements; for these large stars further reactions can occur producing several elements, until the heaviest and most stable element is formed: iron. This is the beginning of the agony of the star, as in a short time, a catastrophic collapse takes place in the form of a supernova; a disastrous event which leads to the formation of heavier elements. In the case of some little stars, supernova do not occur; in this case only medium-sized elements are formed and these can be expelled to the interstellar medium and be detected from the Earth, using spectroscopic techniques.

A certain type of star is labeled "carbon rich" and these can expel large quantities of carbon grains or in other instances, the devastated region, where a supernova has occurred, transforms to become a new stellar body named planetary nebula, a type of stellar body which is formed from shells of ionized gases that are ejected from the old red giant stars, and which are very important in stellar chemistry because the processes of fusion for helium inside these can produce heavier elements, making them objects rich in carbon. These contributions are sufficiently important to increase the quantity of carbon by 50 percent for this element when present in a galaxy.

The polyynes i.e. are the most common carbon chain radicals found in the interstellar medium, the description of these strange molecules is very difficult from a terrestrial point of view because the normal rules for bonding are no longer complied with, saturation by hydrogen and other atoms does not occur and it is possible to find chains containing from two up to ten concatenated carbon atoms (in some cases one or two atoms from different elements can be found joined to these chains).

Figure 14. a). Polyyne b). Polyene.

These chains can collide among themselves giving place to more complicated molecules, such as polyaromatic hydrocarbons (PAH) or fullerenes.

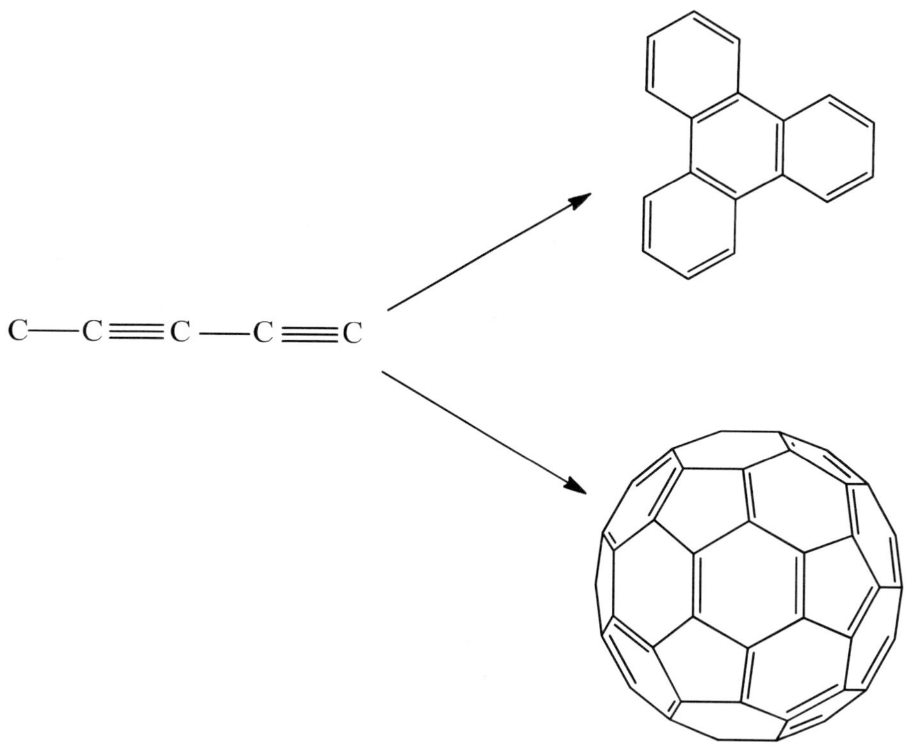

Figure 15. Formation reaction of PAH´s and Fullerenes from Polyynes.

Stability is the most important factor to consider in this context; in the case of fullerenes; these can be very stable species in spite of the thermodynamic conditions of the interstellar medium, but the thermodynamic conditions on Earth are not appropriate enough to provide a natural origin for these types of substances.

Various types of evidence exist to suggest the existence of fullerene in outer space, firstly studies on the R Coronae Borealis stars [33] have indicated that it is possible to find both, PAH and fullerene in any medium rich in carbon dust and furthermore C_{60} [34] and C_{70} [35] have been indisputably identified in two different planetary nebulae: Tc1 and Magellanic cloud [36]; the red giant named Gema in this constellation is the very best known carbon producer, while R Coronae Borealis produces 50% of the carbon found in the Milky Way.

Likewise, the possibility of finding graphene in the same stellar body also exists, confirming that fullerenes and its derivatives can indeed be found in outer space.

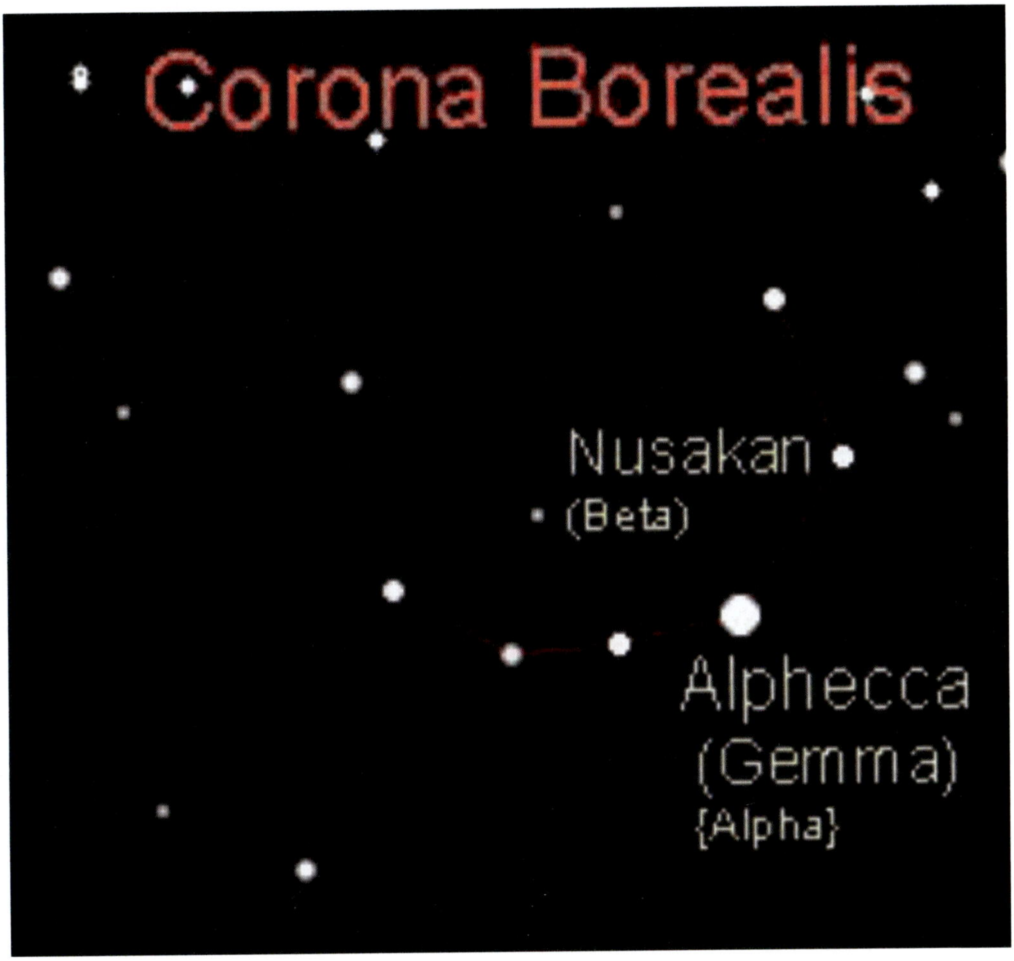

Figure 16. R Coronae Borealis.

4.2. From Carbon Chains to Carbon Cages

4.2.1. The Terrestrial Production of Fullerenes

Harry Kroto visited the University of Rice by invitation of Smalley and his colleagues because ahead of its time, this institution developed a device that could simulate the conditions in outer space, for the purpose of studying novel organic chains. This device was a pulsed supersonic nozzle–laser vaporization apparatus. The original idea was to discover different polyynes and study their particular chemistry [3], however, the unexpected result was C_{60}.

This production process was prohibitive, in terms of producing a large amount of product, however in a relatively short time an appropriate and very simple process was developed by Krätschmer, this required placing two graphite electrodes in a helium atmosphere and passing a voltaic arc through the electrodes, resulting in vaporization and the formation of fullerenes in this medium; therefore, the only precursor of these substances is carbon itself, whereas fullerene and other carbon nanostructures are considered to be allotropic forms of carbon, such as diamond or graphite [4].

4.2.2. Interstellar Production of Fullerenes

The preparation of fullerenes in this medium seems to differ from the Krätschmer synthesis because in spite of the lack of definitive proof, it is believed that carbon can exist in the form of graphite in outer space, however, it is very important to specify the location where it can be found.

As mentioned previously, red giants constitute a good carbon source in outer space, indeed it has been demonstrated that R Coronae Borealis stars represent even better carbon producers than any other star in our galaxy.

It is not clear which process is involved in forming allotropic forms of carbon; firstly, it seems that depending on the environment, graphite can provide a good source of polyynes and possibly polyenes. These families of excited molecules should be the precursors of large stable molecules, as indicated in the following.

Polyynes are considered to be relatively stable molecules in the aggressive environment of a planet nebulae or the far aura of a star and they constitute linear molecules, rich in triple bonds that produce polyaromatic hydrocarbons by interaction with hydrogen. Here the importance of aromaticity is made clear, because large polyaromatic hydrocarbons are even stable enough to resist explosions (one excellent example is the asphaltene species found in oil wells), however large polyynes or polyenes can develop a curved shape. These molecules can be placed in an environment where radiation pressure is very high, the combination of the curved shape molecules, the relatively low concentration of hydrogen [35] (or possibly its total absence) with this high pressure enables the species to join, forming rings or spheres (such as fullerenes); considering the high stability of fullerenes, this should be the final product. A thermochemical result has shown that a fullerene C_{60} is energetically more stable than a large polyaromatic hydrocarbon at 1000 K [31]; thus it is possible to devise competitive reactions for producing fullerenes and polyaromatic hydrocarbons. However, it seems that the conditions in space promote fullerene rather than PAH formation.

The scheme as described suggests the coexistence of both fullerenes and PAHs in the same regions, resulting from similar procedures. At this point, a reader may question how it is possible for both types of substances to be present, as PAHs require the presence of hydrogen, whereas fullerenes do not. In the following, a possible explanation is suggested.

The chemical environment of a planet nebulae and cool stars are rich in carbon grains and manifest a small amount of hydrogen, all of which is consumed in the formation of PAHs; however it is probable that in this situation large quantities of carbon remain, which can form polynne and polyenne chains that are subject to radiation pressure coming from nearby stars, causing a reaction which forms fullerene cages.

4.3. Other Interstellar Derivatives of Fullerene

C_{60} and C_{70} constitute common products from the Krätschmer synthesis, indeed they are commercial products and many laboratories and industries have access to them, however large fullerenes also exist and are notorious because they can play the host to lanthanide metal atoms or even little clusters, which are endohedral complexes. The interesting fact is that these complexes may be detectable in outer space [32] because there are well known isotope anomalies that can be detected in astronomical spectroscopic studies that only can be explained assuming some kind of doping of the fullerene molecules.

A wide variety of large fullerenes, exohedral or endohedral complexes have been prepared in several research laboratories [37, 38]. The goal of a new molecule always makes a good news item and certainly merits international publication, however it seems that the special environments previously mentioned provide the conditions for producing a carbon nanostructure, without the corresponding research work, thus it will be possible in the short term future to find a reservoir of these species in outer space that can be studied with the aim of obtaining specific compounds for different applications.

This work is focused on the presence of fullerenes in the stellar environment, but there are many other carbon nanostructures such as nanotubes or graphene that can be prepared under similar conditions; so it is likely that these types of compounds will be found to reside in the same regions as fullerene.

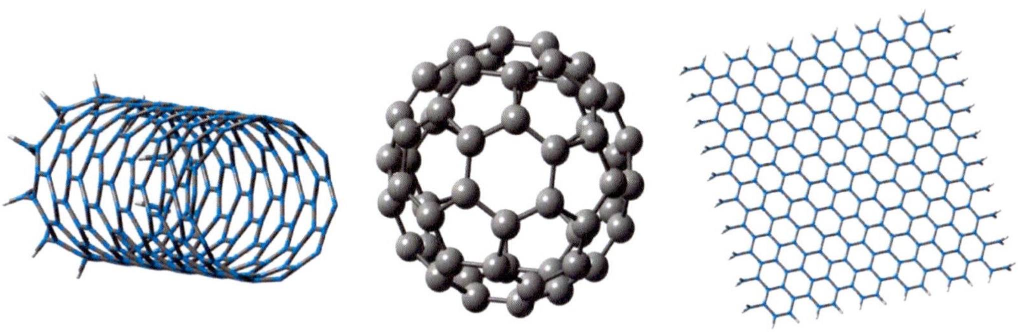

Figure 17. Carbon nanostructures, from left to right: nanotube, fullerene and graphene.

Conclusion

Fullerenes were novel substances when they were discovered in the XX century, whose esthetic shapes and large symmetry gave place to a large quantity of studies solely aimed at explaining their structure. However, this species has been demonstrated to have unique physical and chemical characteristics making them important for industrial, pharmaceutical and energy production applications. They are usually prepared in chemical laboratories following known procedures and exploratory research is taking place in order to find a natural source of fullerene on our planet. Until now all these projects have failed, but there is a very large natural source outside our planet, as the vicinity of cold stars and planetary nebulae have been demonstrated to offer good sources for any kind of carbon nanostructure, so that these places can be considered as large reserves for accessing fullerene and its related compounds. The physical and chemical conditions of regions able to produce these types of substances have been analyzed in this work.

Fullerenes are in the sky; all we have to do is go and get them.

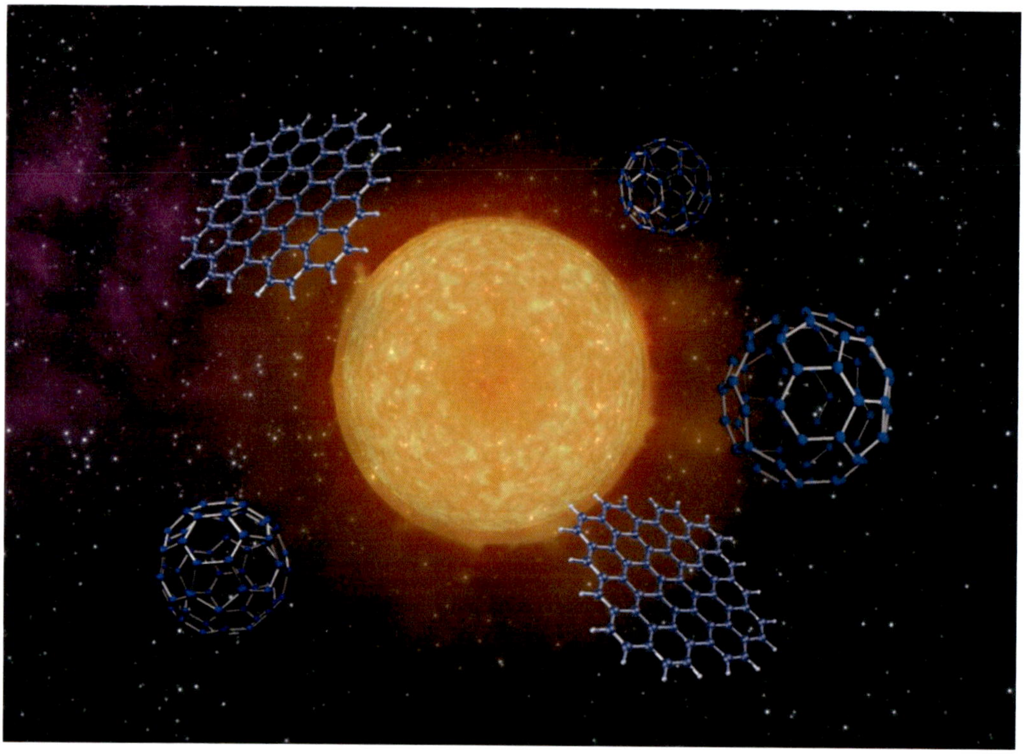

ACKNOWLEDGMENT

We are indebted with Alberto Álvarez and Marco Antonio Salcedo for helping with the art. Also we want to thank Caroline Karslake, Teresa Vázquez and Oralia Jiménez for technical help.

REFERENCES

[1] Ehrenfreund, P. & Sephton, M. A. (2006). Carbon molecules in space: from astrochemistry to astrobiology. *Faraday Discussions,* 133, 277-288.

[2] Ehrenfreund, P. & Foing, B. H. (1995). Search for fullerenes and PAHs in the diffuse interstellar medium. *Planet Space Sci.*, 43, 1183-1187.

[3] Kroto, H. (1997). Symmetry, space, stars and C_{60}. *Reviews of Modern Physics*, 69, 703-722.

[4] Krätschmer, W., Lamb, L.D., Fostiropoulos, K. & Huffman, D. R. (1990), Solid C_{60}: a new form of carbon. *Nature,* 347, 354-357.

[5] Ettl, R., Chao, I., Diederich, F. & Whetten, R. L. (1991), Isolation of C_{76} a chiral (D_2) allotrope of carbon, *Nature*, 353, 149-153.

[6] Diederich, F., Whetten, R. L., Thilgen, C., Ettl, R., Chao, I. & Alvarez, M.M. (1991), Fullerene. Isomerism, isolation of C_{2v}-C_{78} and D_3-C_{78} , *Science*, 254, 1768-1770.

[7] Kikuchi, K., Nakahara, N., Wakabayashi, T., Suzuki, S., Shiromaro H., Miyake, Y., Saito, K., Ikemoto, I., Kainosho, M. & Achiba, Y. (1992) , NMR characterization of isomers of C_{78}, C_{82} and C_{84} fullerenes, *Nature,* 357, 142-145.

[8] Manolopoulos, D. E., Fowler, P.W., Taylor, R., Kroto, H. W. & Walton, D. R. M. (1992), An end to the search of the ground state of C_{84}, *J. Chem. Soc. Faraday Trans.*, 88, 3117-3118.

[9] Bing-she, XU (2008), Prospects and research progress in nano onion-like fullerenes. , *New Carbon Materials.* , 23, 289-301.

[10] Kroto, H. W. (1987), The stability of the fullerenes C_n with n=24, 28, 32, 36, 50, 60, 70, *Nature*, 329, 529-530.

[11] Schmalz, T.G., Seitz, W. A., Klein, D. J. & Hite, G. E.(1988), Elemental carbon cages, *J. Am. Chem. Soc.*, 110, 1113-1127.

[12] Chen, Z. & and King, R. B. (2005). Spherical Aromaticity: Recent Work on Fullerenes, Polyhedral Boranes, and Related Structures. *Chem. Rev.*, 105, 3613-3642.

[13] Rondeau-Gagne, S. & Morin, J. F. (2014), Preparation of carbon nanomaterials from molecular precursors., *Chem. Soc. Rev.*, 43, 85-98.

[14] Mülliken, R.S. (1978), Chemical Bonding., *Ann. Rev. Phys. Chem.*,29, 1-30.

[15] Fukui, K. (1952), A Molecular Orbital theory of reactivity in aromatic Hydrocarbons, *J. Chem. Phys.*, 20, 722-725.

[16] Cioslowski, J. (1995), *Electronic Structure Calculations on Fullerene and their Derivatives*, Oxford University Press, Oxford.

[17] Salcedo, R. (2009), Fullerenocene, *Polyhedron,* 28, 431-436.

[18] Yu, G., Gao, J., Hummelen J. C., Wudil, F., & Heeger, A.J. (1995), Polymer Photovoltaic Cells: Enhanced Efficiencies via a Network of Internal Donor-Acceptor Heterojunctions, *Science*, 270, 1789-1791.

[19] Chaur, M. N., Melin, F., Ortiz, A. L. & Echegoyen, L. (2009), Chemical, Electrochemical, and Structural Properties of Endohedral Metallofullerenes, *Angew. Chem. Int. Ed.,* 48, 7514-7538.

[20] Beavers C. N., Chaur, M. N., Olmstead, M.M, Echegoyen, L. & Balch, A. L. (2009), Large Metal Ions in a Relatively Small Fullerene Cage: The Structure of $Gd_3N@C_2(22010)$-C_{78} Departs from the Isolated Pentagon Rule, *J. Am. Chem. Soc.*,131, 11519-11524.

[21] Mercado, B. Q., Chen, N., Rodríguez-Fortea, A., Mackey, M. A., Stevenson, S., Echegoyen, L., Poblet, J. M., Olmstead, M.M. & Balch, A.L. (2011), The Shape of the $Sc_2(\mu_2$-S) Unit Trapped in C_{82}: Crystallographic, Computational, and Electrochemical Studies of the Isomers, $Sc_2(\mu_2$-S)@Cs(6)-C_{82} and $Sc_2(\mu_2$-S)@C_{3v}(8)-C_{82} *J. Am. Chem. Soc.*, 133, 6752-6760.

[22] Wang, C. R., Kai, T., Tomiyama, T., Yoshida, T., Kobayashi, Y., Nishibori, E., Takata, M., Sakata, M. & Shinohara, H. (2001), A Scandium Carbide Endohedral Metallofullerene: $(Sc_2C_2)@C_{84}$, *Angew. Chem. Int. Ed.*, 40, 397-399.

[23] Wudl, F. (2002), Fullerene materials., *J. Mater. Chem.*,12, 1959–1963.

[24] Yang, S., Liu, F., Chen, C., Jiao, M. & Wei T. (2011), Fullerenes encaging metal clusters—clusterfullerenes., *Chem. Commun.*, 47, 11822–11839.

[25] Hacyan, S. Relatividad para principiantes. (2002). Colección La Ciencia para todos., 78, 3rd ed. México. *Fondo de cultura económica.*

[26] Puget JL, Léger A, Boulanger F (1985). Contribution of large polycyclic aromatic molecules to the infrared emission of the interstellar medium. *A&A* 142:L19

[27] Léger A, d'Hendecourt L (1985) Are polycyclic aromatic hydrocarbons the carriers of the diffuse interstellar bands in the visible?. *A&A* 146:81

[28] Cataldo, F; Iglesias-Grothc, S.; Manchado, A. (2010), Low and High Temperature Infrared Spectroscopy of C_{60} and C_{70} Fullerenes., *Fullerenes, Nanotubes and Carbon Nanostructures.*, 18, 224-235.

[29] Tissot, B.P. & Welte, D.H., (1998), *Petroleum Formation and Occurrence,* Springer-Verlag, New York.

[30] Cami, J; Bernard-Salas, J; Peeters, E.; Malek, S. E. (2010), Detection of C_{60} and C_{70} in a Young Planetary Nebula., *Science*, 329 , 1180-1182.

[31] Salcedo, R. & Fomina, L. (2011) ; Aromaticity under star dust conditions., *Struc. Chem.*, 22, 971-975.

[32] Hare, J.P.; Kroto, H.W. (1992); A postbuckminsterfullerene view of carbon in the galaxy, *Acc. Chem. Res.*, 25, 106-112.

[33] Goeres, A.& Sedlmayr E.(1992), The envelopes of R Coronae Borealis stars I. A physical model of the decline events due to dust formation., *Astron. Astrophys*, 265, 216-236.

[34] Cami J., Bernard-Salas J., Peeters E. &Malek S. E. (2010), Detection of C_{60} and C_{70} in a Young Planetary Nebula, *Science*, 329, 1180-1182.

[35] García-Hernández, D.A., Kameswara-Rao, N. & Lambert, D.L., (2013), Dust around R Coronae Borealis stars. II. Infrared emission features in an H-poor environment., *Astroph. J.*, 773:107, 107-123.

[36] García-Hernández D. A., Iglesias-Groth S., Acosta-Pulido J., Manchado A., García-Lario P., St anghellini L., Villaver E., Shaw R. A. & Cataldo F. (2011), The formation of fullerenes: clues from new C_{60}, C_{70}, and (possible) planar C_{24} detections in magellanic cloud planetary nebulae., *Astroph. J. Lett.*, 737:L30.

[37] Rivera-Nazario D.M, Pinzón J.R., Stevenson S. & Echegoyen L. A. (2012), Buckyball maracas:exploring the inside and outside propierties of endohedral fullerenes., *J. Phys. Org. Chem.*, 26, 194-205.

[38] Akasaka T., Wudl F. & Nagase S. (2010), *Chemistry of Nanocarbons*, Wiley-Blackwell, London.

In: Fullerenes
Editor: Shannon B. Ellis
ISBN: 978-1-63321-385-2

Chapter 9

BIOMEDICAL APPLICATION OF FULLERENES

Ljiljana Djekic and Marija Primorac
University of Belgrade – Faculty of Pharmacy, Belgrade, Serbia

ABSTRACT

The unique physical, chemical and biological properties of fullerenes have opened an intensive research on their possible applications in nanomedicine in the last few decades. The highly unsaturated structure of the hydrophobic pristine fullerene (C_{60}) is suitable for: 1) functionalization of molecule with different functional groups and production of water-soluble derivatives, such as polyhydroxyfullerenes (fullerols or fullerenols) and carboxyfullerenes; 2) conjugation with cyclodextrins, poly(2-oxazolines), isostearic acid, polyethyleneglycol or amino acids; 3) encapsulation into carriers with hydrosoluble surface such as liposomes and nanoparticles of polyvinylpyrrolidone (PVP). Such strategies enable dispersion in aquous vehicles, prevent formation of fullerene aggregates and may increase fullerene capacity to interact with biological environment. Fullerenes as well as their derivatives and C_{60} based composites, demonstrated to have antioxidant activity *in vitro* and *in vivo*. Therefore, they are known as "free-radical sponges". Several studies have demonstrated that fullerene induce suppression of human skin cell injuries caused by ultraviolet (UV) light irradiation and/or peroxides, brain cells damage by peroxide, and oxidative stress induced damage of lung. Thus, they are promissing agents for prophylaxis or therapy of free radical-related skin diseases and other serious disorders such as Alzheimer`s disease, Parkinson`s disease, neutrophilic lung inflammation. PVP-entrapped C_{60} submicron particles (Radical Sponge) are already employed as an ingredient of the „anti-aging" cosmetic products on the current market. Fullerenes are potentially useful for acne treatment, due to their properties as free radical scavengers. On the other hand, fullerenes may become reactive oxygen species (ROS) generators, under visible or UV light exposure. The cytotoxicity of fullerenes is perspective for photo- and radio-therapy of cancers and as antimicrobial agents. The role of fullerenes as antioxidants or prooxidants is dependent on applied dose/concentration, molecule/particle size, surface morphology, chemical modification/conjugation/encapsulation, and the presence of light or ionizing radiation. HIV-1 protease inhibition is a new application that has been proposed recently for fullerenes and ~~its~~ their derivatives. The future of fullerene family, including novel nanomaterials, such as PEGylated fullerene/iron oxide nanocomposites and pH-activated glycol chitosan/fullerene nanogels, is focused on development of carriers suitable for targeted drug delivery and/or diagnostic strategies. This chapter highlights both benefits and potential health risks of fullerens as active agents of nanopharmaceuticals and cosmetic products.

INTRODUCTION

Fullerenes are family of molecules composed of carbon atoms arranged in hexagonal and pentagonal rings to form hollow spheres with icosohedral symmetry. C_{60} (also known as *pristine fullerene, Buckminsterfullerene, Buckyball*) is the first fullerene molecule discovered in soot in 1985. by Kroto`s group at Rice University (Houston, USA) [1]. The molecule consists of 60 carbon atoms arranged in 12 pentagons and 20 hexagons with icosohedral symmetry resembling the soccer ball design. Currently, major higher fullerenes, C_{60} and C_{70}, are the most massive produced carbon nanomaterials [2]. Furhermore, various C_{60}-based derivatives have been prepared by covalent and non-covalent derivatisation of fullerenes. The size of the individual molecules of fullerenes and their derivatives is at the nanoscale (1 - 100 nm), and they are classified as nanomaterials, according to Recommendation 2011/696/EU [3]. Like other nanotechnology products, fullerenes poses unique physico-chemical characteristics. During the last three decades, the biological activity of C_{60} and its derivatives has been extensively investigated [4]. In biological environment, some fullerenes showed excellent antioxidant acitivity against reactive oxygen species (ROS), that made them cytoprotective substances suitable for therapeutical and cosmetic purposes. Pristine fullerene-PVP complex and fullerene-containing vegetable squalane, available under trade names Radical Sponge® and LipoFullerene® (Vitamin C60 BioResearch Corp., Japan), respectively, have been approved as the antioxidant ingredients in several marketed cosmetic products [5, 6]. On the other hand, the cytotoxic activity of some fullerene family members was observed, likely caused by induction of oxidative stress processes. Such compounds are potential therapeutics for cancer and deseases caused by microorganisms (e.g., AIDS, influenza, acne) [7 - 10]. Recently, a biomedical application of fullerenes spreads in the area of drug delivery and diagnostic [11 - 14].

PHYSICOCHEMICAL PROPERTIES OF FULLERENES

Fullerenes are black to brown odourless thermostabile solids (for example C_{60} sublimes at 800 K). They are sparingly soluble in common non-polar solvents such as benzene, toluene, chloroform and carbon disulfide, at room temperature [15]. Lipofullerene® is a solution of highly purified and organic solvent-free C_{60}, at nearly saturated concentration of 278 ppm, in squalane prepared from olive oil [6]. Pristine fullerene is highly hydrophobic and insoluble in water [15], that is assumed as a drawback for its biological application. Aquous suspensions of C_{60} can be prepared: 1) by prolonged (2-8 weeks) stirring of fullerene powder with pure water [16, 17] or water phase containing suitable solibilizers (e.g., polysorbate 80 [18], PEG-hydrogenated castor oil [19]); 2) by solvents exchange method, using non-polar solvent which is miscible with water to dissolve fullerene and subsequently to form suspension after water addition, under evaporation or sonication [20]. In such colloid systems fullerene molecules self-assembly into crystals of the nanoscale size. For example, $C_{60/70}$ colloid, reffered as nano-

C_{60} (nC_{60}), was produced by evaporating tetrahydrofuran from a solution after addition of water [21], using the procedure first described by Deguchi et al. (2001) [22] and modified by Fortner et al. (2005) [23] (Figure 1).

The concentration of nC_{60} in water suspension was adjusted to 5 μg/ml. The average size of nC_{60}, determined by transmission electron microscopy and dynamic light scattering, was approximately 100 nm [21]. The concentration of the dispersed phase in such nanosuspensions should be below 100 μg/ml, in order to avoid the formation of aggregates. At the higher concentrations aggregation sholud be suitably prevented or minimised by adition of surfactants and hydrophilic polymers [22, 24].

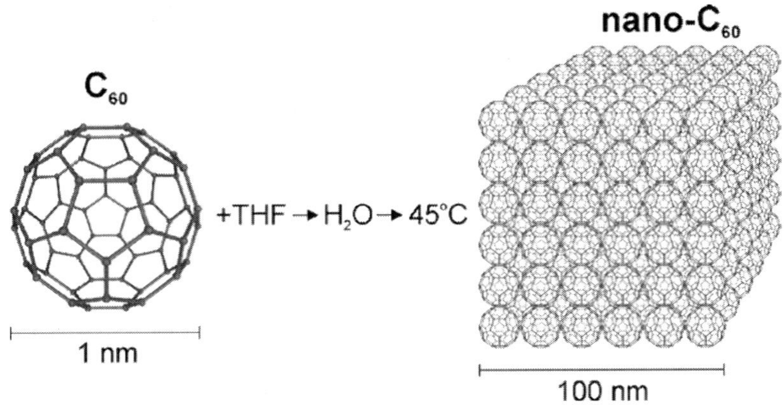

Figure 1. The preparation and structure of nC_{60} [21].

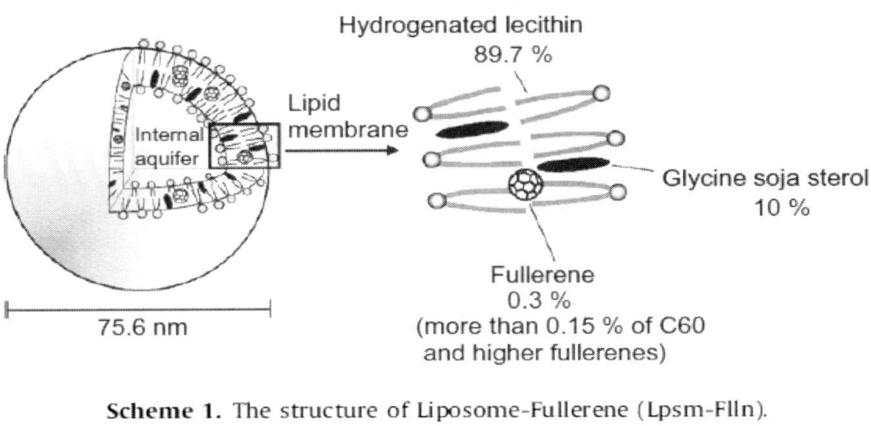

Figure 2. The structure of Liposome-Fullerene [40].

Each carbon atom within the unmodified fullerenes molecules is bonded to three others and is sp^2 hybridised. Thus they may behave like alkenes with electron accepting properties (e.g., C_{60} and C_{70}) or both electron acceptor/donor character (e.g., C_{76}) [25, 26]. They are susceptible for chemical additions at double bonds. Such reactivity make them attractive for various derivatisations. Although the non-polar character of fullerenes is preffered for their penetration/permeation through biological membranes [27], however, sufficient water solubility is essential for interactions with biomolecules [28]. Relevant literature describes approaches to achieve high water solubility for fullerene molecules in order to increase their reactivity within the biological environment by covalent functionalization with different hydrophilic functional groups [29, 30]. For example, fullerols and carboxyfullerenes are derivatives of the C_{60} molecule with multiple hydroxyl or carboxy groups, respectively, and high water solubility [31]. Other hydrosoluble or amphipathic derivatives of fullerenes are obtained by conjugation with cyclodextrins [32], poly(2-oxazoline) [33], isostearic acid [34], polyethyleneglycol (PEGylated fullerene) [35], amino acids [36], dendrimers [37]. Such conjugates in water tend to mutualy connect *via* hydrogen bonds and form micelles with fullerene part coated with hydrophilic fragments embaded in sorrounding water. Complexes with polymers containing electron-donor groups (e.g., PVP, poly(2-alkyl-2-oxazolines) (PO_x), usually have high hydrophilicity and colloidal stability and low toxicity [33, 38]. Such characteristics are preferable for development of pharmaceutical preparations and cosmetic products. Radical Sponge® is produced by entrappement of fullerene purified by sublimation (BioFullerene, Vitamin C60 BioResearch, Japan) in PVP of 60-80 kDa at a molar ratio range of 0.42–0.67:1, resulting in a formation of nanoparticles with a mean diameter of 150–250 nm (as powder) and approximately 688 nm in the aqueous vehicle. The difference in nanoparticles size in dry and hydrated state was attributed to swelling and shrinkage of the polymer. PVP-fullerene submicron particles are hydrophilic and stabile cosmetic ingredient that was classified as safe because their size is to large for permeation thorough the human skin [38].

Fullerenes are also suitable for encapsulation into vesicular nanocarriers or different nanoparticles that isolate them from sorrounding water phase [39, 40]. Moreover, nanoencapsulation may increase fullerenes capacity to interact with the biological environment. Kato et al., (2010) [40] designed liposomes consisting of hydrogenated lecithin and soybean sterol and fullerene in the weight ratio of 89.7:10:0.3 (containing more than 0.15% of C_{60} and higher fullerenes), designated as Liposome-Fullerene. C_{60} molecules are embedded in the phospholipid bilayer of the liposomes (Figure 2).

The average size of Liposome-Fullerene in 0.5% aqueous phase dispersion is approximately 76 nm. Liposome-Fullerene did not exert significant toxic effects in biological systems, such as photocytotoxicity and bacterial reverse mutagenicity.

BIOLOGICAL ACTIVITY OF FULLERENES

During the last two decades, the biological activity of C_{60} and its hydrophilic derivatives has been extensively investigated [30]. They possess a high capacity for scavenging of biological ROS (e.g., superoxide, hydroxyl, singlet oxygen and nitrogen-based radicals and peroxides) [33, 41 - 43]. Compared to other antioxidants, fullerene shows higher thermostability, photostability, and antioxidant activity that is up to hundred-fold higher than that of other antioxidants [25]. Antioxidants suppress the production of the ROS in a concentration dependent manner. Xia et al., (2010) [44] reported a linear correlation between nC_{60} and α-tocoferol, regarding antioxidant activity, up to the concentration of 200 μg/ml nC_{60}. Hydrophile derivatives also demonstrated antioxidant activity *in vitro* and *in vivo* acting as potent free radical scavengers as good or better than commonly used antioxidants. Therefore, fullerenes are considered "free-radical sponges" which may protect human cells against damage by ROS. Suspensions of nC_{60} protected the rodent liver in a dose-dependent manner against free-radical damage [45]. The production of ROS in the skin, promoted under the UV irradiation, leads to the DNA damage, cell membrane disruption through a pathway of intracellular oxidative stress, resulting in cellular apoptosis. Different fullerene derivatives showed antioxidative activity and suppressive effects against ROS generation induced by UV irradiation [34, 38, 47, 48]. Fullerol $C_{60}(OH)_{24}$ demonstrated an acceptor activity for superoxide radical in xanthine/xanthine oxidase system and cytoprotective properties in a wide range of concentrations [49]. Different fullerols ($C_{60}(OH)_{44}$ x $8H_2O$, $C_{60}(OH)_{6-12}$, and $C_{60}(OH)_{32-34}$ x $7H_2O$), demonstrated ROS acceptor activity as well as cytoprotective effect for skin keratinocytes under UVA and UVB irradiation. It was observed that protective effect of $C_{60}(OH)_{44}$ x $8H_2O$ was higher for UVB-induced damages, in comparison with UVA irradiation effects [50]. There are also evidences demonstrating the antioxidative effect of water-soluble fullerenes *in vivo*. Ito et al. (2010) [48] have investigated a prooxidant or an antioxidant role of fullerene in UV light-induced disorders in mouse skin. Aqueous vehicle containing various concentrations of ascorbic acid (2 % *w/w*) and/or fullerene (0.04 – 5 % *w/w*), applied to the skin of mice at 100 mg/cm^2, gave no phototoxic effect under UVB irradiation, in comparison with the control (aquous solution of carboxymethyl cellulose sodium 1 % *w/w*), while erythema, the ROS index, and the apoptosis index were significanly decreased. The results of this study suggested that fullerene has a ROS-reducing effect and this effect was enhanced by the co-application with ascorbic acid. In another study [34], chemically generated hydroxyl radicals were scavenged by PEG-modified fullerene, hydroxyfullerene and isostearate-mixed fullerene as efficiently as ascorbic acid or its 2-*O*-phosphorylated derivative. Enzymatically generated superoxide anion radicals were also scavenged by PEG-modified or PVP-entrapped fullerene similarly as by ascorbic acid or 2-*O*-phosphorylated ascorbic acid. Several fullerene derivatives, especially PVP-fullerene, were shown to diminish the ROS amounts in terms of molecular and cellular levels against either UVB or *tert*-butyl hydroperoxide (t-BuOOH), which has been used as a representative radical source in biochemical studies [51, 52]. Fullerols inhibit production of matrix metalloproteinases MMP-1, MMP-3 and MMP-13 in human chondrocytes [53].

In contrast to antioxidant activity, different authors reported that C_{60} and fullerols become ROS generators, under visible or UV light exposure, causing cyto(photo)toxic effects in representative cell lines or tissues [54 - 57]. C_{60} nanocrystals prepared by solvents exchange method, produced large amounts of ROS in reaction with water, which lead to lipid peroxidation-mediated cell membrane damage and necrotic cell death for various types of normal and transformed human and rodent cells [53, 58]. Sayes et al., (2004) [59] showed that C_{60} is cytotoxic to human liver carcinoma cells and fibroblasts. Therefore, fullerenes have characteristics of both antioxidants and prooxidants. The mechanisms of the two opposite effects are different and the origin and intesity of the effects depend on numeorus factors such as chemical structure and size of a fullerene molecule, applied dose, duration and intensity of irradiation, size and surface morphology of fullerene nanocrystals. Fullerene derivatives with nonionic or anionic groups consider non-cytotoxic, while those with cationic moieties have shown toxicity toward various types of mammalian cells [30]. Water-soluble C_{60} derivatives have displayed either cytoprotective or cytotoxic effects. For example, polyhydroxylated and malonic acid C_{60} derivatives behave as "free radical sponges" and protect cells from oxidative stress-mediated cell damage *in vitro* and *in vivo* [60 - 62], while at higher concentrations or upon exposure to UV light, they could kill cells through ROS-dependent or ROS-independent mechanisms [63 - 65]. Several types of water-soluble fullerenes cause oxidative damage to cellular membranes resulting in cell death with or without visible light. Wielgus et al., [66] observed cytotoxycity of fullerols for pigment epithelial cells of human retina at concentrations higher than 10 μmol/l. Rouse et al., (2006) [67] were evaluated the influence of fullerene-amino acid derivatives (Baa), ranging in concentrations of 0.4–0.00004 mg/ml, on human epidermal keratinocytes (HEK) in a humidified 5 % CO_2 atmosphere at 37 °C. Additionally, human cytokines IL-6, IL-8, TNF-a, IL-1b, and IL-10 were assessed for fullerene-amino acid concentrations ranging from 0.4–0.004 mg/ml. Media was harvested at 1, 4, 8, 12, 24 and 48 h for cytokine analysis. TEM analysis revealed that at 24 h, the HEK treated with 0.00004 mg/ml Baa, the lowest concentration, did not show uptake of the Baa. The 0.0004 mg/ml concentration rarely depicted particles in the cytoplasm of the HEK, while few particles were found in the cytoplasm of HEK treated with 0.004 mg/ml. Unstained TEM of the HEK treated with 0.04 mg/ml at 24 h demonstrates several cells containing a few Baa particles in many vacuoles (Figure 3). The obtained results indicate that concentrations lower than 0.04 mg/ml initiate less cytokine activity and maintain cell viability. For fullerene-amino acid concentrations of 0.4 mg/ml and 0.04 mg/ml cell viability was decreased and a pro-inflammatory response was induced.

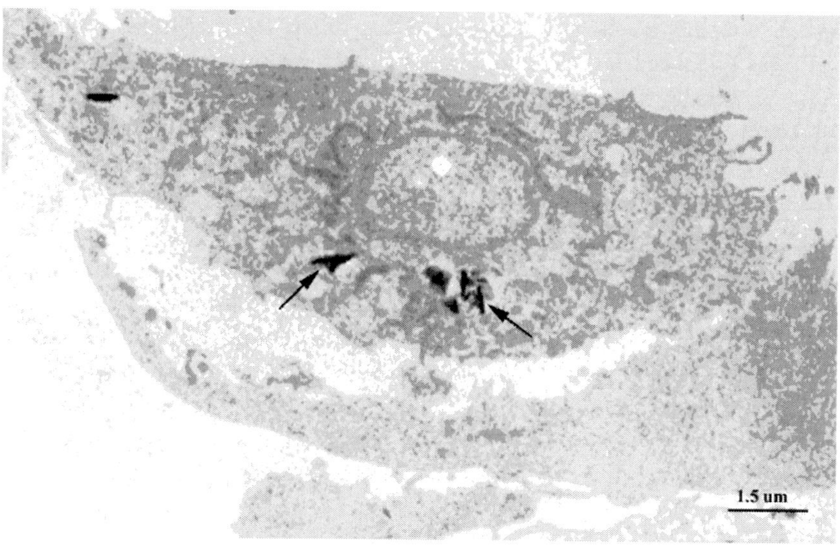

Figure 3. TEM micrograph depicting intracytoplasmic vacuole localization of 0.04 mg/ml Baa (arrows) after 24 h (unstained) [67].

Prooxidant activity of fullerenes probably is not the only mechanism related with their cytotoxicity. Literature describes the influence of fullerenols, carboxyfullerenes, C_{60}-(glucosamine)$_6$, and C_{60}-(cystine)$_5$ derivatives on proteins that leads to modulation of apoptosis in different types of cells, including neurons, via ROS-independent mechanisms [68, 69]. Fullerenes may interact with functional proteins, such as enzymes, and induce an apoptosis modulating effect. *In silico* analysis suggested that C_{60} fits in the hydrophobic binding pocket of HIV-1 protease with potentially consequent inhibition of the enzyme and blocking the HIV replication cycle [70]. Also, ther are reports on antiviral effect, even against drug-resistant HIV-variants [71-73], and antibacterial properties of fullerenes [74-76].

Current knowledge on complex and likely diverse antiapoptotic and proapoptotic mechanisms of fullerenes is not sufficient to derive conclusions which derivatives under which conditions may enhance proliferation processes or apoptosis [4]. In order to elucidate a significance of the functionalization type, size and surface topology of fullerene nanoparticles for promoting of either cytoprotective or cytotoxic effects, more intensive research is required.

POTENTIAL APPLICATIONS OF FULLERENES IN BIOMEDICINE

Fullerenes are strong antioxidants which can reduce cellular damage and inflammation caused by increased production of ROS, that is typical in pathogenesis of neurodegenerative disorders, cancer, arthritis, cardiovascular deseases [33, 41, 43, 60, 77 – 79]. Therefore, fullerenes are potentially useful in their medical treatment. C_{60} and some fullerols and carboxyfullerenes are able to penetrate through the blood–brain barrier and provide neuroprotective acitivity by the oxidative stress supression [41, 60]. Several biological studies have demonstrated that some fullerenols and carboxyfullerenes prevent neurotoxicity caused by levodopa [80] and alcohol [81]. Solution of fullerenes, prepared with addition of 1 % of polysorbate 80, suppress toxicity of polynarcotics and decrease a lethal effect caused by simultaneous action of methamphetamine and morphine [82]. Cai et al. (2008) [83] showed that fullerol $C_{60}(OH)_{24}$ was able to protect human neuroblastoma cells against deficits in cell viability and mitochondrial activity induced by 1-methyl-4-phenylpyridinium, a toxic molecule that interferes with the process of oxidative phosphorylation. Fullerene derivatives have been studied regarding their possible usage in therapy of Alzheimer`s disease and Parkinson`s disease [41]. Tong et al., (2011) [33] observed that nanoparticles of C_{60}-PVP and $C_{60}(PO_x)$ are non-toxic, neuronal cell permeable, superoxide scavenging antioxidants that might be promising candidates for the treatment of brain-related diseases associated with increased levels of superoxide. One type of C_{60}-PVP (PVP 10 kDa) and two types of C_{60}-(PO_x) complexes ($PEtO_x$ 5 kDa, and $P(EtO_x$-co-$BuO_x)$ 8 kDa) were prepared and their superoxide scavenging activities were quantitatively measured and compared in cultured neurons stimulated with angiotensin II. Interestingly, the obtained results show that C_{60}-$(PEtO)_x$ exhibited greater cellular uptake and significant intracellular superoxide scavenging activity, that is not found in C_{60}-PVP complex. Efficient scavenging of intracellular superoxide suggests the therapeutic potential of C_{60}-(PO_x) complexes in angiotensin II-related cardiovascular diseases like hypertension and heart failure. However, in order to achieve clinical relevance of these nanomaterials, it is necessary to exclude any risk of neurotoxicity.

In a recent review on pulmonary exposure to fullerenes was inferred that they are capable of eliciting localized antiinflammatory responses [27]. Several studies have demonstrated a fullerene induced suppression of damage of lung caused by oxidative stress [84-88]. Some C_{60} derivatives (1,2-methanofulleren-$C_{60})_{61}$-carbolacid and fullerols) are known to possess an inhibitory effect in the bronchial asthma model [84], oxidative stress in both the RAW 264.7 macrophage cell line, ischemia-reperfused rat lungs [85, 86], and neutrophilic lung inflammation in mice [87]. For example, in RAW 264.7 cells, sodium nitroprusside (1 mM) and H_2O_2 (400 μM) caused a marked (90 %) decrease in cell viability and mitochondrial membrane potential. These parameters were dose dependently reversed by pretreatment with $C_{60}(OH)_{5-9}$ (10-50 μM), although high concentration of $C_{60}(OH)_{5-9}$ (1 and 1.5 mM) lead to apoptosis or cell necrosis. In the isolated rat lung, the increases in pulmonary artery pressure and capillary filtration pressure induced by sodium nitroprusside were reversed significantly by $C_{60}(OH)_{5-9}$ (10 mg/kg). C_{70}-tetraglycolate (TGA) inhibit mast cell function *in vitro* and *in vivo* and provide therapeutic effect on asthma induced in mice. The animals were treated intranasally with TGA either simultaneously with treatment or after induction of pathogenesis. TGA-treated mice have significantly reduced airway inflammation, eosinophilia, and bronchoconstriction. The TGA treatments are effective, even when given after disease is established. Moreover, a novel inhibitory mechanism was reported, because TGA stimulates the production of antiinflammatory P-450 eicosanoid metabolites in the lung. The reported inhibitory capabilities of TGA were promising platform for consideration of fullerene-based treatments for asthma [88].

Hydroxyl radical, hydrogen peroxyde, superoxyde anion radical, and lipid peroxides, generated in human skin cells under UV rays exposure, may cause photodamage of cell membrane, proteins and DNA, that may promote further development of inflammation, immunosuppression, photoaging, pigmentaton, and/or skin cancer [50]. Usually, the progression of UV-induced skin damage may be reduced by administration of antioxidants such as ascorbic acid, tocoferol, and polyphenols. These substances are common active substances of „anti-aging" cosmetic products and pharmaceuticals for prevention and therapy of skin deseases. Fullerenes applied to the skin also may protect keratinocyte against apoptosis caused by ROS [5, 34]. Thus, they may be highly useful in preventing many skin problems related to oxidative stress. Important advantages of fullerenes over the other antioxidant agents are superior physico-chemical stability (it is well known that ascorbic acid undergo fast oxidative decomposition in the aqueous solution) and enhanced penetration into the deeper segments of the human skin epidermis. Several fullerene-enriched cosmetic products are already available on the market (e.g., Dr. Wu Anti-Oxidant Whitening Lotion With Fullerene RS™, Dr. Wu Singapore, Singapore; Zelens® Fullerene C-60 Night Cream, Zelens, London, UK). Extensive investigations were conducted on suppression of UV irradiation-induced cell injuries and melanogenesis in human skin cells and/or human skin model induced by different fullerenes. Fumelli et al. (2000) [47] have demonstrated that carboxyfullerenes protected human keratinocytes from UV-induced apoptosis. Xiao et al., (2005) [38] evaluated the free-radical scavenging activities of five water-soluble fullerene derivatives using electron spin resonance (ESR) spectrometry. Additionaly, intracellular inhibitory potential of these fullerenes in UV irradiation-induced and t-BuOOH-induced oxidation in human skin keratinocytes was assessed. Although, the current study confirmed the antioxydant activity of the investigayed molecules, this study pointed that such activity of different fullerenes may be derived by different pathways in accordance with the nature of side groups or modifier substances. Saitoh et al., (2011) [50] evaluated antioxidative activities and suppressive effects against UVA and UVB irradiation-induced intracellular ROS generation and cell injuries of three novel fullerenols ($C_{60}(OH)_{6-12}$, $C_{60}(OH)_{32-34}$ x $7H_2O$, and $C_{60}(OH)_{44}$ x $8H_2O$) in human skin keratinocytes. The main goal of this study was to elucidate fullerene effectiveness levels regarding the differences in number of intramolecular hydroxyl groups. The cytoprotective effect of $C_{60}(OH)_{44}$ x $8H_2O$ for human skin keratinoytes had a tendency to be superior to the two other fullerols. Furthermore, the cytoprotective effects of

$C_{60}(OH)_{44}$ x $8H_2O$ against UVB-induced injuries, prevailed in comparison with the harmful effects of UVA irradiation. $C_{60}(OH)_{44}$ x $8H_2O$ is suggested to be the most attractive candidate for maintaining of skin homeostasis, and to be protectant against skin disorders and photoaging. Xiao et al. (2006) [5] investigated photo-biological activity of Radical Sponge® in human skin keratinocytes. The cells were repeatedly irradiated with a visible light of wavelengths of 400–2000 nm. The results showed that Radical Sponge® prevents both UV-induced skin injuries and skin aging, within a defined concentration range (10 – 40 μM), while photosensitization and cytotoxicity were not observed. Also, it was concluded that Radical Sponge® caused a more marked cytoprotection when it was pre-irradiationally administered and rinsed out immediately before the irradiation, than when administered only during or after the irradiation, indicating the preventive rather than therapeutic effect on UVA injuries toward human skin keratinocytes. Fujimoto et al., (2010) [90] described ability of C_{60}-PVP for inhibition of UVA-induced melanogenesis in human melanocytes and whitening effects in human volunteers. PVP-fullerene significantly reduced UVA-induced melanogenesis in a dose dependent manner and in a larger extent than in cells treated with ascorbic acide and arbutin. Clinical studies were performed on 32 women volunteers. A gel formulation containing 2 ppm of C_{60}-PVP, was prepared and applied on the face twice daily. After 6 weeks of application, the skin of 94 % subjects became fairer than before treatment without any inflammation or irritation. The obtained results implicated that C_{60}-PVP has a whitening effect because of its ability to scaveng UV-induced free radicals. LipoFullerene® is also developed for usage as a cosmetic ingredient with antioxidant ability and anti-wrinkle effect. Kato et al. (2010) [6] assessed anti-wrinkle formation efficacy of LipoFullerene® enriched cream within 23 Japanese women (group I: age 38.9±3.8, n=11, group II: age 39.4±4.3, n=12) in a randomized, matched pair double-blind 8-week study. The cream was applied twice daily on the right or left half of the face, and C_{60}-unloaded cream enriched with squalane, was applied as the control on another half of the face. As clinical evaluations of wrinkle grades, visual observation and photographs, and silicone replicas of both crow's feet areas were taken at 0 week and after 4 and 8 weeks. Subsequently, trans-epidermal water loss (TEWL), moisture levels of the stratum corneum, and visco-elasticity were measured on cheeks by instrumental analysis. LipoFullerene®-loaded cream enhanced the skin hydration and showed the anti-wrinkle effect significantly higher than the control at 8th week ($p < 0.05$) without severe side effects. The anti-acne activity of LipoFullerene® was evaluated in an open trial on the study population comprised of 5 men and 6 women with acne vulgaris [91]. Their ages ranged from 23 to 39 years. During the trial, the subjects applied on their entire face, except on erosions, 0.4 ml of 1 % LipoFullerene® in a gel vehicle comprising squalene (1%), phenoxyethanol (0.5%), I.T.O. gel base (I.T.O. Co. Ltd., Tokyo, Japan) (consisting of glycerin, carbomer, sodium polyacrylate, methylparaben, propylparaben, water) (49.25%), and water (49.25%), twice a day (immediately after washing the face in the morning and night). The concentration of C_{60} in the gel was 75 μmol/l. At 4 and 8 weeks, the mean number of inflammatory lesions (erythematous papules and pustules) significantly decreased for 23.2% and 37.8%, respectively. The number of pustules, consisting of accumulation of neutrophils, was significantly decreased for 87.6%. Fullerene suppresses acne through suppression of neutrophil infiltration and sebum production, possibly through its antioxidant effect on skin cells. After treatment for 8 weeks, the water content of the skin significantly. Therefore, the LipoFullerene® enriched gel may help in controlling acne vulgaris with skin care benefit. Fullerenol $C_{60}(OH)_{44}$ was evaluated as inhibitory agent for sebum production in hamster sebocytes and for *Propionbacterium acnes* lipase activity *in vitro*. Sebum production was significantly reduced by 1.5 μM of fullerenol in cells that had been irradiated with 10 mJ/cm^2 of UVB, although it was not altered in the non-irradiated cells, indicating that fullerene is a sebum suppressor for sebocytes under oxidative stress, such as that induced by UVB light. It was also found that fullerenol has inhibitory activity against *P. acnes* lipase [10]. Important advantage of fullerene in acne treatment is ability to penetrate deep into the epidermis without causing skin irritation [92].

Liposome-Fullerene has a potential to be safely utilized as a cosmetic ingredient for UVA-protection. It may help to maintain the integrity of the skin barrier by protection against UVA-induced oxidative stress when administered on the surface of three-dimensional human skin tissue model, rinsed out before each UVA-irradiation at 4 J/cm^2, and thereafter added again, followed by 19-cycle-repetition for 4 days. UVA-caused corneum scaling and disruption of epidermis layer, breakdown of collagen type I/IV, DNA strand cleavage and pycnosis/karyorrhexis were scarcely repressed by liposome alone, but repressed by 250 ppm of Liposome-Fullerene complex. Upon administration of Liposome-Fullerene (in concentration equivalent to 12 ppm of C_{60}) on human abdomen skin biopsies mounted in Franz diffusion cells, C_{60} permeated after 24 h into the epidermis, but it was not detected in the basal lamina and dermis [40].

Fullerenes, due to their ability to modulate cell death, are promising drug candidates for therapy of bacterial and viral infections and cancer [4, 21, 30, 71]. Fullerene derivatives bearing S-triazine fragment demonstrated a moderate to good level of inhibition of bacteria [9]. Fenetylaminosuccinate-C_{60} possess anti-HIV activity with K_{inh}=5.3 μmol/l [7]. Dendrite derivatives of fullerene [93] showed EC50 equal to 0.22 μmol/l in lymphocytes of a patient with HIV-1 acute phase. PVP-fullerene possesses antiviral activity for influenza A and B viruses [94]. It has been demonstrated antibacterial activity against *Escherichia coli* and *P. acnes* [76]. It was observed cytotoxicity of several water-soluble fullerenes for human dermal fibroblast and liver carcinoma cell lines [59]. The nano-C_{60} water suspension make large amounts of ROS in reaction with water, which leads to lipid peroxidation-mediated cell membrane damage and necrotic cell death [54, 58]. A very high capacity for ROS generation could either stem from some unusual physico-chemical characteristics of underivatized C_{60} nanocrystals or possibly involve a synergistic interaction of crystalline C_{60} and a relatively small amount (10 %) of otherwise non-toxic residual organic solvent (tetrahydrofuran) incorporated into its lattice [58]. However, the mechanisms other than oxidative stress-mediated necrosis might be involved in nano-C60 antitumor action. Harhaji et al., (2007) [21] demonstrated the multiple mechanisms in the *in vitro* anticancer effects of the nC_{60} using the rat glioma cell line C6 and the human glioma cell line U251. Nano-C_{60} in a dose-dependent manner reduced the tumor cell numbers after 24 h of incubation. The observed antiglioma action of nano-C_{60} at high concentration (1 μg/ml) was due to a reactive oxygen species-mediated necrotic cell damage that was partly dependent on oxidative stress-induced activation of extracellular signal-regulated kinase (ERK). On the other hand, low-dose nano-C_{60} (0.25 μg/ml) did not induce either necrotic or apoptotic cell death, but caused oxidative stress/ERK independent cell cycle block in G2/M phase and subsequent inhibition of tumor cell proliferation. Treatment with either high-dose or low-dose nC_{60} caused the appearance of acidified intracytoplasmic vesicles indicative of autophagy, but only the antiglioma effect of low-dose nC_{60} was significantly attenuated by inhibiting autophagy with bafilomycin A1. Importantly, primary rat astrocytes were less sensitive than their transformed counterparts to a cytostatic action of low-dose nC_{60}. Regardless of the mechanisms underlying the production of reactive oxygen intermediates by nano-C_{60}, the ability to rapidly kill tumor cells qualifies it for a potential anticancer drug. Some forms of fullerols demonstrated antitumor activity *in vivo* in relation of hepatocarcinoma H22 [95] and *in vitro* cytoxicity against human liver carcinoma cells (HepG2) [54, 96]. The effect depended on the dose and size of particles. Daily doses of 0.08 and 0.4 mg/ml were used during investigation of fullerol $C_{60}(OH)_{20}$ for antitumor and antimetastatic activity on EMT-6, metastatic model of breast cancer the average size of nanoparticles at pH 7 varied from 223 - 364 nm, correspondingly [97]. In addition to expected modulation of oxidative stress, decreasing of expression of a few angiogenic factors in tumor tissues was observed. In particular, expression of CD-31 (PECAM-1 – molecule of adhesion of endothelial cells thrombocytes) and vascular density was decreased in tumors treated by fullerol. By this mode $C_{60}(OH)_{20}$ inhibits a tumor growth and suppresses metastasizing through signaling pathway linked with tyrosine kinase. Experiments on transplantation of malignant tumors to group of mice demonstrated efficiency of C_{60} usage to inhibit growth of such tumors [98]. This is linked not only with antioxidant activity of C_{60}, but also with blockage of specific cell receptors, in particular, endothelia growth factor.

Fullerenes have been considered as a promising nanomaterial in the field of targeted tumor therapy [99-101]. In some animal studies, the cytotoxicity of PEG-modified C_{60} was successfully targeted to tumor tissue, causing tumor regression without discernible damage to healthy tissues [21]. It is known that the C_{60} molecule and some of its derivatives, photo-activated by UV incidence, induced lipid peroxidation in hepatic rat microsomes [102]. Wielgus et al. (2010) [103] reported that fullerol was toxic to retinal pigment epithelial cells both in the presence of visible light and in the dark. Vileno et al., (2010) also demonstrated phototoxic effects of fullerol via ROS generation under visible light incidence in a monkey fibroblast cell line. Fullerenes which absorbe UV or visible light or X-rays and generate ROS, are potential photo- and radiotherapeutic agents, respectively [104-106]. This may be employed for targeted damage of pathogenic microorganisms or malignant tumors, with minimisation of damage of a healthy tissue [4]. Very recently Kato et al., (2014) [108] have investigated proliferation inhibition of human malignant melanoma HMV-II cells under X-ray irradiation by using C_{60} dissolved in squalene as radiosensitizer. Radiosensitization is attributed to peroxidation of unsaturated bonds of squalene by X-ray-excited C_{60}. Use of fullerenes in photo- and radio-therapy of cancer seems to be very promising.

Some fullerene derivatives attract a serious attention as carriers in targeted drug delivery and theranostic techniques [14, 110]. Conjugates of fullerenes with known antineoplastics may enhance cancer distruction employing fullerene as generator of singlet oxygen. However, uptake of such conjugates and their biodistribution may differ from those for free drug molecules. For example, doxorubicin is localized in a cell nucleus, but its conjugate with fullerene is in a cytoplasm [109]. Shi et al., (2013) [14] prepared nanocomposites of fullerene C_{60} nanoparticles with iron oxide nanoparticles (IONPs) onto the surface. The nanocoposite was PEGylated in order to improve hydrophilicity and biocompatibility and conjugated with hematoporphyrin monomethyl ether (HMME), a new photodynamic anti-cancer drug. Such nanomaterial posesses strong superparamagnetism and high capacity for targeted photodynamic cancer therapy. Compared with free HMME, the nanocomposite remarkably enhanced photodynamic cancer cell killing effect in a cultured B16-F10 cells *in vitro* and murine tumor model *in vivo,* due to 23-fold higher drug uptake into tumor and strong photodynamic activity of the encapsulated drug. Moreover, the nanocomposite could be used as a T2-contrast agent for *in vivo* magnetic resonance imaging of cancer.

Kim et al. (2014) [12] developed a conjugate of glycol chitosan (GC) grafted with 2,3-dimethylmaleic acid (DMA) and fullerene (C_{60}) (GC-g-DMA-g-C_{60}) for use as a photosensitizer prodrug. This conjugate, consisting of a hydrophilic block (GC and DMA) and a hydrophobic block (C_{60}), was self-assembled to form polysaccharidic nanogels. Such nanogels also formed multi-nanogel aggregates at pH 7.4 that can be disintegrated at pH 5.0 due to the reduction of electrostatic interaction that emerged in acidic environment. Upon illumination by 670 nm light, photo-responsive properties of the nanogel aggregates allowed generation of different singlet oxygen in accordance with the pH condition: the singlet oxygen generation was reduced at pH 7.4, while it was enhanced at pH 5.0. Therefore, GC-g-DMA-g-C_{60} nanogel aggregates can be a perspective candidate for the selective pH targeting of endosome and photodynamic therapy in various malignant tumor cells.

Cationic fullerenes offer great potential for gene delivery because they can form complexes with DNA penetrable into the cell [111]. Gene delivery efficacy of tetra(piperazino)fullerene epoxide (TPFE) was evaluated *in vitro* and *in vivo* [13]. Delivery of enhanced green fluorescent protein gene (EGFP) by TPFE on pregnant female ICR mice showed distinct organ selectivity compared with a cationic liposome formulation (Lipofectin Reagent, Life Technologies, USA). Moreover, higher gene expression by TPFE was found in liver and spleen, but not in the lung. No acute toxicity of TPFE was found for the liver and kidney, although Lipofectin significantly increased liver enzymes and blood urea nitrogen. In mice fetal tissues, neither TPFE nor Lipofectin induced EGFP gene expression. Delivery of insulin 2 gene to female C57/BL6 mice increased plasma insulin levels and reduced blood glucose concentrations, as efficiently as a cationic liposome based system, indicating the potential of TPFE-based gene delivery for clinical application. TPFE can induce gene expression by releasing DNA inside the cell because of its resistance to digestion by nucleases.

Safety Considerations on Biomedical Application of Fullerenes

Therapeutic applications of fullerenes are still a controversy. Current knowledge indicate that fullerenes show a double-edged effect and it is essential to elucidate all aspects important for achievement of their beneficial functions with avoidance of harmful manifestations. Pharmacokinetic studies, which will be necessary for human health risk assessments, are limited. On the other hand, fullerenes family consists of diverse members regarding molecular size, surface chemistry, solubility, and aggregation/agglomeration tendency. Available results of the studies related to potential toxicity of different types of fullerenes are still poor for establishment of general conclusions. Sayes and associates [54, 59] suggest that the underivatized C_{60} is cytotoxic to human liver carcinoma cells and fibroblasts and can induce oxidative stress in juvenile largemouth bass [55]. However, it was reported that C_{60} were not affected mitochondrial activity in human keratinocyte HaCaT and lung carcinoma A549 cells [17, 113]. On the other hand, C_{70} dispersion inhibited colony forming ability at the concentration of 25.2 μg/ml. Exposure to C_{70} dispersion caused an increase in intracellular

ROS and lipid peroxidation levels. The induction of intracellular ROS level was inhibited by pre-treatment of the cells by antioxidants. TEM observations of C_{70} exposed cells showed cellular uptake. Although C_{60} and C_{70} were not caused cell death, they caused the induction of intracellular oxidative stress, at a level 2–3 times higher than in unexposed cells, however milder than for some metal oxide nanoparticles [113-115]. The review of different reports [116] pointed that colloidal fullerene nanoparticles nC_{60} was very toxic to human cells and microorganisms when prepared using tetrahydrofuran (THF) as solvent (so-called THF method). Other studies pointed that C_{60} and C_{60}-containing composites at low concentrations do not manifest hemolytic or cytotoxic effects [98]. Also, the aqueous suspension of nC_{60} showed no acute or subacute toxicity in rodents, particularly after X-ray irradiation [58, 114], or when prepared via non-THF methods [116]. Several studies confirmed that the toxic effects observed with nC_{60} suspension could be attributed to the formation of THF/nC_{60} complex [118, 119] or the oxidative by-products of THF, formed during the preparation processes [120, 121]. No toxicity was observed for 16 h to *E. coli* or *Bacillus subtilis* by nC_{60} prepared with non-THF method at concentrations up to 342 μg/ml, which was 100 times higher than the reported minimum inhibitory concentrations of nC_{60} prepared using THF method for these two bacteria. When *E. coli* was exposed to 85.5 μg/ml nC_{60} with daily passage for 4 days, the stationary phase populations at different passages were not statistically different from the control without nC_{60} nanoparticles. No mortality to human epidermal keratinocytes was observed at a concentration 170 times higher than the reported LC_{50} values for other human cell lines [17]. A large amount of nC_{60} after solvent evaporation showed no toxicity in human epithelial cells and liver carcinoma cells [122]. These new results reveal that the intrinsic biological property of nC_{60} is most likely non-toxic. It is important to point out that chemical derivatization of C_{60} and applied dose increase may significantly influence the mode of interaction of fullerenes with biological systems and decrease apparent toxicity [17]. No extrapolation to all fullerene types can be made. Currently available studies do not indicate a chronic fullerene exposure risk in relation to the consequences for health. Although the toxicity of fullerenes may be small, further metabolic transformations in the human body has been scarcely investigated. Finaly, nanoparticle size is an important factor related with their potential cytotoxicity or safety. A recent study on risk assessment for humans by reviewing the limited animal toxicity data available revealed that a period-limited acceptable exposure level of C_{60} particles with average size of 96 nm was estimated to be 0.39 mg/m^3 for healthy workers and 1.4×10^{-2} mg/m^3 for the general human population [124]. In spite of the rapid research involving usage of fullerenes as potential active ingredients of pharmaceutical preparations and cosmetic products, it is necessary an exhaustive characterization of their effects on human health, including adverse biological responses and potential risks for occurence of systemic toxicity, by applying a case-to-case approach.

The consumers of the cosmetic products with fullerenes are exposed *via* direct dermal application. Current state of art suggests that fullerenes-enriched cosmetic products are harmless to human skin or health. Xia et al., (2010) [44] observed that C_{60} (~1 nm diameter) applied as solution in mineral oil, a common ingredient of numerous cosmetic products, do not penetrate into the skin of Yorkshire weanling pigs, for four days. The solution was applied each day and the content of the permeated fullerene was determined in the samples collected by tape-stripping at the site of the solution application. The abscence of C_{60} penetration was confirmed *in vitro* using isolated pig skin *stratum corneum* in flow-through diffusion cell. The results of few other studies demonstrated that fullerenes penetrate intact healthy epidermis and provide potential enhancement in cosmetic effects, however, without potential risks for absorption and systemic toxicity. Derivatized fullerenes are localized within intercellular spaces of the epidermis and do not reach dermis. The external mechanical forces applied to the skin while flexing proves to have a significant effect on both the rate and extent of fullerene penetration. A study of Rouse et al., (2007) [125] confirms that fullerene-based peptides can penetrate intact skin and that mechanical stressors, such as those associated with a repetitive flexing motion, increase the rate at which these particles permeate through epidermis. The mechanical stress regimes used in this study attempt to mimic physiological forces that can occur during conditions involved in consumer use. The worst-case scenario is a permeation of fullerenes into subdermal vascular network. Leavens et al., (2010) [126] assessed tissue distribution of nC_{60} by quantifying nanoparticle extraction from the vascular space into tissues in an isolated perfused porcine skin flap (IPPSF). IPPSFs were perfused for 4 h with 0.885 μg/ml nC_{60} in media with immunoglobulin G present (IgG+) or absent (IgG−), followed by a 4 h perfusion with media only during a washout phase. The steady state differences in the arterial and venous concentrations in the IPPSF were small with extraction percentages (mean±sd) of 8.2±5.7% and 4.2±6.7% for IgG+ and IgG− media, respectively, and were not significantly different between the types of media. The apparent volumes of distribution at steady state were 0.12±0.047 and 0.10±0.034 L/kg, for IgG+ and IgG− media, respectively. By 4 h following infusion of nC_{60}, the recovery of nC60 in the venous effluent was 94±5.5% and 97±6.8% of the infused nC_{60} for IgG+ and IgG− media, respectively. Based on the apparent volume of distribution, the low extraction during the perfusion, and the high percentage recovery following the washout phase, there was limited distribution of nC_{60} from the vascular space into the extracellular space and negligible intracellular uptake of C_{60} nanoparticles in this system. Although the current facts suggests that fullerene used presently in cosmetic preparations are harmless to human skin or health, potential side effects or toxic activities of other fullerene derivatives with the intention to use in cosmetic products have not been fully assesed yet.

Conclusion

The production of fullerenes and their use in consumer products is expected to increase in future. Biological activities of fullerenes range from cytoprotection to cytotoxicity. Although many fullerenes are able to protect cells from UV irradiation, some fullerenes under certain conditions may absorbe X-ray, UV or visible light and generate ROS. Fullerene has beneficial features for usage as active ingredients of nanopharmaceuticals and cosmetics. Their current application include anti-aging, skin whitening and anti-acne cosmetic products. Also, over the last years a great attention is paid to use different fullerenes as cytoprotective agents, especially important for prevention and therapy of neurodegenerative diseases, or cytotoxic agents for various therapeutic purposes, antitumor drugs in targeted drug delivery and theranostic strategies for cancer therapy. The current literature implicated that the intrinsic biological activities of fullerenes family is not fully elucidated. Correlation between the fullerene structures and biological activities needs further investigation for better understanding of their interactions with the constituents of the biological environment. The potential risks for the fullerene based side effects or toxicity have not been fully assessed yet, and it is of great importance to intensify such evaluation in pre-clinical and clinical trials. This will enhance estimation of the toxicological and safety profiles and further spreading of biomedical application of fullerenes.

REFERENCES

[1] Kroto, HW; Heath, JR; O'Brien, SC; Curl, RF; Smalley, RE. C60: Buckminsterfullerene. *Nature*, 1985, 318, 162-163.

[2] Kratschmer, W; Lamb, LD; Fostiropoulos, K; Huffman, DR; Solid C60: a new form of carbon. *Nature*, 1990, 347, 354–358.

[3] Commission Recommendation of 18 October 2011 on the definition of nanomaterial (Text with EEA relevance). Official Journal of the European Union (2011/696/EU) L275/38 - L275/40 (Available from: URL: http://eur-lex.europa.eu/LexUriServ/LexUriServ.do?uri=OJ:L:2011:275:0038:0040:EN:PDF

[4] Orlova, MA; Trofimova, TP; Shatalov, OA. Fullerene nanoparticles operating the apoptosis and cell proliferation processes in normal and malignant cells. *Der Pharmacia Lettre*, 2013, 5(3), 99-139.

[5] Xiao, L; Takada, H; Gana, X; Miwa, N. The water-soluble fullerene derivative 'Radical Sponge®' exerts cytoprotective action against UVA irradiation but not visible-light-catalyzed cytotoxicity in human skin keratinocytes. *Bioorganic & Medicinal Chemistry Letters* 2006, 16, 1590–1595.

[6] Kato, S; Taira, H; Aoshima, H; Saitoh, Y; Miwa, N. Clinical Evaluation of Fullerene-C_{60} Dissolved in Squalane for Anti-Wrinkle Cosmetics. *Journal of Nanoscience and Nanotechnology*, 2010, 10(10), 6769-6774.

[7] Friedman, SH; De Camp, DL, Sijbesma, RP; Srdanov, G; Wudl, F; Kenyon, GL. Inhibition of the HIV-1 protease by fullerene derivatives: model building studies and experimental verification. *J. Am. Chem. Soc.*, 1993, 115, 6506-6509.

[8] Aquino, A; Chan, J; Giolma, K; Loh, M. The Effect of a Fullerene Water Suspension on the Growth, Cell Viability, and Membrane Integrity of Escherichia coli B23. *Journal of Experimental Microbiology and Immunology* (*JEMI*), 2010, 14, 13-20.

[9] Ji, H; Yang, Z; Jiang, W; Geng, C; Gong, M; Xiao, H; Wang, Z; Cheng, L. Antiviral activity of nano carbon fullerene liposome against influenza virus in vitro. *J. Huazhong Univ. Sci. Technol. Med. Sci.*, 2008, 28, 243–246.

[10] Inui, S; Aoshima, H; Ito, M; Kobuko, K; Itami, S. Inhibition of sebum production and Propionibacterium acnes lipase activity by fullerenol, a novel polyhydroxylated fullerene: potential as a therapeutic reagent for acne. *J. Cosmet. Sci.*, 2012, 63(4), 259-265.

[11] Montellano, A; Da Ros, T; Bianco, A; Prato, M. Fullerene C60 as a multifunctional system for drug and gene delivery. *Nanoscale*, 2011, 5;3(10), 4035-4041.

[12] Kim, S; Lee, DJ; Kwag, DS; Lee,UY; Youn, YS; Lee, ES. Acid pH-activated glycol chitosan/fullerene nanogels for efficienttumor therapy. *Carbohydrate Polymers*, 2014, 101, 692– 698.

[13] Maeda-Mamiya, R; Noiri, E; Isobe, H; Nakanishi, W; Okamoto, K; Doi, K; Sugaya, T; Izumi, T; Homma, T; Nakamura, E. In vivo gene delivery by cationic tetraamino fullerene. *PNAS*, 2010, 107, 5339–5344.

[14] Shi, J; Yu, X; Wang, L; Liu, Y; Gao, J; Zhang, J; Ma, R; Liu, R; Zhang, Z. PEGylated fullerene/iron oxide nanocomposites for photodynamic therapy, targeted drug delivery and MR imaging. *Biomaterials*, 2013, 34, 9666-9677.

[15] Chen, HS; Kortan, AR; Haddon, RC; Fleming, DA.Thermodynamics of fullerene (C60) in pure oxygen, nitrogen and argon. *J. Phys. Chem.*, 1992, 96(3), 1016–1018.

[16] Lyon,D; Adams,L; Falkner, JC; Alvarez, PJJ. Water Suspensions: Effects of Preparation Method and Particle Size. *Environ. Sci. Technol.* 2006, 40, 4360-4366.

[17] Horie, M; Nishio, K; Kato, H; Shinohara, N; Nakamura, A; Fujita, K; Kinugasa, S; Endoh, S; Yoshida, Y; Hagihara, Y; Iwahashi, H. In vitro evaluation of cellular influences induced by stable fullerene C70 medium dispersion: Induction of cellular oxidative stress. *Chemosphere*, 2013, 93, 1182–1188.

[18] Shinohara, N; Nakazato, T; Tamura, M; Endoh, S; Fukui, H; Morimoto, Y; Myojo, T; Shimada, M; Yamamoto, KT; Hiroaki, YY; Nakanishi, Junko. Clearance Kinetics of Fullerene C60 Nanoparticles from Rat Lungs after Intratracheal C60 Instillation and Inhalation C60 Exposure. *Toxicological Science*, 2010, 18(2), 564–573.

[19] Seki, M; Fujishima, S; Gondo, Y; Inoue, Y; Nozaka, T; Suemura, K; Takatsuki, M. Acute toxicity of fullerene C60 in aquatic organism. *Environ. Sci.*, 2008, 21, 53–62.

[20] Hilburn, ME; Murdianti, BS; Maples, RD; Williams, JS; Damron, JT; Kuriyavar, SI; Ausman, KD. Synthesizing aqueous fullerene colloidal suspensions by new solvent-exchange methods. *Colloids and Surfaces A: Physicochemical and Engineering Aspects*, 2012, 401, 48–53.

[21] Harhaji, Lj; Isakovic, A; Raicevic, N; Markovic, Z; Todorovic-Markovic, B; Nikolic, N; Vranjes-Djuric, S; Markovic, I; Trajkovic, V. Multiple mechanisms underlying the anticancer action of nanocrystalline fullerene. *European Journal of Pharmacology*, 2007, 568, 89–98.

[22] Deguchi, S; Rossitza, GA; Tsujii, K. Stable dispersions of fullerenes, C60 andC70 in water. Preparation and characterization. *Langmuir*, 2001, 17, 6013–6017.

[23] Fortner, JD; Lyon, DY; Sayes, CM; Boyd, AM; Falkner, JC; Hotze, EM; Alemany, LB; Tao, YJ; Guo, W; Ausman, KD; Colvin, VL; Hughes, JB. C60 in water: nanocrystal formation and microbial response. *Environ. Sci. Technol.*, 2005, 39, 4307–4316.

[24] Dhawan, A; Taurozzi, JS; Pandey, AK; Shan, WQ; Miller, SM; Hashsham, SA; Tarabara, VV. Stable colloidal dispersions of C60 fullerenes in water: evidence for genotoxicity. *Environ. Sci. Technol.*, 2006, 40, 7394–7401.

[25] Krusic, PJ; Wasserman, E; Keizer, PN; Morton, JR; Preston, KF. Radical reaction of C60. *Science*, 1991, 254, 1183–1185.

[26] Fukuzumi, S; Ohkubo, K; Imahori, H,; Guldi, DM. Driving force dependence of intermolecular electron-transfer reactions of fullerenes. *Chemistry* (Easton), 2003, 9, 1585–1593.

[27] Johnston, HJ; Hutchison, GR; Christensen, FM; Aschberger, K; Stone, V. The biological mechanisms and physicochemical characteristics responsible for driving fullerene toxicity. *Toxicol. Sci.*, 2010, 114, 162-182.

[28] Tabata, Y. & Ikada Y. Biological functions of fullerene. *Pure Appl. Chem.*, 1999, 71(11), 2047-2053.

[29] Yang, J. & Barron, A. R. A new route to fullerene substituted phenylalanine derivatives. *Chem. Commun.* (Camb), 2004, 21(24), 2884-2885.

[30] Bosi, S; Da Ros, T; Spalluto, G; Prato, M. Fullerene derivatives: an attractive tool for biological applications. *European Journal of Medicinal Chemistry*, 2003, 38, 913-923.

[31] Tsao, N; Luh, T; Chou, C; Chang, T; Wu, J; Liu, C; Lei, H. In vitro action of carboxyfullerene. *Antimicrob. Agents Chemother.* 2002, 49, 641-649.

[32] Ikeda, A; Iizuka, T; Maekubo, N; Aono, R; Kikuchi, J; Akiyama, M; Konishi, T; Ogawa, T; Ishida-Kitagawa, N; Tatebe, H; Shiozaki, K. Cyclodextrin Complexed [60]Fullerene Derivatives with High Levels of Photodynamic Activity by Long Wavelength Excitation. *ACS Med. Chem. Lett.*, 2013, 4(8), 752–756.

[33] Tong, J; Zimmerman,MC; Li,S; Yi, X; Luxenhofer, R; Jordan, R; Kabanov AV. Neuronal uptake and intracellular superoxide scavenging of a fullerene (C60)-poly(2-oxazoline)s nanoformulation. Biomaterials 2011, 32, 3654-3665.

[34] Xiao, L; Matsubayashi, K; Miwa, N. Inhibitory effect of the water-soluble polymer-wrapped derivative of fullerene on UVA-induced melanogenesis via downregulation of tyrosinase expression in human melanocytes and skin tissues. *Arch. Dermatol. Res.* 2007, 299, 245-257.

[35] Asada, R; Liao, F; Saitoh, Y; Miwa, N. Photodynamic anti-cancer effects of fullerene [C_{60}]–PEG complex on fibrosarcomas preferentially over normal fibroblasts in terms of fullerene uptake and cytotoxicity. *Molecular and Cellular Biochemistry* 2014, 390(1-2), 175-184.

[36] Bianco, A; Da Ros, T; Prato, M; Toniolo, C. Fullerene-based amino acids and peptides. *J. Pept. Sci.*, 2001, 7(4), 208-219.

[37] Hahn, U; Vögtle, F; Nierengarten, J-F. Synthetic Strategies towards Fullerene-Rich Dendrimer Assemblies. *Polymers* 2012, 4(1), 501-538.

[38] Xiao, L; Takada, H; Maeda, K; Haramoto, M; Miwa, N. Antioxidant effects of water-soluble fullerene derivatives against ultraviolet ray or peroxylipid through their action of scavenging the reactive oxygen species in human skin keratinocytes. *Biomed. Pharmacother.* 2005, 59(7), 351-358.

[39] Zhou, Z. Liposome Formulation of Fullerene-Based Molecular Diagnostic and Therapeutic Agents Pharmaceutics, 2013, 5, 525-541.

[40] Kato, S; Aoshima, H; Saitoh, Y; Miwa, N. Fullerene-C60/liposome complex: Defensive effects against UVA-induced damages in skin structure, nucleus and collagen type I/IV fibrils, and the permeability into human skin tissue. *Journal of Photochemistry and Photobiology B: Biology*, 2010, 98, 99–105.

[41] Dugan, LL; Lovett, E; Cuddihy, S; Ma, B-W; Lin, T-S; Choi, DW. Carboxyfullerenes as neuroprotective antioxidants. In: Kadish KM, Ruoff RS, editors. Fullerenes: Chemistry, Physics, and Technology. Wiley, New York, 2000; 467–479.

[42] Ali, SS; Hardt, JI; Quick, KL; Kim-Han, JS; Erlanger, BF; Huang, TT; Epstein, CJ; Dugan, LL. A biologically effective fullerene (C60) derivative with superoxide dismutase mimetic properties. *Free Rad. Biol. Med.* 2004, 37, 1191–1202.

[43] Quick, K. & Dugan, L. Fullerene derivative (C3) functions as a SOD mimetic by reducing agerelated increase in superoxide levels and prevention of age-related loss of mitochondrial membrane potential in brain. *Free Rad. Biol. Med.* 2004, 37, S163–S163.

[44] Xia, XR; Monteiro-Riviere, NA; Riviere, JE. Intrinsic biological property of colloidal fullerene nanoparticles (nC60): Lack of lethality after high dose exposure to human epidermal and bacterial cells. *Toxicology Letters* 2010, 197, 128–134.

[45] Gharbi, N; Pressac, M; Hadchouel, M; Szwarc, H; Wilson, SR; Moussa, F. Fullerene is a powerful antioxidant in vivo with no acute or subacute toxicity. *Nano Lett.* 2005, 5, 2578–2585.

[46] Zhao, QF; Zhu, Y; Ran, TC; Li, JG; Li, QN; Li, WX. Cytotoxicity of fullerenols on Tetrahymena pyriformis. *Nucl. Sci. Technol./Hewuli* 2006, 17, 280–284.

[47] Fumelli, C; Marconi, A; Salvioli, S; Straface, E; Malorni, W; Offidani, AM; Pellicciari, R; Schettini, G; Giannetti, A; Monti, D; Franceschi, C; Pincelli, Carlo. Carboxyfullerenes Protect Human Keratinocytes from Ultraviolet-B-Induced Apoptosis. *Journal of Investigative Dermatology*, 2000, 115, 835–841.

[48] Ito, S; Itoga, K; Yamato, M; Akamatsu, H; Okanoa, T. The co-application effects of fullerene and ascorbic acid on UV-B irradiated mouse skin. *Toxicology2010*, 267, 27–38.

[49] Mirkov, SM; Djordjevic, AN; Andric, NL; Andric, SA; Kostic, TS; Bogdanovic, GM; Vojinovic-Miloradov, MB; Kovacevic, RZ. Nitric oxide-scavenging activity of polyhydroxylated fullerenol, C60(OH)24. *Nitric Oxide*, 2004, 11, 201-207.

[50] Saitoh, Y; Miyanishi, A; Mizuno, H; Kato, S; Aoshima, H; Kokubo, K; Miwa, N. Super-highly hydroxylated fullerene derivative protects human keratinocytes from UV-induced cell injuries together with the decreases in intracellular ROS generation and DNA damages (2011), *J. Photochem. Photobiol. B*, 2011, 102, 69-76.

[51] Gan, L; Huang, S; Zhang, X; Zhang, A; Cheng, B; Cheng, H. Fullerenes as a tert-butylperoxy radical trap, metal catalyzed reaction of tert-butyl hydroperoxide with fullerenes, and formation of the first fullerene mixed peroxides C60(O)(OOtBu)4 and C70(OOtBu)10. *Jam. Chem. Soc.* 2002, 124, 13384–13385.

[52] Huang, S; Xiao, Z; Wang, F; Gan, L; Zhang, X; Hu, X. Selective preparation of oxygen-rich [60]fullerene derivatives by stepwise addition of tert-butylperoxy radical and further functionalization of the fullerene mixed peroxides. *J. Org. Chem.* 2004, 69, 2442–2453.

[53] Alcaraz, MJ; Megıras, J; Garcıra-Arnandis, I; Clerrigues, V; Guillern, MI. New molecular targets for the treatment of osteoarthritis. *Biochem. Pharmacol.*, 2010, 80, 13–21.

[54] Sayes, CM; Gobin, AM; Ausman, KD; Mendez, J; West, J; Colvin, VL. Nano-C60 cytotoxicity is due to lipid peroxidation. *Biomaterials*, 2005, 26, 7587-7595.

[55] Oberdörster, E. Manufactured nanomaterials (fullerenes, C60) induce oxidative stress in the brain of juvenile largemouth bass. *Environ. Health Perspect.*, 2004, 112, 1058-1062.

[56] Vileno, B; Jeney, S; Sienkiewicz, A; Marcoux, PR; Miller, LM; Forró, L. Evidence of lipid peroxidation and protein phosphorylation in cells upon oxidative stress photo-generated by fullerols, *Biophys. Chem.*, 2010, 152, 164-169.

[57] Socoowski Britto, R; Garcia, ML; Martins da Rocha, A; Flores, JA; Pinheiro, MV; Monserrat, JM; Ferreira, JL.. Effects of carbon nanomaterials fullerene C60 and fullerol C60(OH)18–22 on gills of fish Cyprinus carpio (Cyprinidae) exposed to ultraviolet radiation. *Aquatic Toxicology* 2012, 114–115, 80– 87.

[58] Isakovic, A; Markovic, Z; Todorovic-Markovic, B; Nikolic, N; Vranjes-Djuric, S; Mirkovic, M; Dramicanin, M; Harhaji, L; Raicevic, N; Nikolic, Z; Trajkovic, V. Distinct cytotoxic mechanisms of pristine versus hydroxylated fullerene. *Toxicol. Sci.*, 2006, 91, 173–183.

[59] Sayes, C; Fortner, J; Guo, W; Lyon, D; Boyd, A; Ausman, K; Tao, Y; Sitharaman, B; Wilson, J; Hughes, J; West, J; Colvin, V. The differential cytotoxicity of water-soluble fullerenes. *Nano Letters* 2004, 4(10), 1881–1887.

[60] Dugan, LL; Lovett, EG; Quick, KL; Lotharius, J; Lin, TT; O'Malley, KL. Fullerene-based antioxidants and neurodegenerative disorders. *Parkinson Relat. Disord.*, 2001, 7, 243-246.

[61] Lin, J. & Wu, C. Surface characterization and platelet adhesion studies on polyurethane surface immobilized with C60. *Biomaterials*, 1999, 20, 1613-1620.

[62] Monti, D; Moretti, L; Salvioli, S; Strafolacince, E; Malorni, W; Pellicciari, R; Schettini, G; Bisaglia, M; Pincelli, C; Fumelli, C; Bonafe, M; Franceschi, C. C60 carboxyfullerene exerts a protective activity against oxidative stress-induced apoptosis in human peripheral blood mononuclear cells. *Biochem. Biophys. Res. Commun*, 2000, 277, 711-717.

[63] Yamawaki, H. & Iwai, N. Cytotoxicity of water-soluble fullerene in vascular endothelial cells. *Am. J. Physiol. Cell Physiol.*, 2006, 290, C1495–C1502.

[64] Yang, XL; Fan, CH; Zhu, HS. Photo-induced cytotoxicity of malonic acid [C60]fullerene derivatives and its mechanism. *Toxicol. In Vitro*, 2002, 16, 41–46.

[65] Rancan, F; Rosan, S; Boehm, F; Cantrell, A; Brellreich, M; Schoenberger, H; Hirsch, A; Moussa, F. Cytotoxicity and photocytotoxicity of a dendritic C60 mono-adduct and a malonic acid C60 tris-adduct on Jurkat cells. *J. Photochem. Photobiol.*, B, 2002, 67, 157–162.

[66] Wielgus, AR; Zhao, B; Chignell, CF; Hu, DN; Roberts, JE. Phototoxicity and cytotoxicity of fullerol in human retinal pigment epithelial cells. *Toxicol. Appl. Pharmacol.*, 2010, 242, 79-90.

[67] Rouse, JG; Yang, J; Barron, AR; Monteiro-Riviere, NA. Fullerene-based amino acid nanoparticle interactions with human epidermal keratinocytes. *Toxicology in Vitro*, 2006, 20, 1313–1320.

[68] Isakovic, A; Markovic, Z; Todorovic-Markovic, B; Nikolic, N; Vranjes-Djuric, S; Mirkovic, M; Dramicanin, M; Harhaji, L; Raicevic, N; Nikolic, Z; Trajkovic, V. Distinct cytotoxic mechanisms of pristine versus hydroxylated fullerene. *Toxicol. Sci.*, 2006, 91, 173–183.

[69] Yang, XL; Fan, CH; Zhu, HS. Photo-induced cytotoxicity of malonic acid [C60]fullerene derivatives and its mechanism. *Toxicol. In Vitro*, 2002, 16, 41–46.

[70] Tzoupis, H; Leonis, G; Durdagi, S; Mouchlis, V; Mavromoustakos, T; Papadopoulos, MG. Binding of novel fullerene inhibitors to HIV-1 protease: insight through molecular dynamics and molecular mechanics Poisson–Boltzmann surface area calculations. *J. Comput. Aided Mol. Des.*, 2011, 25, 959–976.

[71] Bosi, S; Da Ros, T; Spalluto, G; Balzarini, J; Prato, M. Synthesis and anti-HIV properties of new watersoluble bis-functionalized [60]fullerene derivatives. *Bioorg. Med. Chem. Lett.*, 2003, 13, 4437-4440.

[72] Marchesan, S; Da Ros, T; Spalluto, G; Balzarini, J; Prato, M. Anti-HIV properties of cationic fullerene derivatives. *Bioorg. Med. Chem. Lett.*, 2005, 15, 3615–3618.

[73] Marcorin, GL; Da Ros, T; Castellano, S; Stefancich, G; Bonin, I; Miertus, S; Prato, M. Design and synthesis of novel [60]fullerene derivatives as potential HIV aspartic protease inhibitors. *Org. Lett.*, 2000, 2, 3955-3958.

[74] Schuster, DI; Wilson, SR; Schinazi, RF. Anti-human immunodeficiency virus activity and cytotoxicity of derivatized Buckminsterfullerenes. *Bioorg. Med. Chem. Lett.*, 1996, 6, 1253-1256.

[75] Tang, YJ; Ashcroft, JM; Chen, D; Min, G; Kim, CH; Murkhejee, B; Larabell, C; Keasling, JD; Chen, FF. Charge-associated effects of fullerene derivatives on microbial structural integrity and central metabolism. *Nano Lett.*, 2007, 7, 754–760.

[76] Aoshima, H; Kokubo, K.; Shirakawa, S; Ito, M; Yamana, S; Oshima, T. Antimicrovial activity of fullerenes and their hydroxylated derivatives, *Biocontrol. Sci.*, 2009, 14, 69–72.

[77] Yudoh, K; Karasawa, R; Masuko, K,; Kato, T. Water-soluble fullerene (C60) inhibits the osteoclast differentiation and bone destruction in arthritis. *Int. J. Nanomed.*, 2009, 4, 233-239.

[78] Yudoh, K; Karasawa, R; Masuko, K; Kato, T. Water-soluble fullerene (C60) inhibits the development of arthritis in the rat model of arthritis. *Int. J. Nanomed.*, 2009, 4, 217-225.

[79] Yudoh, K; Shishido, K; Murayama, H; Yano, M; Matsubayashi, K; Takada, H; et al. Water-soluble C60 fullerene prevents degeneration of articular cartilage in osteoarthritis via down-regulation of chondrocyte catabolic activity and inhibition of cartilage degeneration during disease development. *Arthritis Rheum.*, 2007, 56, 3307-3318.

[80] Corona-Morales, AA; Castell, A; Escober, A; Drucker-Colin, R; Zhang, L. Fullerene C60 and ascorbic acid protect cultured chromaffin cells against levodopa toxicity. *J. Neurosci. Res.*, 2003, 71(1), 121–126.

[81] Tykhomyrov, AA; Nedzvetsky, VS; Klochkov, VK; Andrievsky,GV. Nanostructures of hydrated C60 fullerene (C60HyFn) protect rat brain against alcohol impact and attenuate behavioral impairments of alcoholized animals. *Toxicology*, 2008, 246, 158-165.

[82] Mori, T; Ito, S; Namiki, M; Suzuki, T; Kobayashi, S; Matsubayashi, K; Sawaguchi, T. Involvement of free radicals followed by the activation of phospholipase A2 in the mechanism that underlies the combined effects of methamphetamine and morphine on subacute toxicity or lethality in mice: Comparison of the therapeutic potential of fullerene, mepacrine, and cooling, *Toxicology*, 2007, 236, 149-157.

[83] Cai, X; Jai, H, Liu, Z; Hou, B, Luo, C; Feng, Z; Li, W; Liu, J. Polyhydroxylated fullerene derivative C(60)(OH)(24) prevents mitochondrial dysfunction and oxidative damage in an MPP(+) -induced cellular model of Parkinson's disease. *J. Neurosci. Res.*, 2008, 86(16), 3622-3634.

[84] Bobrova, NA; Mikitiuk, MV; Kutsenko, LA; Kaĭdashev, IP. [Effect of fullerene C60 on free-radical induced lipid peroxidation processes in bronchial asthma]. *Patol. Fiziol. Eksp. Ter.*, 2012, 3, 109-114.

[85] Lai, YL; Murugan, P; Hwang, KC. Fullerene derivative attenuates ischemia-reperfusion-induced lung injury. *Life Sci.*, 2003, 72, 1271-1278.

[86] Chen, YW; Hwang, KCH; Yen, CC; Lai, YL. Fullerene derivatives protect against oxidative stress in RAW 264.7 cells and ischemia-reperfused lungs. *Am. J. Physiol.*, 2004, 287, 21-26.

[87] Roursgaard, M; Poulsen, SS; Kepley, CL; Hammer, M; Nielsen, GD; Larsen, ST. Polyhydroxylated C60 Fullerene (Fullerenol) Attenuates Neutrophilic Lung Inflammation in Mice. *Basic & Clinical Pharmacology & Toxicology*, 2008, 103(4), 386–388.

[88] Norton, SK; Wijesinghe, DS; Dellinger, A; Sturgill, J; Zhou, Z; Barbour, S; Chalfant, C; Conrad, DH; Kepley, CL. Epoxyeicosatrienoic acids are involved in the C(70) fullerene derivative-induced control of allergic asthma. *J. Allergy Clin. Immunol.*, 2012, 130(3), 761-769.

[89] Xiao, L; Matsubayashi, K; Miwa, N. Inhibitory effect of the water-soluble polymer-wrapped derivative of fullerene on UVA-induced melanogenesis via downregulation of tyrosinase expression in human melanocytes and skin tissues. *Arch. Dermatol. Res.*, 2007, 299, 245-257.

[90] Fujimoto, T; Aoshima, H; Kokubo, K; Ito, M; Miwa, N. Antioxidant property of fullerene is effective in skin whitening. AB54 *J. Am. Acad. Dermatol.* March 2010. P1600.

[91] Inui, S; Aoshima, H; Nishiyama, A; Itami, S. Improvement of acne vulgaris by topical fullerene application: unique impact on skin care. *Nanomed. Nanotechnol. Biol. Med.*, 2011, 7, 238-241.

[92] Huczko, A. & Lange, H. Fullerenes: experimental evidence for a null risk of skin irritation and allergy. *Fullerene Sci. Technol.* 1999, 7, 935-939.

[93] Brettreich, M. & Hirsch, A. A highly water-soluble dendro[60]fullerene. *Tetrahedron Lett.*, 1998, 39, 2731-2734.

[94] Shimanovsky, NS. Nanotechnologies in contemporary pharmacology. *Int. Med. J.* (Russ.), 2009, 1, 131-135.

[95] Zhu, J; Ji, Z; Wang, J; Sun, R; Zhang, X; Gao, Y; Sun, H; Liu, Y; Wang, Z; Li, A; Ma, J; Wang, T; Jia, G; Gu, Y. Tumor-inhibitory effect and immunomodulatory activity of fullerol C60(OH)x. *Small*, 2008, 4, 1168–1175.

[96] Rebecca, M; Hsing-Lin, W; Jun, G; Srinivas, I; Gabriel, MA; Jennifer, M; Andrew, SP; Yuping, B; Chun-Chih, W; Zhong, C; Yuan, G; Rashi, I. Impact of physicochemical properties of engineered fullerenes on key biological responses. *Toxicol. Appl. Pharmacol.*, 2009, 234, 58-67.

[97] Jiao, F; Liu, Y; Qu, Y; Li, W; Zhou, G; Ge, C; Li, Y; Sun, B; Chen, C. Studies on anti-tumor and antimetastatic activities of fullerenol in a mouse breast cancer model, *Carbon*, 2010, 48, 2231-2243.

[98] Prylutska, S; Burlaka, AP; Klymenko, PP; Grynyuk, II; Prylutskyy, YuI; Schütze, C; Ritter, U. Using water-soluble C60 fullerenes in anticancer therapy, *Cancer Nanotechnol.*, 2011, 2, 105-110.

[99] Braun, K; Dunsch, L; Pipkorn, R; Bock, M; Baeuerle, T; Waldeck, W; Wiessler, M. Gain of 500-fold sensitivity on an intravital MR contrast agent based on an endohedral gadolinium-cluster-fullerene-conjugate: a new chance in cancer diagnostics. *Int. J. Med. Sci.*, 2010, 7, 136–146.

[100] Chen, Z; Mao, R; Liu, Y. Fullerenes for cancer diagnosis and therapy: preparation, biological and clinical perspectives. *Curr. Drug Metab.*, 2012, 13, 1035–1045.

[101] Dou, Z; Xu, Y; Sun, H; Liu, Y. Synthesis of PEGylated fullerene-5-fluorouracil conjugates to enhance the antitumor effect of 5-fluorouracil. *Nanoscale* 2012, 4, 4624–4630.

[102] Kamat, JP; Devasagayam, TP; Priyadarsini, KI; Mohan, H; Mittal, JP. Oxidative damage induced by the fullerene C60 on photosensitization in rat liver microsomes. *Chem. Biol. Interact.*, 1998, 114(3), 145-159.

[103] Wielgus, AR; Zhao, B; Chignell, CF; Hu, DN; Roberts, JE. Phototoxicity and cytotoxicity of fullerol in human retinal pigment epithelial cells. *Toxicol. Appl. Pharmacol.*, 2010, 242, 79-90.

[104] Vileno, B; Jeney, S; Sienkiewicz, A; Marcoux, PR; Miller, LM; Forró, L. Evidence of lipid peroxidation and protein phosphorylation in cells upon oxidative stress photo-generated by fullerols. *Biophys. Chem.*, 2010, 152(1-3), 164-169.

[105] Sharma, SK; Chiang, LY; Hamblin, MR. Photodynamic therapy with fullerenes in vivo: reality or a dream? *Nanomedicine* (Lond). 2011, 6(10), 813-825.

[106] Liao, F; Saitoh, Y; Miwa, N. Anticancer effects of fullerene [C60] included in polyethylene glycol combined with visible light irradiation through ROS generation and DNA fragmentation on fibrosarcoma cells with scarce cytotoxicity to normal fibroblasts. *Oncol. Res.*, 2011, 19(5), 203-216.

[107] Yu, JG; Jiao, FP; Chen, XQ; Jiang, XY; Peng, ZG; Zeng, DM; Huang, DS. Irradiation-mediated carbon nanotubes' use in cancer therapy. *J. Cancer Res. Ther.*, 2012, 8(3), 348-354.

[108] Kato, S; Kimura, M; Miwa, N. Radiosensitization by fullerene-C60 dissolved in squalene on human malignant melanoma through lipid peroxidation and enhanced mitochondrial membrane potential. *Radiation Physics and Chemistry*, 2014, 97, 134–141.

[109] Liu, JH; Cao, L; Luo, PG; Yang, ST; Lu, F; Wang, H; Meziani, MJ; Haque, SA; Liu, Y; Lacher, S; Sun, YP. Fullerene-conjugated doxorubicin in cells, *Acs Appl. Mater. Interf.*, 2010, 2, 1384-1389.

[110] Prabhu, P. & Patravale V. The upcoming field of theranostic nanomedicine: an overview. *J. Biomed. Nanotechnol.*, 2012, 8(6), 859-882.

[111] Ramos-Perez, V; Cifuentes, A; Coronas, N; de Pablo, A; Borrós, S. Modification of carbon nanotubes for gene delivery vectors. *Methods Mol. Biol.*, 2013, 1025, 261-268.

[112] Oberdörster E. Manufactured nanomaterials (fullerenes, C60) induce oxidative stress in the brain of juvenile largemouth bass. *Environ. Health Perspect.*, 2004, 112(10), 1058-1062.

[113] Horie, M; Nishio, K,; Kato, H; Shinohara, N; Nakamura, A; Fujita, K; Kinugasa, S; Endoh, S; Yamamoto, K; Yamamoto, O; Niki, E; Yoshida, Y; Iwahashi, H. In vitro evaluation of cellular responses induced by stable fullerene C60 medium dispersion. *J. Biochem.* 2010, 148(3), 289-298.

[114] Nishizawa, C; Hashimoto, N; Yokoo, S; Funakoshi-Tago, M; Kasahara, T; Takahashi, K; Nakamura, S; Mashino, T. Pyrrolidinium-type fullerene derivative-induced apoptosis by the generation of reactive oxygen species in HL-60 cells. *Free Radic. Res.*, 2009, 43(12), 1240-1247.

[115] Aschberger, K; Johnston, HJ; Stone, V; Aitken, RJ; Hankin, SM; Peters, SA; Tran, CL; Christensen, FM. Review of fullerene toxicity and exposure – Appraisal of a human

health risk assessment, based on open literature. *Regulatory Toxicology and Pharmacology*, 2010, 58, 455–473.

[116] Xia, XR; Monteiro-Riviere, NA; Riviere, JE. Intrinsic biological property of colloidal fullerene nanoparticles (nC60): Lack of lethality after high dose exposure to human epidermal and bacterial cells. *Toxicology Letters*, 2010, 197, 128–134.

[117] Andrievsky, GV; Bruskov, VI; Tykhomyrov, AA; Gudkov, SV. Peculiarities of the antioxidant and radioprotective effects of hydrated C60 fullerene nanostuctures in vitro and in vivo. *Free Radic. Biol. Med.*, 2009, 47(6), 786-793.

[118] Brant, JA; Labille, J; Bottero, JY; Wiesner, MR. Characterizing the impact of preparation method on fullerene cluster structure and chemistry. *Langmuir*, 2006, 22(8), 878-885.

[119] Avdeev, MV; Khokhryakov, AA; Tropin, TV; Andrievsky, GV; Klochkov, VK; Derevyanchenko, LI; Rosta, L; Garamus, VM; Priezzhev, VB; Korobov, MV; Aksenov, VL. Structural features of molecular-colloidal solutions of C60 fullerenes in water by small-angle neutron scattering. *Langmuir*, 2004, 20(11), 4363-4368.

[120] Henry, TB; Menn, FM; Fleming, JT; Wilgus, J; Compton, RN; Sayler, GS. Attributing effects of aqueous C60 nano-aggregates to tetrahydrofuran decomposition products in larval zebrafish by assessment of gene expression. *Environ. Health Perspect.*, 2007, 115(7), 1059-1065.

[121] Kovochich, M; Espinasse, B; Auffan, M; Hotze, EM; Wessel, L; Xia, T; Nel, AE; Wiesner, MR. Comparative toxicity of C60 aggregates toward mammalian cells: role of tetrahydrofuran (THF) decomposition. *Environ. Sci. Technol.*, 2009, 43(16), 6378-6384.

[122] Levi, N; Hantgan, RR; Lively, MO; Carroll, DL; Prasad, GL. C60-fullerenes: detection of intracellular photoluminescence and lack of cytotoxic effects. *J. Nanobiotechnology*, 2006, 4, 14 (doi:10.1186/1477-3155-4-14).

[123] Gao, J; Wang, HL; Shreve, A; Iyer, R. Fullerene derivatives induce premature senescence: A new toxicity paradigm or novel biomedical applications, *Toxicol. Appl. Pharmacol.*, 2010, 244, 130–143.

[124] Shinohara, N; Gamo, M; Nakanishi, J. Fullerene C60: inhalation hazard assessment and derivation of a period limited acceptable exposure level. *Toxicol. Sci.*, 2011, 123, 576-589.

[125] Rouse, JG; Yang, J; Ryman-Rasmussen, JP; Barron, AR; Monteiro-Riviere, NA. Effects of mechanical flexion on the penetration of fullerene amino acid-derivatized peptide nanoparticles through skin. *Nano Lett.*, 2007, 7(1), 155-160.

[126] Leavens, TL; Xia, XR; Lee, HA; Monteiro-Riviere, NA; Brooks, JD; Riviere, JE. Evaluation of perfused porcine skin as a model system to quantitate tissue distribution of fullerene nanoparticles. *Toxicology Letters*, 2010, 197, 1–6.

In: Fullerenes
Editor: Shannon B. Ellis

ISBN: 978-1-63321-385-2

Chapter 10

SYNTHESIS OF FULLERENES IN FLAMES: A REVIEW

Z. A. Mansurov**, M. Nazhipkyzy and N. G. Prikhodko
The Institute of Combustion Problems, Almaty, Republic of Kazakhstan

ABSTRACT

A fullerene is any molecule composed entirely of carbon, in the form of a hollow sphere, ellipsoid, tube, and many other shapes. In this paper were considered works on formation and the synthesis of fullerenes in flames in the following sections, such as, the structure and properties of fullerenes; applications of fullerenes; methods for identification of fullerenes; methods for producing fullerenes; formation and synthesis of fullerenes in flame; synthesis of fullerenes in flames with applied electrical field; mechanism of fullerenes formation. Also was presented modified scheme for soot and fullerenes formation process in flames.

It has been found that if the peripheral zone of the benzene flame is heated by some external source, such as a laser beam, which not just burns the soot but also creates the same conditions as in the middle of the flame, the concentration of fullerenes increases.

The increase in the yield of fullerenes in the case where a ring electrode is positioned above the peripheral part of the reaction zone of a flame is due to the glow-discharge conditions providing an effective synthesis of fullerenes.

Thus, it has been established that the yield of fullerenes increases under the action of a glow discharge in the case where an electrode (a needle or a ring) is positioned directly above a flame, and a maximum yield of fullerenes is attained with the use of a ring electrode positioned above the central region of the flame front. The maximum yield of the fullerene C_{60} was β=15% of the soot formed.

INTRODUCTION

The discovery of fullerenes representing a new form of existence of one of the most abundant elements on the Earth - carbon - has been recognized as one of the most significant achievements in 20th century science. Despite the fact that it has long been known that carbon

* Email: zmansurov@kaznu.kz.

atoms have a unique ability to bind into complex three-dimensional branched structures, which forms the basis of the whole subject of organic chemistry, the possibility of only one carbon forming stable skeleton.

The C_{60} and C_{70} fullerene ions were detected in premixed acethylene/oxygen/argon and benzene/oxygen/argon flames in 1987, identified by the mass-spectrometry method [1]. Howard et al. [2] have obtained large amounts of C_{60} and C_{70} in laminar premixed soot forming flames of benzene and oxygen at low pressures. Unlike the evaporation of graphite, in the fullerenes formed in flames the ratio C_{70}/C_{60} changes from 0.26 to 8.8 (in the case of evaporation of graphite, this ratio changes from 0.02 to 0.18).

The synthesis of fullerenes in nature has been a subject of intense discussion for many years. It is unclear whether they form through assembly of small carbon bearing molecules (bottom-up) or fragmentations of large compounds (top-down). The discovery of fullerenes in rocks (Buseck et al., 1992) suggests that fullerenes can be generated through a solid-state process. Many interesting scenarios, such as cataclysmic impact, extensive wildfires, chondritic impactor, vaporization of carbon by lightning strike, and pyrolysis of organic matter, have been proposed to explain the formation of fullerenes in geological environments (Buseck, 2002, and references therein). The hydrocarbon combustion process yields a C70/C60 ratio ranging from 0.26 to 0.57 (Howard et al., 1991), which tends to increase with increasing pressure. This ratio is much larger than that obtained through graphite evaporation methods (0.02–0.18, e.g., Ajie et al., 1990). One cannot simply argue that the low C70/C60 ratio in PNs seems not to favor the top-down scenario, because the circumstellar envelopes have a significantly lower pressure than laboratories and the fullerene formation in combustion experiments is not completely understood. Moreover, this C70/C60 ratio range is very uncertain since C70 has been measured only in a very few sources. The abundance of generated higher fullerenes beyond C70 is relatively low, which cannot be theoretically explained. In addition, we cannot rule out the possibility that the C70/C60 ratio can be modified by chemical processes in circumstellar envelopes. Circumstellar envelopes around evolved stars, therefore, provide a unique laboratory to investigate fullerene formation and processes under low pressure environments.

The knowledge of fullerenes formation in flames has grown significantly during the short time since this process was first established. In this paper we present a review of the formation and the synthesis of fullerenes in flames in the following sections:

- The structure and properties of fullerenes;
- Applications of fullerenes;
- Methods for identification of fullerenes;
- Methods for producing fullerenes;
- Formation and synthesis of fullerenes in flame;
- Synthesis of fullerenes in flames with applied electrical field;
- Mechanism of fullerenes formation.

Fullerenes are closed hollow spherical shells or more complex 3D shapes, constructed from pentagonal and hexagonal carbon cells (C_{2n}, $n \geq 15$), each of which has 60 or more carbon atoms. Molecular group known as fullerenes, contain other three-dimensional carbon molecules, including C_{70}, C_{76}, C_{78}, C_{84}, etc. Among these C_{60} is the most common fullerene consisting of 20 hexagons and 12 pentagons. C_{60} is a stable molecule crystal spheroid with a radius of 0.357 nm and the third allotropic form of carbon after diamond, graphite and carbon.

THE STRUCTURE AND PROPERTIES OF FULLERENES

The structure of fullerene C_{60} proposed by R. Smalley resembles a football, so it is sometimes called footballene, and that of C_{70} resembles a rugby ball. Fullerenes C_{60} and C_{70} were identified in 1985 and prepared in macroscopic quantities in 1990, both by evaporation of graphite in an arc discharge [3]. In flames, fullerene ions were found in 1987, and in 1991, C_{60} and C_{70} were extracted in large quantities from laminar soot flames of premixed mixtures of benzene and oxygen at low pressures [4] and then was identified spectroscopically. Kr¨atschmer et al. (1990b) developed a method to efficiently produce C_{60} in the laboratory, making it possible to study in detail its electronic and vibrational properties. Because of its high stability and symmetry, C60 can be taken as a benchmark to study other fullerenes. C_{60} and its derivatives have long been suspected to be present in the universe. The search for this compound in nature started soon after its synthesis in the laboratory, and it has been detected in geological materials and meteorites and reported by several groups (e.g., Buseck et al., 1992; Becker and Bada, 1994; Becker et al., 1994; Heymann et al., 1996).

Formation Mechanism of C_{60}

I

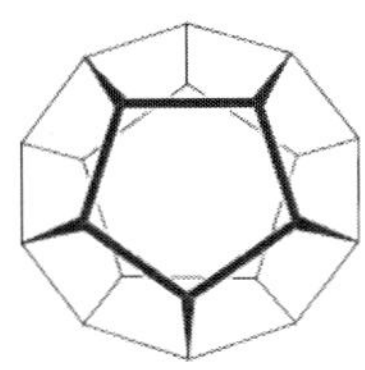

II (H omitted)

Fullerene C_{70} is formed in an amount of 5-6 times less than C_{60}, in the electric method. In form it is close to an ellipsoid, whose axes size - 0,788 and 0,682 nm.

All the fullerenes relate into the class of carbons with a polyhedral-closed shell. They were detected as ionized particles in the fuel-rich plane of premixed acetylene and benzene-oxygen flames at low pressures with the use of molecular-beam sampling in combination with mass-spectrometry analysis [1].

In C_{70} there are five types of carbon atoms, i.e., there are five crystallochemical nonequivalent positions of carbon atoms (for example, there is the top where only six-membered rings converge). The length of the C-C take values 0,138-0,146 nm. The C_{70} molecule has a lower symmetry than C_{60}. Fullerene C_{70} at room temperature forms a hexagonal close packing. Fullerene molecules perform a rotational motion around the equilibrium position with a frequency of 1012 Hz. With decreasing temperature, the fcc lattice becomes a simple cubic.

The solubility of fullerenes. Fullerenes are practically insoluble in polar solvents such as alcohols, acetone, tetrahydrofuran, etc. They are poorly soluble in alkanes (acyclic saturated hydrocarbons) of normal structure (pentane, hexane, decane, etc.), and the increasing number of carbon atoms increases the solubility in alkanes. Fullerenes with the highest solubility are characterized by aromatic hydrocarbons and their derivatives, among which naphthalene derivatives are the highest. The analysis shows that fullerenes are best dissolved in solvents, for which the value of the specific enthalpy of vaporization is divided by the volume of the solvent molecule close to the corresponding value for the fullerene molecule (around 100 kal/sm^3).

Fullerenes have appreciable solubility in a broad class of organic solvents. This feature is explained by the unusual structure of fullerenes, with increased reactivity. Interest in investigating the behavior of fullerenes in solutions is due to the fact that fullerenes are the only soluble form of carbon that has both applied and fundamental importance. The uniqueness of the fullerene is also evident since they form clusters of several molecules in the solutions. Solvent interaction with fullerenes is due to van-der-Waals forces. Thus solutions of fullerenes begin to show unusual optical, thermodynamic, kinetic, and other properties. This is manifested in a nonmonotonic temperature dependence of the solubility of fullerenes in some solutions, and with increasing temperature the solubility, for instance C_{60} decreases. However, it should be noted that nonmonotonic temperature dependence of solubility is characteristic only in fullerenes C_{60} and it is not observed in other fullerenes. This characterized complex of solubility of structurally related molecules indicates the presence of several distinct competing mechanisms in the solubility of fullerenes, that in turn reflect on the shape of the temperature dependence. Solutions of fullerenes also have non-linear concentration dependence of the optical susceptibility, which can be used to block the light when it reaches a certain critical value (as a nonlinear optical shutter for protecting optical devices from intensive optical radiation).

Of the fullerene family, C_{60} and C_{70} are the major isomers obtained with 73 and 23% yield, respectively, a variety of molecules larger than C_{60} and C_{70} representing a total amount of 3 to 4% by weight. The colors of the fullerene family vary according to their molecular weight and symmetry: their colors in solution are magenta (C_{60}), port-wine red (C_{70}), brown (C_{76}, C_{78}) and yellow-green (C_{84}).

The physical properties. Due to the high reactivity of higher fullerenes and a small percentage of their output in the synthesis of fullerenes, their properties in literature are very poorly portrayed. The properties of fullerenes C_{60} and C_{70} are more widely studied and reported. Electron affinity in C_{60} is 2.65 eV, the ionization potential of - 7.61 eV binding energy per carbon atom - 7 eV, the polarizability is close to 80 Å3. The molecule can take up to 12 electrons and release one electron, i.e. the charge on the C_{60} can vary from 1 to 12. In the condensed state is a fullerene with molecular crystals - fullerites, in which the binding energy between molecules is 1.86 eV and is carried out by van der Waals forces. Fullerites have an orientational disorder of characterizing the ability of the thermal motion of molecules as to the relative change in the conditions of the spatial orientation of the crystal environment, which is typical for low energy barriers between the different positions of the molecules of the order of 0.2-0.3 eV.

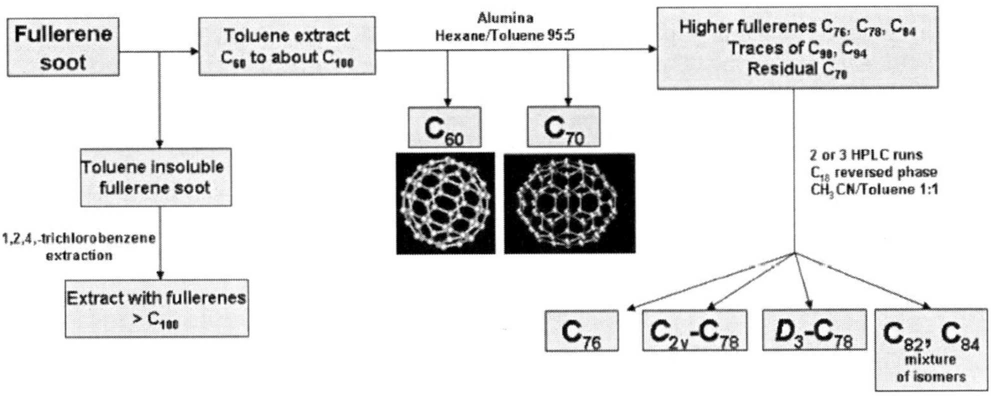

Fullerenes purification and isolation protocol [36].

Applications of Fullerenes

The uniqueness of the fullerene structure determines their unusual physical and chemical properties that have great promise for practical use. Fullerenes are similar in their properties to graphite, and therefore inherit its high heat resistance and extraordinary electrical properties. Also, it is widely known for its graphite anti-friction properties. The specific three-dimensional structure of fullerene molecules makes it possible to obtain a previously unknown class of heteroatomic compounds - doped fullerenes, and these compounds have begun a new turn in chemical (biochemical) science. Here are the most promising areas of application of fullerenes and their derivatives, some of which received a practical embodiment. The greatest number of patents from abroad dedicated the application of fullerenes in semiconductor technology and nanoelectronics as photodiodes, transistors and solar cells. C_{60}, having high electro negativity, readily forms charge-transfer complexes, and acts as an electron acceptor. In diodes consisting of a fullerene as an acceptor covalently bound to a photoactive donor, and exposed to light, a photo induced electron transition occurs. High efficiency of light absorption and charge separation is a fundamental principle of solar panels. In the near future, this property of the fullerenes may be used in the creation of energy solar cells and other devices using photoconductivity. Application of fullerenes for growing diamond films also proves promising. In the process of destruction by microwave radiation of C_{60}, a pair of carbon C-C is formed. Deposited on the surface of SiO_2, they form diamond microcrystals forming the smooth diamond film. There are developments on the use of endohedral fullerenes and their derivatives, including inside the carcass half-integer spin atoms such as hydrogen or phosphorus, as the quantum computer parts [6].

Application of fullerenes promise anti-friction coatings, solid lubricants and lubricant additives. Experiments have shown that the introduction of fullerene additives to liquid oils and greases form fullerene coatings that can significantly improve the parameters of friction and wear parts and mechanisms, in some cases - repeatedly. This makes their application promising for a broad range of machinery and industrial equipment [7]. A small percentage of additives by weight are good enough in many cases. One result is shown using fullerene soot, which provides a real absorptive effect. Materials with improved tribological properties are also prepared by introducing the polymeric binder in carbon Nanostructures polyhedral fulleroid multilayer type and high-temperature pressing of powder mixtures of iron and cobalt and fullerites [8]. In the second case, due to the formation of superhard carbon particles in a metal matrix, the wear resistance of the resulting composite is much higher than the high-tool-steels. Composites are versatile: they can be used in a variety of friction conditions.

Fullerenes are able to take over a large number of free electrons, so they are strong oxidizing agents and form many new chemical compounds. The possible applications of fullerenes in pharmacology are perhaps more numerous. At present fullerenes based on antiviral and neurotropic drugs, and hemosorption adsorbents are being developed, as well as methods of treatment using them.

Fullerene structures are promising catalysts of different transformations and provide high selectivity [9]. Dehydrogenate activity of fullerenic catalysts with respect to alkanes, including methane, are manifested in a high temperature range. Fullerenic catalysts can be considered as functional analogs of noble metals.

Fullerenes are able to connect a large number of free electrons, so they are strong oxidizing agents and form many new chemical compounds. Possible applications of fullerenes in pharmacology are perhaps the most numerous. Antiviral fullerenes and neurotropic drugs, adsorbents for hemosorption, as well as methods of treatment using them are currently being developed.

It was created a series of derivatives of fullerene C_{60} with pronounced membranotropic, antioxidant, antiviral and immunomodulatory properties, including inhibition of HIV and cytomegalovirus infections [10]. Organic derivatives of fullerenes have potentially wide applications in biology, in particular for photodisintegration DNA, inhibiting protease of a HIV, neuroprotection and apoptosis (natural cell death). Their use is planned for the treatment of Alzheimer's and Parkinson's, stroke, atherosclerosis. Their use also looks promising for particular compounds grafted to amino acid fullerenes. Fullerenes can be used for modification of proteins. The diameter of the molecule of C_{60} is close to the steroid hormone or protein alpha-helices, so fullerene derivatives are ideal ligand enzymes and receptors. C_{60} has proved itself as a good antioxidant and is 10-100 times more effective than vitamin E. The damage oxidative by free radicals refers to the main factors causing cell death and disease, which is the cause of many diseases (degeneration of the central nervous system, diabetes, atherosclerosis, coronary artery disease) and physical aging. Free radicals occur during of normal oxygen uptake and by chain reactions, which can disrupt chemical bonds in hundreds of enzymes, cell membranes, and DNA molecules. The healthy body produces endogenous antioxidants - enzymes (superoxide dismutase, catalase) and small molecules (glutathione, vitamins C and E), which deactivate free radicals. However, with age and with certain diseases the concentration of the antioxidants produced by the body is reduced. Fullerene C_{60} is unique in that it contains 30 conjugated double bonds, each of which can neutralize free radicals, and does so with a very high speed. With the interaction of free radicals, C_{60} will form very stable compounds. It is assumed that this interaction has catalytic character. Once in the body, the fullerene molecules accumulate in the cell mitochondria, where disease is created with the highest concentration of free radicals. Effective antioxidants in vivo (in a living organism) and in vitro (in artificial media) showed itself as a fullerene derivative trimalon of acid.

Water-soluble derivatives of C60 have cytotoxic effects on cancer cells [11], and have antitumor activity [12].

Another potential use of fullerene derivatives is their use as materials for nonlinear optics to create optical fiber telecommunications, which do not require converting the optical signal into an electric signal and vice versa [13].

Recently, more and more attention has been paid to the use of such materials. Different organic molecules are formed by the donor and acceptor moieties are interconnected by a bridge of unsaturated bonds. The fullerene core is a powerful acceptor of electrons, so its derivatives may have a high hyperpolarizability, which are the main factors characterizing nonlinear optical properties of the molecule. Another method is the preparation of compounds with nonlinear fullerene-based optical properties, including derivatives of fullerenes, into a polymeric carrier.

Endo fullerenes (containing, for example, 3He or 129Xe) can be used as a label for the investigation of biological processes in the living body by NMR spectroscopy. Endohedral fullerenes containing Gd (for example, water-soluble polyhydroxylated Gd@C_{82} and Gd@C_{60}, and Gd@C_{60} [C (COONa) 2] 10) is also promising for MRI studies of the human body. The uniqueness of the Gd3 + ion is that it has 7 unpaired electrons. It chelates with ion Gd3 + and acid dietiltetraminpenta, which has the commercial name "Magnevist", and are already used in medical practice. The compound enhances signals of water protons and produces higher contrast images of internal organs.

Methods for Identification of Fullerenes

One of the most common and reliable methods for the identification of fullerenes is spectral analysis. This method serves as a source of the most complete and reliable information on the structure and distinctive features of this class of compounds.

Table 1. Results of Identification of Fullerenes by the Mass-Spectra of Benzene Soot Extracts

Formula of a substance	Calculated mass, a.m.u	Mass determined by the spectrum, a.m.u
C_{24}	288,26	288,3
$C_{16}H_{10}$ (pyrene, fluoranthene)	202,26	202,4
$C_{18}H_{10}$ (cyclopentapyrene benzofluoranthene)	226,28	226,4
$C_{20}H_{10}$ (corannulene)	250,3	250,3
$C_{60}O$	736,66	736,5
$C_{60}O_3$	768,66	768,5
$C_{60}H_2$	722,68	722,5
$C_{70}O$	856,77	856,7
$C_{70}H_2$	842,78	842,6
C_{74}	888,81	888,6
C_{78}	936,8	936,8
C_{90}	1080,99	1081,7
C_{94}	1129,03	1128,9
$C_{94}O$	1145,59	1145,5

Studying the time-flight mass spectrometer with a laser ionization of the analyte is frequently used. Characteristic mass spectrum is generated in the flame of positively charged ions of fullerenes in acetylene - oxygen flame [2].

The mass-spectrometric investigations of benzene extracts of soot were carried out with the use of an ion-cyclotron-resonance Fourier spectrometer of the FT ICR MS Bruker "Apex Qe" type with an electric-spray ionization. They have shown that the fullerene oxides $C_{60}O$, $C_{60}O_2$, $C_{60}O_3$, $C_{70}O$, and $C_{94}O$ as well as the higher fullerenes C_{74}, C_{78}, C_{90}, and C_{94} are present in the indicated extracts [14]. The results of identification of fullerenes by their mass spectra are presented in Table 1.

Thus, it has been established that the yield of fullerenes increases under the action of a glow discharge in the case where an electrode (a needle or a ring) is positioned directly above a flame, and the maximum yield of fullerenes is attained with the use of a ring electrode positioned above the central region of the flame front. The maximum yield of the fullerene C_{60} was β=15% of the soot formed.

METHODS FOR PRODUCING FULLERENES

Methods of producing fullerenes are quite varied, and in particular, include an arc discharge in an inert gas atmosphere using a graphite electrode, laser vaporization of a graphite surface, the laser pyrolysis of hydrocarbons, the combustion of aliphatic and aromatic hydrocarbons at higher equivalence ratio, the pyrolysis of hydrocarbons in a reactor, and an arc plasma jet at atmospheric pressure. Analysis methods for obtaining fullerenes on an industrial scale have shown that not all of them are suitable for industrial production because of high energy costs and low productivity. Recently, however, they have all been developed to some extent. At the present time the predominant method of industrial production of fullerenes is arc evaporation of graphite proposed in 1990 by B. Krechmer. Optimization of the production process of the arc is a certain relation between the carbon concentrations, temperature and gas flow rate flowing out of the electrode gap. However, to obtain such optimum ratio in the arc is almost impossible, since the arc current changes the three parameters simultaneously. The most comprehensive study was carried out in [6], where the rate of erosion of the graphite anode, the growth rate of carbon formation on the cathode and the content of fullerenes in the soot on the arc current, type and pressure of the gas, and the inter electrode gap were obtained. For a graphite anode with a diameter of 6 mm, maximum values of the fullerene yield of 10-15% are obtained at a current of 80 A, a helium pressure of 100 Torr, and the gap between the electrodes is 3-5 mm. With the decrease and increase of current and pressure at the optimal electrode spacing a drop in the yield of fullerenes was observed.

FORMATION AND SYNTHESIS OF FULLERENES IN THE FLAME

Evidence that fullerenes can be formed in flames was at first elusive but progress was eventually made. In 1991 significant quantities of C60 and C70 were found in samples collected from low pressure premixed benzene/oxygen flames. Further investigations showed that fullerenes could be produced in sustainable quantities by sub-atmospheric pressure, laminar, and premixed flames of benzene in an oxygen-deficient atmosphere with or without the presence of an inert gas. The evolution of fullerene production technology in 2009 is shown below.

The Evolution of Fullerene Production Technology

	Carbon Arc	**1st Generation Combustion**	**2nd Generation Combustion**
Inventors	Kratchmer-Huffman/UA THE UNIVERSITY OF ARIZONA.	Howard/MIT	nano-c
Synthesis	Individual graphite rods are vaporized with electrical currents in low pressure inert gas. 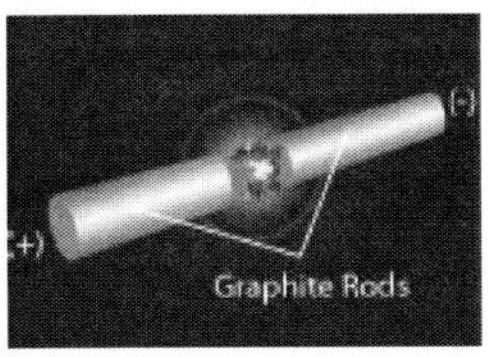	A continuous low flow of hydrocarbon fuel is burned at low pressure in a "flat" flame. 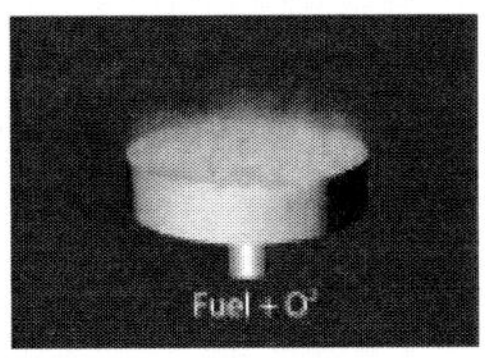	A continuous high-flow of hydrocarbon is burned at low pressure in a 3-D chamber. 
Fullerene Extraction	Solvents to separate fullerenes from other combustion byproducts.	Solvents to separate fullerenes from other combustion byproducts.	Not Necessary
Projected Price of Mixtures	**>$10,000/kg**	**<$1,000/kg**	**<$200/kg**
Fullerene Separations	High Pressure Liquid Chromatography (HPLC) using solvents.	High Pressure Liquid Chromatography (HPLC) using solvents.	Not Necessary
Projected Price of Purified C_{60}	**>$25,000/kg**	**<$16,000/kg**	**<$200/kg**

The experimental results of Ugarte (1992) presented evidence of the spontaneous tendency of electron irradiated graphite to include pentagons in its hexagonal network, and hence form curved structures. This has been further confirmed by Chuvilin et al. (2010), who have shown that fullerenes can form directly from grapheme in a similar fashion. In the experiment of Chuvilin and co-workers, a sheet of graphene was exposed to a beam of energetic 80 keV electrons. The energy transferred to the graphene results in the loss of carbon atoms from the edges of the sheet. If the size of the graphene arc is not too big (less than a few hundreds of carbon atoms), the loss of carbon atoms at the edge will lead to the formation of pentagons, which triggers the curving of graphene into a bowl-shaped structure. Further carbon removal from the edges using the electron beam will reduce the size of the curved structure until it is sufficiently small to zip up its open edges and isomerize into a closed fullerene structure.

All the fullerenes fall into the class of carbons with a polyhedral-closed shell. They were detected as ionized particles in the fuel-rich plane of premixed acetylene and benzene-oxygen flames at low pressures with the use of molecular-beam sampling in combination with mass-spectrometry analysis [15].

Dunk et al. (2012) claim to have identified the fundamental processes leading to the formation of fullerenes in a recent experimental work, based on laser vaporization of pure C_{60} in the presence of carbon vapor. According to this study, fullerenes would self-assemble bottom up through a closed network growth (CNG) mechanism based on the ingestion of atomic carbon and C_2 clusters.

It should be noted that the work from Dunk et al. (2012) provides experimental evidence of the growth of larger fullerenes from pre-existing C_{60} only. Because the experiments show that the CNG of larger fullerenes does not result in the production of C_{60}, it is deduced that C_{60} formation must be a result of CNG from smaller fullerenes. However, the initial formation mechanism of such smaller fullerenes is still under debate.

More recently, macroscopic amounts of not only C_{60} and C_{70} fullerenes [16], but also of larger ones, such as C_{116} [17], were extracted from the low-pressure laminar premixed benzene flames with the use of soot solvents. The reactions of formation of fullerenes and soot in an arc discharge can have much in common with the reactions occurring in the fuel-rich flames. The Bulb-like fullerene nanostructures, formed along with fullerenes and soot, were detected in soot with the use of high-resolution transmission electron microscopy (HRTEM).

It has been shown [3] that the maximum fullerene formation is shifted to the right relative to the maximum soot formation. In a detailed study of fullerene formation from benzene flames, the second maximum at 70 mm from the burner matrix was detected. This data has provided the basis for the development of an alternative method of producing fullerenes in the hydrocarbon combustion regime (Figure 2).

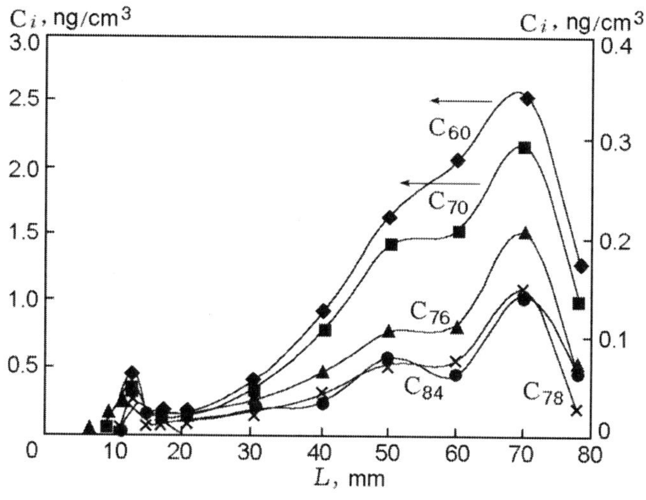

Figure 1. Concentration profiles of fullerenes on the axis of a premixed benzene–oxygen–argon flame [3].

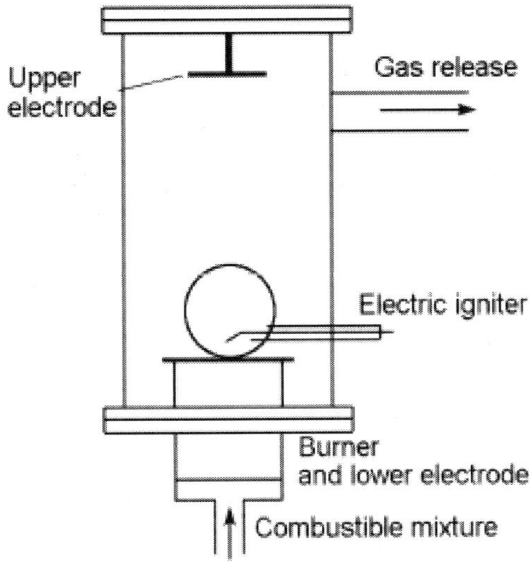

Figure 2. The scheme of the setup.

The influence of gas discharge, the type of electrode, and the inter electrode spacing on the yield of fullerenes in combustion was studied. The experimental setup is shown in Figure 2. It was established that polycyclic aromatic hydrocarbons were positively and negatively charged; therefore the process could be controlled with an electric field.

The experimental conditions were as follows: C/O=1, p=40 torr, flow rate of benzene 250 cm^3/min, oxygen 758 cm^3/min, argon 101 cm^3/min, v=18.4 cm/s, T=1200 K, and δ=0.5–0.8 cm; electrode systems are the needle–plane, plane–plane, and ring–plane with an inter-electrode spacing of 4–21 cm; voltage U = 0.5–20 kV. A premixed $C_6H_6/O_2/Ar$ flame under conditions corresponding to the maximum yield of fullerenes was studied [18]. Processing of the experimental data revealed the advantage of the ring electrode over the electrode in the form of a needle and showed that the yield of fullerene C_{60} ($\approx$15%) was the highest when the electrode was placed in the middle of the flame (L = 4 cm).

It has been found that if the peripheral zone of the benzene flame is heated by some external source, such as a laser beam, which not just burns the soot but also creates the same conditions as in the middle of the flame, the concentration of fullerenes increases.

The effect of acetylene–oxygen flames on the different zones along the height of the benzene–oxygen flames have been studied using a ring burner. It has been found that fullerene formation is activated at the boundary of contact of the acetylene–oxygen and benzene–oxygen flames. The positive effect of the acetylene flame consists of increasing the temperature in the upper part of the reaction zone of the main benzene–oxygen flame by 50 ^{0}C, increasing the degree of ionization, and intense conversion of PAHs present in the benzene–oxygen flame to fullerenes.

It has been established that in the soot at the front of a flame the number of structures with a closed shell as well as the concentration of the fullerene molecules in the gas phase increases with their time of stay in this region. In the region located at a height of 70 mm above a burner the concentrations of the fullerenes and structures with a closed shell are lower than those in regions located at heights of 60–120 mm. The highly ordered nanostructures, such as nanotubes and fullerene bulbs, were detected in solid samples taken from the walls and the upper part of the combustion chamber, which points to the fact that they are formed in the internal-redistribution processes occurring in the solid carbon during a time not longer than 100 ms [4].

The photographs of soot samples obtained with the use of a high-resolution JEM-3010 electron microscope of the JEOL firm with a magnifying power of up to 1,500,000 times at an accelerating voltage of 300 keV, show that these samples are highly structured, which is characteristic of fullerene formations. It is known that the fullerenes formed in a flame are adsorbed, along with the polycyclic hydrocarbons, on the soot particles. The indicated photographs are presented in Figure 3 [19, 20].

The soot samples shown on them comprise agglomerates consisting of spherical and nonspherical particles and an amorphous soot (Figure 12a). In these samples, spherical nanostructures with a diameter of up to 2 nm are also present (Figure 12b). In (Figure 12c), straight and bent carbon structures are seen. In photographs obtained with high magnifying power (Figure 12d), one can see a layered carbonaceous material.

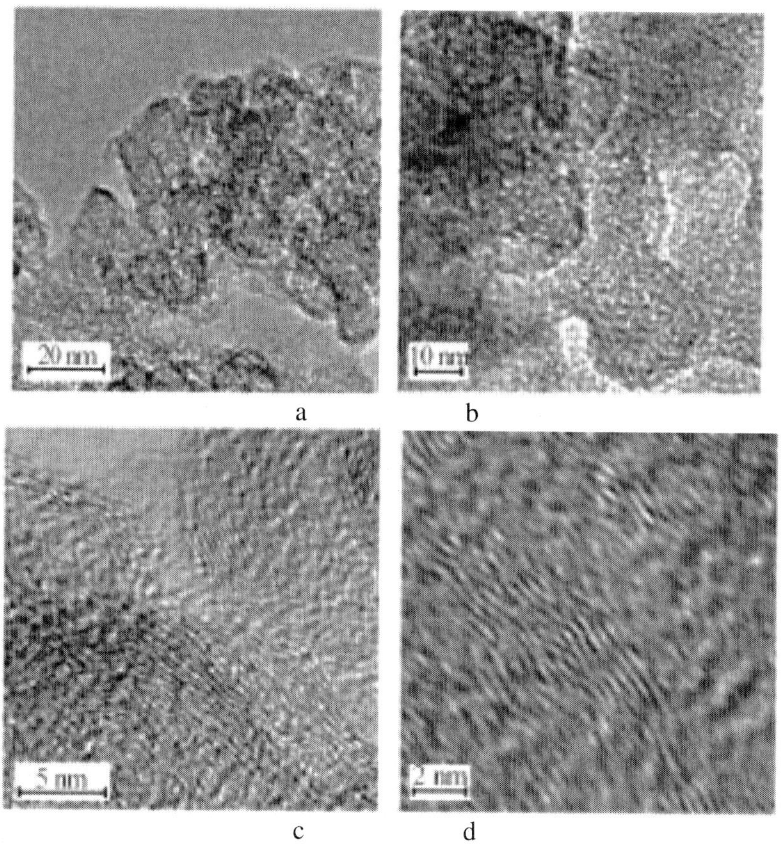

Figure 3. Microscopic photographs of fullerene containing soot samples obtained with different magnifying powers.

Contrary to the supposition that the fullerene structures are the precursors of soot [17], the results of Grieco point to the fact that a gas-phase growth of fullerene molecules along their vertical path in a flame takes place simultaneously with the nucleation of soot particles and fullerenes are deposited on the growing soot particles.

Two ways of formation of fullerenes in flames were discussed. Frenklach and Ebert suggest that the number of the bent structures, i.e., PAHs containing five- and six-membered rings, increase successively, which leads to the formation of fullerenes. They believe that the "bent" PAHs are formed more rarely than the plane PAHs in flames, which is in agreement with measurements of the concentrations of the PAHs (including corannulene - a bent molecule) in the fullerene-forming flames [21].

Corannulene represents a subsystem for almost all fullerenes and, therefore, is of special interest as a fullerene precursor. Corannulene was detected in certain concentrations in ethylene-air flames at atmospheric pressure, in which fullerenes are absent, and it was detected in large amounts in the low-pressure fullerene-forming benzene flames.

The kinetic reliability of the successive formation of fullerenes in a homogeneous gas-phase reaction was theoretically proved in [22], where calculations for a piston reactor were carried out on the basis of the experimentally measured concentrations of different-type particles. The authors of the indicated work described the formation of the C_{60} and C_{70} fullerenes, beginning with fluoranthene, and their growth provided by the successive emission of hydrogen, the addition of C_2H_2, or the use of corannulene as an intermediate compound.

The formation of the C_{60} and C_{70} fullerenes through the splitting out of hydrogen or the addition of C_2H_2 was also tested in the process of simulation of a fullerene-forming flat premixed benzene–oxygen–argon flame at low pressure with an equivalent ratio of 2.4 [22].

The concentration profiles of fullerenes are characterized by two local maxima. Based on the fact that the concentrations of fullerenes, PAHs, and soot change with increase in the distance above a burner, Grieco et al. arrived at the conclusion that the first maximum can be due to the active coagulation of the PAHs, accompanied by their intramolecular condensation, including the dehydrogenation, rearrangement, and formation of rings.

Most of the PAHs are evidently expended for the soot formation and a much smaller part of them is expended for the formation of fullerenes. Grieco et al. concluded that the second, largest concentration maximum, corresponding to the region of the most intensive fullerene formation, cannot be explained by only the reaction of the PAHs, the concentrations of which in this region are at the detection limit of the analytical equipment or are lower than it; this maximum can be explained by the successive addition of acetylene to the PAHs.

Detailed investigations of the ways of formation of fullerenes in flames were carried out by Homman's group who studied the structure of a flame mainly by the direct introduction of ions and neutral particles into a mass spectrometer [19]. They investigated the premixed flames of acetylene, benzene, butadiene, and naphthalene with oxygen at a low pressure with the use of molecular-beam sampling in combination with the mass-spectrometric analysis. It has been established that the bimolecular reaction between two PAHs with a coordinated detachment of hydrogen — the so-called zipper mechanism - plays an important role in the growth of fullerenes.

SYNTHESIS OF FULLERENES IN FLAMES WITH APPLIED ELECTRICAL FIELD

It is practically impossible to increase the yield of fullerenes in combustion using a traditional method. The authors of [18] have shown that the synthesis of fullerenes formed in the plasma of an electric arc increases with the increase in the electron density and in the ionization instability, causing a change in the electron concentration from 10^{10} to 10^{11} cm^{-3}. Despite the fact that the conditions of formation of fullerenes in a flame differ from those in an electric arc, they share a common trait: the initial products are ionized. An electric field applied to a flame changes the electron concentration and the flame temperature, with the result that conditions necessary for the formation of fullerenes are established.

A series of experiments on the study of the yield of fullerenes in a premixed benzene-argon-oxygen flame exposed to a longitudinal electric field in the region of a dark corona discharge and a glow discharge at C/O = 1.0, P = 40 torr, benzene-flow rate Q_1 = 250 cm^3/min, oxygen-flow rate Q_2 = 758 cm^3/min, argon-flow rate Q_3 = 101 cm^3/min (10% of the combustion-mixture volume), and velocity of the cold mixture inflow from a perforated stabilizer v = 18.4 cm/s have been carried out [18]. Under these conditions, created with a margin of steady burning flame front from the burner δ = 0,5 ÷ 0,8 cm, studies were conducted in a voltage range of 0.5 - 20 kV with needle electrode systems. The ring plane-plane at different inter-electrode distances varying ranging from 1 to 18 cm length of one experiment was τ = 20 min. After completion of the experiment soot was collected from the combustion chamber and filter. After that samples of the soot were weighed and were extracted with benzene and examined using physico-chemical methods. Identification of C_{60} and C_{70} was carried out by absorption spectra obtained in the infrared wavelength range in the study sample of soot extracts using a FTIR spectrometer from Perkin Elmer.

It has been established in [24] that a negative-polarity upper electrode exerts a large influence on the fullerene formation as compared to a positive-polarity one. However, in experiments conducted with a negative-polarity, the upper electrode in the region of dark, corona, and glow discharges at an inter-electrode distance H = 18 cm with the use of a needle–plane electrode system, a significant increase in the yield of fullerenes was not detected. At an inter-electrode distance of H = 18 cm, the upper electrode was positioned at a height of 10–11 cm above the upper edge of a flame. Under the indicated conditions, a discharge acted on the upper part of the flame, where a smaller number of ions are found [25] and the formation of the output combustion products is practically absent. In this case, the calculated electron density did not exceed ~10^5–10^7 cm^{-3}. It has been established in earlier investigations that the optimum density of the free electrons, providing a maximum yield of fullerenes, is ~10^9 cm^{-3}. It is known [26] that, if the density of the charged particles in a gas is very small, they interact mainly with neutral particles. This interaction is short-lived and the pair collisions are of main importance in such an ionized gas. When the density of the charged particles increases, the role of their interactions with each other increases gradually too. It is apparent that, under the above-described experimental conditions, the density of the charged particles is insufficient to effectively influence the fullerene formation at an interelectrode distance of 18 cm.

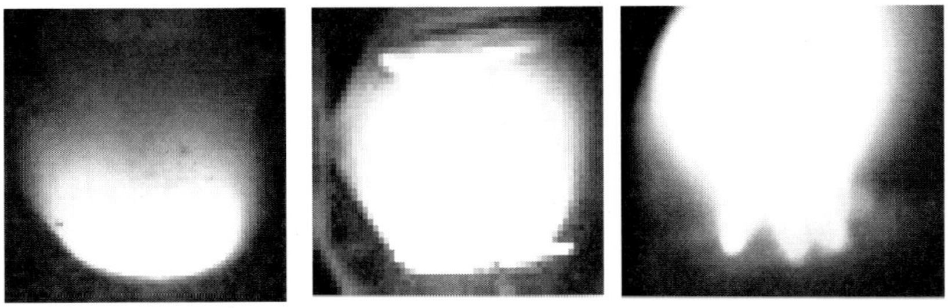

Figure 4. Action of a glow discharge on the yield of fullerenes in the case where a ring cathode is positioned above the central part of a flame: a) a flame without a discharge; b) the flame exposed to a discharge (U = 7 kV); c) strimer.

Table 2. Influence of an Electric Field on the Formation of Fullerenes in the Combustion of the Benzene–Oxygen–Argon Mixture

Electrode	Soot yield, mg	Soluble part yield, %	Yield of C_{60}, %
$H = 6$ cm, $I = 6,6$ mA, $U_н = 7$ kV			
Needle	0,5879	65,64	5
Ring	0,3825	47,54	13
$H = 4$ cm, $I = 7$ mA, $U_н = 7$ kV			
Needle	0,3421	42,15	10
Ring	0,2245	62,56	15
$H = 6$ cm, $I = 4,5$ mA, $U_н = 12,5$ kV			
Ring	0,4909	36,68	2
$H = 4$ cm, $I = 4$ mA, $U_н=7$ kV			
Ring	0,3623	40,81	5
$H = 1$ cm, $I = 3,8$ mA, $U_н = 5$ kV			
Ring	0,5284	24,79	<1
Without field			
No electrodes	0,5958	22,55	<1

Investigations carried out under conditions where the upper electrode was positioned directly above a flame have shown that, in this case, the yield of fullerenes increases. The influence of the type of electrode and the height of its position above a flame on the formation of fullerenes was investigated to determine the conditions providing their maximum yield. Electrodes in the form of a needle and a ring were used. The investigations were carried out at a negative polarity of the upper electrode under conditions where $U = 7$ kV and $H = 1$–9 cm (with a step of 1 cm). A glow discharge appeared independently of the type of electrode at different heights of its disposition above a flame; this discharge caused an intense glow of the whole flame (Figure 4). In this case, the average temperature of the flame increased to 1200°C (without a field, this temperature was equal to 950°C).

Thus, in the case where the upper electrode is placed directly above a flame, the near-cathode region of the glow discharge exerts a stronger influence on the processes occurring in the flame. In this case, in the IR spectra of soot extracts, all the spectral lines corresponding to the fullerenes C_{60} ($\lambda = 528, 577, 1183$, and $1429\ cm^{-1}$) appeared.

It has been established that a ring electrode offers an advantage over a needle one, and a maximum yield of the C_{60} fullerene (β=15% of the soot formed) was obtained in the case where this electrode was positioned above the peripheral part of the central reaction zone of the flame (Table 2) [27].

The increase in the yield of fullerenes, in the case where a ring electrode is positioned above the peripheral part of the reaction zone of a flame, is due to the glow-discharge conditions providing an effective synthesis of fullerenes.

This supposition is based on the results of earlier investigations [23], showing that, if the peripheral zone of a benzene flame is heated by any external heat source, e.g., a laser beam, which does not simply burn the soot but forms conditions identical to those at the center of the flame, the concentration of fullerenes increase.

THE MECHANISM OF FULLERENES FORMATION

The combustion method is by far more important than the carbon vapor condensation method [28], giving C60/C70 yields of 20% and higher based on the weight of extractable and solid products. The mechanism of C60/C70 formation in combustion has long been considered to be different from that of the carbon vapor method because the combustion method involves hydrogen atoms in addition to carbon and leads to polyaromatic hydrocarbon (PAHs) (or nanographenes) as the major products, with C60/C70 as very minor by-products. This view was held by Homann who constructed his own formation mechanism for C60 based on the intermediacy of PAHs [29].

However, the situation changed when Howard used benzene premixed with oxygen flame. Although he never clearly described the flame temperature, he once mentioned about 1,800 K [16], which is close to 2,000 K used in the size-up stage by Irle/Morokuma.

Howard initially achieved only 0.003–9.0% yields [16], but they increased to 20% and, shortly before his untimely death in 2007, to quantitative yield [28]. Although his latest flaming conditions were never disclosed, we can guess if the combustion method follows the SHG road. This last assumption is likely in view of the following reasons:

1. Combustion methods give soot as the major by-product, which is an aggregate of defective multishell fullerenes [29]. In the case of the carbon vapor method, the major by-product is carbon nano-onions (synonym to multishell fullerenes). By-products are essentially the same.
2. In high-temperature flames, most of the bonded H atoms on the surface of feedstock molecules will be preferentially burnt to water. H atoms and H-containing intermediates will also burn out quickly in the early stage of combustion. In the inner region of the flame, some of the H atoms might remain on the edges of small PAH. However, these flat molecules tend to move slower than C_2, small polyynes and other typical dissipative structures in the stream of flame; therefore, they tend to transfer themselves from the inner to the outer flame zone and burn out.

As long as the flame of the combustion method is long enough and in sufficiently high temperature, the composition of the downstream inner flame will be the same carbon “soup” as in the carbon vapor method, but medium gas is different, consisting of H_2O, CO_2 and CO. Among these, only CO is reactive. One likely phenomenon that would happen in the size-down process of abstraction of C_2 from the abutting pentagon unit (III, see Figure 5) in giant fullerene by carbon mono-oxide (IV) from the gas by way of stable cyclopropenone intermediate (V):

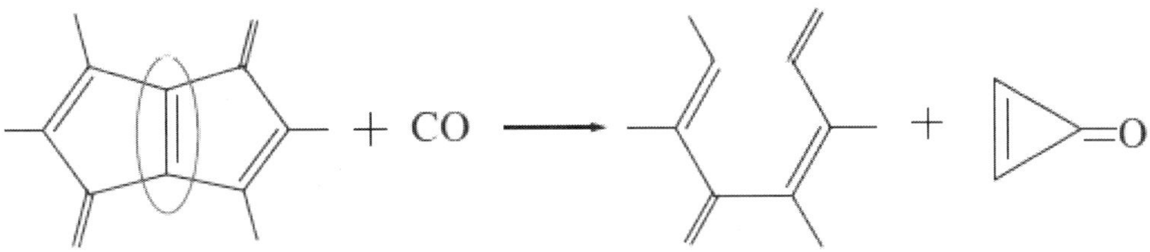

III IV V

Figure 5. Model of formation of five- and six-member rings as a result of the binding of two PAHs by the zipper mechanism [29].

The well-known aromaticity in V would contribute to reduce the transition state of C_2 popping-out reaction from III. Once the abstract reaction is complete, V will be quickly decomposed to give back carbon-monoxide IV; hence, equation (1) represents an ideal autocatalysis reaction that reduces the barrier of the C_2-popping reaction, promotes the shrink step and increases the yields of C_{60}/C_{70}.

A definite advantage of the combustion method is the possibility to adjust the conditions to optimize the flame structure, temperature distribution, flow rates of feedstock and air in order to maximize the yield of C_{60} and combustive removal of by-products. By all accounts, the combustion method is the best candidate for quantitative production of C_{60}/C_{70}.

Fullerenes formation in flames can be viewed as a molecule weight growth process, analogous to the formation of polycyclic aromatic hydrocarbons and soot. An oxygen deficiency leads to the formation of soot, and PAHs are nuclei of soot particles. From Figure 6, it is evident that PCAHs can be converted to soot or fullerenes, depending on combustion conditions, among which the determining parameter is pressure.

Fullerenes are formed in the process of step-by-step monomolecular reactions, such as the cyclization of C_{30} cells, the rupture of the C–H bonds, and the intramolecular rearrangement; therefore, a high-temperature process is preferred. Fullerenes can also be formed by a combined mechanism. Since the majority of aromers are geometrically incompatible for the formation of fullerenes as the finite products of the reaction between different PAHs, it is suggested that reactions with acetylene can lead to further growth of the molecules as long as they become geometrically compatible for the cell closure.

It is known that the formation and synthesis of fullerenes in the traditional method of arc evaporation of graphite is performed at pressures $p < 40$ torr.

A necessary condition for this mechanism is low pressure. As the pressure increases, i.e., with transition to atmospheric pressure or above, where triple collisions dominate, PCAHs coagulate to form soot clusters.

As noted in [30], nano- and subnano size particles are formed by gas-phase condensation in asymptotic giant branched stars. The experiments were carried out at pressures of 0.1-2.6

and 7 mbar, close to the values of p in the astrophysical atmosphere at temperatures T<1700 K, at which the formation of fullerenes was observed. As fullerenes are formed at low pressures, the corresponding spatial orientation is important, which requires that the steric factor should be considered. For a molecule of C_{60} to form, a spatial orientation of two molecules of C_{30} is required. There are various models of formation of fullerenes C_{60}, one of which is the zip mechanism [1, 31].

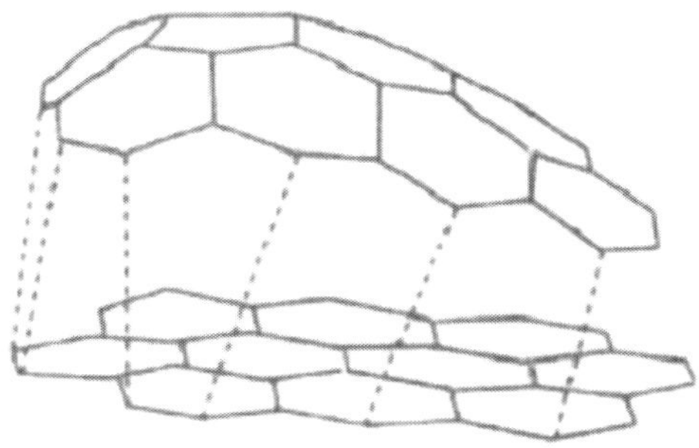

Figure 6. Model of formation of five- and six-member rings as a result of the binding of two PAHs by the zipper mechanism.

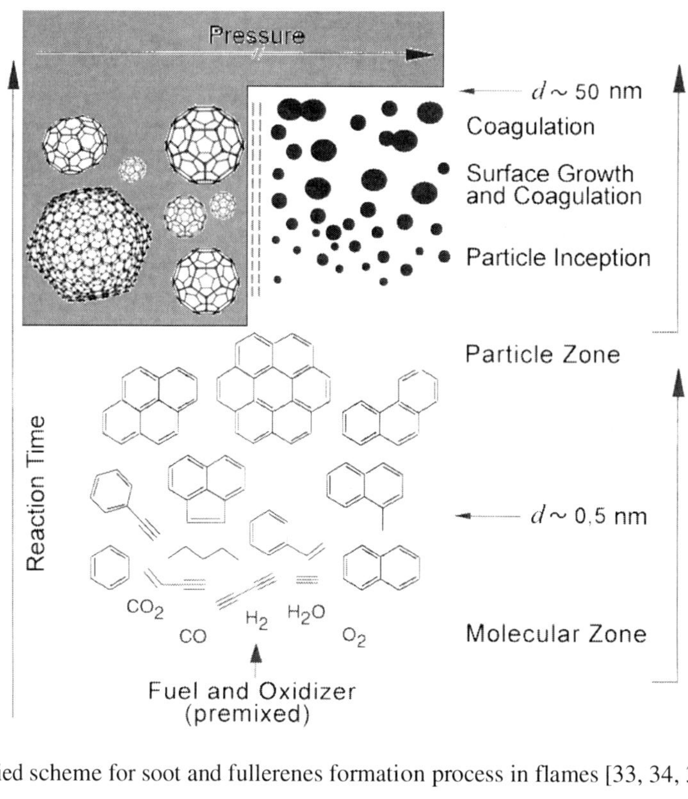

Figure 7. Modified scheme for soot and fullerenes formation process in flames [33, 34, 35].

The flame front structure of soot formation flames can be considered in close connection with the mechanism of fuel conversion. A scheme that was suggested by Bockhorn [32] in

1994 is known, where polycyclic aromatic hydrocarbons (PAH) are precursors of soot particles. Since then, in rich fuel flames the nano-sized particles such as fullerenes and graphenes are found. The fullerenes formation is generally observed at pressures below 60 Torr, because for fullerenes formation an adherence of steric spatial factor is necessary, but in flames at atmospheric pressure this factor is prevented by triple collisions. Therefore, the pressure coordinate can be introduced to the general scheme for soot formation. At low pressure the formation of fullerenes from polycyclic aromatic hydrocarbon (PAH) occurs, but with pressure increase the polycyclic aromatic hydrocarbons (PAH) are coagulates to soot particles. On the basis of data on the synthesis of fullerenes in the flame it is possible to modify the general scheme proposed by H. Bockhorn for rich fuel flames, namely to make a pressure-coordinate, which allows for the formation of fullerenes at low pressures, and soot at high pressures.

The fuel flames lacks of oxygen leads to soot formation, at that PAH are the nucleus of soot particles. From Figure 7 we can see that PAH can be converted to soot or fullerenes, depending on combustion conditions, [33, 34].

Conclusion

In review, works on formation and the synthesis of fullerenes in flames were considered in the sections, such as, the structure and properties of fullerenes; applications of fullerenes; methods for identification of fullerenes; methods for producing fullerenes; formation and synthesis of fullerenes in flame; synthesis of fullerenes in flames with applied electrical field; mechanism of fullerenes formation. A modified scheme for soot and fullerenes formation process in flames was also presented.

It was found that if the peripheral zone of the benzene flame is heated by some external source, such as a laser beam that does not just burn soot but also creates the same conditions as in the middle of the flame, the concentration of fullerenes increases.

References

[1] J. Ahrens, M. Bachmann, Th. Baum, J. Griesheimer, R. Kovacs, P. WeilmËunster, K. H. Homann: *Int. J. Mass Spectrom. Ion Proc.* 138, 133–148 (1994).

[2] J. B. Howard, A. L. Lafleur. Fullerenes synthesis in combustion. *Carbon*, 1992; 30 (8):1183-1201.

[3] H. W. Kroto, J. R. Health, S. C. O'Brien, R. F.Curl, R. F. Smalley. *Nature*, 1985. p. 162–164.

[4] Howard J. B. Fullerenes formation in flames. Invited topical review, 24th, Symp. (Int.) on Combustion. Pittsburgh, (Pennsylvania, USA): The combustion institute, 1992; p. 933–946.

[5] W. Krätschmer, L. D. Lamb, K. Fostiropoulos and D. R. Huffman, *Nature*, 1990, 347, 354.

[6] R. C. Huddon, et al. // *Nature* 350, 320, (1991).

[7] Jensen AW., Daniels C. // *J. Org. Chem.* – 2003. – V. 68. – P. 207.

[8] Thakral Seema, Mehta R. M. // *Ind. J. Pharm. Sci.* – 2006. – V. 68. – № 1. – P.13.
[9] Tagmatarchis N., Shinohara H. // *Mini Reviews in Medicinal Chemistry.* – 2001. – V. 1. – № 4. – P. 339.
[10] Da Ros T., Spalluto G., Prato M. // *Croatica Chernica Acta.* – 2001. – V. 74. – №4. – P. 743.
[11] Yang X. L., Fan C. H., Zhu H. S. // *Toxicology in Vitro.* – 2002. – V. 16. – P. 41.
[12] Irie K, Nakamura Y., Ohigashi H. et al. // *Biosci. Biotechnol. Biochem.* – 1996. – V. 60. – №8. – P. 1359.
[13] Sun Y.-P., Lawson G. E., Riggs J. E., Ma B., Wang N., Moton D. K. // *J. Phys. Chem., A*, 102, 5520 (1998).
[14] Z. A. Mansurov, B. T. Lesbaev, D. I. Chenchik, et. al. Synthesis of fullerenes and carbon nanotubes in flames, Book of abstracts, Int. Conf. on Carbon, Nagano (2008), pp. 134–139.
[15] P. Gerhardt, S. Loffler, and K.-H. Homann, The formation of polyhedral carbon ions in fuel-rich acetylene e and benzene flames, *Proc. 22nd Int. Symp. Combust.,* The Combustion Inst., Pittsburgh (1988), pp. 395–401.
[16] J. B. Howard, J. T. McKinnon, Y. Makarovsky, et al., Fullerenes C60 and C70 in flames, *Nature*, 352, 139–141 (1991).
[17] H. Richter, A. J. Labrocca, W. J. Grieco, K. Taghizadeh, A. L. Lafleur, and J. B. Howard, Generation of higher fullerenes in flames, *J. Phys. Chem. B,* 101, 1556–1560 (1997).
[18] Z. A. Mansurov, N. G. Prikhodko, T. T. Mashan, and B. T. Lesbaev, "The Study of Influence of Electric Field on Soot Formation at Low Pressure," *Chem. Physics* 25 (10), 18–22 (2006).
[19] Z. A. Mansurov, N. G. Prikhodko, B. T. Lesbaev, and T. T. Mashan, Combustion of the premixed benzene oxygen mixture in electric field at low pressure, Proc. 31st Int. Symp. Combust., Heidelberg (2006), P. 164.
[20] Z. A. Mansurov, T. A. Shabanova, V. L. Levin, and N. G. Prikhod'ko, A novel characteristic of a C60–C70-fullerene containing substance (according to electronic-microscopy data), Vestn. KazNU, *Ser. Khimicheskaya*, 39, No. 3, 444–448 (2005).
[21] A. L. Lafleur, J. B. Howard, J. A. Marr, and T. Yadav, Proposed fullerene precursor corannulene identified in flames both in the presence and absence of fullerene production, *J. Phys. Chem.*, 97, 13539–13543 (1993).
[22] T. Baum, P. Loffler, P. Weilmunster, and K.-H. Homann, Fullerene ions and their relation to PAH and soot in low-pressure hydrocarbon flames, *Der Bunsenges Phys. Chem.*, 96, 841–857 (1992).
[23] M. Bachmann, W. Wiese, and K.-H. Homann, *Thermal and chemical influences on the soot mass growth*, 25th Symp. (Int.) on Combustion, The Combustion Inst., Pittsburgh (1994), pp. 635–643.
[24] Z. A. Mansurov, N. G. Prikhodko, T. T. Mashan, and B. T. Lesbaev, Formation of PAH, fullerenes, nanoparticles and soot at combustion of hydrocarbons in electric field, Proc. 20th Int. *Colloquium on the Dynamics of Reactive Systems,* July 31–August 5, 2005, Montreal, Canada. CD, P. 5.
[25] H. F. Calcote and R. J. Gill, Comparison of the ionic mechanism of soot formation with a free radical mechanism, in: H. Bockhorn (Ed.), Soot Formation in Combustion.

Mechanisms and Models, *Springer Series in Chem. Phys.*, Springer, Berlin (1994), Vol. 59, pp. 471–484.

[26] Yu. P. Raizer, Gas Discharge Physics [in Russian], *Nauka*, Moscow (1987).

[27] N. G. Prikhodko, B. T. Lesbaev, and Z. A. Mansurov, Formation of fullerenes in a flame subjected to a gas discharge, *Izv. Nats. Akad. Nauk RK, Ser. Khimicheskaya,* No. 2, 15–18 (2006).

[28] J. B. Howard, Winchester, D. F. Kronholm, A. J. Modestino., H. Richter. Method for combustion synthesis of fullerenes. Patent No/. US 7,771,692 B2. Date of Patent: Aug. 10, 2010.

[29] E. Ōsawa. Fullerenes, Nanotubes and Carbon Nanostructures. Formation Mechanism of C60 under Non-equilibrium and Irreversible Conditions. - subscription information: An Annotation. NanoCarbon Research Institute, AREC, Shinshu University, Nagano, Japan, 2012.

[30] C. JËager, F. Huisken, Jansa I. Lamas, Th. Henning: *Astrophys.*, No 696, 706–712 (2009)

[31] Homann, K. H. (1998) Fullerenes and soot formation – new pathways to large particles in flames. *Angew. Chem. Int. Ed.*, 37: 2435–2451.

[32] H.Bockhorn (ed). Soot formation in Combustion. Springer, Berlin / Heidelberg, P.4, 1994.

[33] Z. A. Mansurov. Producing Nanomaterials in Combustion. Combustion, Explosion, and Shock Waves, Vol. 48, No. 5, pp. 561–569, 2012. Original Russian Text. Z. A. Mansurov.

[34] Z. A. Mansurov, N. G. Prikhodko, A. V. Saveliev. Obrazobanie PTSAU, fullerenov, uglerodnyh nanotrubok e sazhi v protsessah gorenyia. [Formation PAH, fullerenes, carbon nanotubes and soot in combustion processes] / - Almaty: Kazakh University, 2012. - 383 p.

[35] Z. Mansurov. Soot and Nanomaterials Synthesis in the Flame. *Journal of Materials Science and Chemical Engineering*. Published Online January 2014. http://dx.doi.org/10.4236/msce.2014.21001

[36] http://www-ipcms.u-strasbg.fr/spip.php?article2166&lang=en. *(Endo)fullerenes: from production to isolation.*

In: Fullerenes
Editor: Shannon B. Ellis

ISBN: 978-1-63321-385-2

Chapter 11

SPHERICALITY OF SOME CLASSES OF FULLERENES MEASURED BY TOPOLOGY

Fatemeh Koorepazan-Moftakhar[1,2]**,**
Ali Reza Ashrafi[1,2,*]**,** ***Ottorino Ori***[3,4,†] ***and Mihai V. Putz***[4,‡]

[1]Department of Nanocomputing, Institute of Nanoscience and Nanotechnology, University of Kashan, Kashan, I. R. Iran
[2]Department of Pure Mathematics, Faculty of Mathematical Sciences, University of Kashan, Kashan, I. R. Iran
[3]Actinium Chemical Research, Rome, Italy
[4]Laboratory of Computational and Structural Physical-Chemistry for Nanosciences and QSAR, Department of Biology-Chemistry, West University of Timişoara, Timişoara, Romania

ABSTRACT

The measure of the sphericality of a fullerene cage C_n with n carbon atoms is obtained by computing the topological invariant $\rho = 2W / n\underline{w}$, being W the topological Wiener index of the fullerene molecular graph F and $\underline{w}$ the contribution to W coming from the minimal vertices. Such an originally model is applied to the infinite families of tubular-like 1D fullerenes with exactly *50 + 10n*, *60 + 12n*, *10n* and *12n* carbon atoms, correlating molecular sphericality and Hückel energy according to a peculiar parabolic behavior. Formally, the upper limit $1 \leq \rho < 4/3$ is also derived.

* Ali Reza Ashrafi: Department of Nanocomputing, Institute of Nanoscience and Nanotechnology, University of Kashan, Kashan 87317–51167, I. R. Iran. Department of Pure Mathematics, Faculty of Mathematical Sciences, University of Kashan, Kashan 87317–51167, I. R. Iran. E-mail: ARA: ashrafi@kashanu.ac.ir.

† Ottorino Ori: Actinium Chemical Research, Via Casilina 1626/A, 00133 Rome, Italy. Laboratory of Computational and Structural Physical-Chemistry for Nanosciences and QSAR, Department of Biology-Chemistry, West University of Timişoara, Pestalozzi Str. No. 16, Timişoara 300115, Romania. E-mail: OO: ottorino.ori@gmail.com.

‡ Mihai V. Putz: Laboratory of Computational and Structural Physical-Chemistry for Nanosciences and QSAR, Department of Biology-Chemistry, West University of Timişoara, Pestalozzi Str. No. 16, Timişoara 300115, Romania. E-mail: MVP: mv_putz@yahoo.com.

Keywords: Topological efficiency index, Topological sphericality, Wiener index, fullerene, Hückel energy

1. INTRODUCTION

An orientable fullerene graph F is a connected cubic graph (or trivalent graph) such that F is embeddable on a sphere with every face being either a pentagon or a hexagon [1-3]. These cubic *chemical graphs* represent a fullerene molecules C_n with n vertices (carbon atoms), n_e edges (chemical bonds) and n_f faces subject to Euler's constraint $n + n_f - n_e = 2$. In particular, the total number of rings covering molecular surface consists of $f_6 = n/2\text{-}10$ hexagons and exactly 12 pentagons. According to a possible description, the surface of a fullerene arises from the overlap of 12 nanocones interconnected by graphenic regions. In the chemical picture, the isomers of a C_n fullerene are hollow molecules entirely composed by n sp^2 carbon connected by n_e bonds with different sphericality assuming the shape of a sphere, a pyramid, a triangular cushion, a nanocone or menhir, a barrel, a tube. The study of these stable allotropic forms of carbon, constantly attracts a large interest due to their potential applications in many different fields.

Since 1985, when Kroto and his team reported their pioneering work on fullerene chemistry, several researchers devoted their time for discovering mathematical properties of this new material. Fowler and Manolopoulos [2], introduced a large class of fullerene graphs together with a complete description of their symmetry and other mathematical properties. Two useful free computer programs for working with fullerenes are reported in [4, 5].

In a series of papers [6-9], Ori and coworkers presented a new important topological parameter named *topological efficiency* ρ for ranking the relative stability of fullerene isomers and determining the *topological sphericality* σ of fullerenes and other nanostructures.

The invariant ρ represents so far the unique topology-rooted quantity proposed in literature for measuring and comparing the sphericality of a given set of C_n molecules and it accomplish this essential task just by exploiting the long-range connectivity properties of the molecular graphs, a truly remarkable answer to the request of a *simple tool* to predict isomeric stabilities for higher fullerenes "such as the measure of non-sphericality" recently raised by Gutman and coauthors [10]. At the end of this chapter, the correlation between the *topological sphericality* σ and the Hückel energy (HE) will be finally presented for the some classes of fullerenes. Taking in to consideration the *topological definition* of HE in which the energies of the orbital which are populated in the ground state) are calculated as the eigenvalues of the adjacency matrix $A(G)$ of the graph, the total π - electron energy E_π of a given molecule is proportional to the summation of the first $n/2$ eigenvalues [11]. According to this approach, the Hückel energy is considered in the present work, like the *topological sphericality,* genuine topological invariants.

2. The Topological Sphericality

To introduce the *topological efficiency index* ρ of a general connected molecular graph Γ having the set of vertices $V(\Gamma)=\{v_1, v_2, ..., v_n\}$ we start from the chemical distance $d_{ij} = d(v_i,v_j)$ corresponding the number of bonds in the minimum path connecting v_i to v_j.

The Wiener index of Γ, named $W(\Gamma)$, is a topological molecular descriptor corresponding to the half-summation of all distances between 2–element subsets of $V(\Gamma)$. If $w_i = \Sigma_j d_{ij}$ then we can easily prove that

$$W(\Gamma) = \frac{1}{2}\sum_{i=1}^{i=n} w_i \tag{1}$$

The graph invariant w_i expresses the contribution to $W(\Gamma)$ coming from the vertex v_i. Suppose $\underline{w} = \mathrm{Min}\{w_i\}$ and $\overline{w} = Max\{w_i\}$. Graph nodes v_i featuring $w_i=\underline{w}$ (or $w_i=\overline{w}$) are called *minimal* (or *maximal*) *vertices* respectively. The topological efficiency $\rho(\Gamma)$ is the defined as

$$\rho(\Gamma) = \frac{2W(\Gamma)}{n\underline{w}} \tag{2}$$

This graph invariant expresses the efficiency of a given graph in filling the space when compared to its *minimal vertex*, the most deeply embedded node in the graph. A variant of ρ has been suggested by Vukičević [12] and inserted in the computer package [5], leading to the so-called *extreme topological efficiency* index:

$$\rho^E(\Gamma) = \frac{\overline{w}}{\underline{w}} \tag{3}$$

By definition, the following inequalities hold: $\rho(\Gamma) \geq 1$ and $\rho^E(\Gamma) \geq 1$ for any graph Γ.

Both topological descriptors (2) and (3) have been applied by the authors in the last few years to investigate and compute the *topological sphericality* on various classes of molecules and infinite lattices by posing:

$$\sigma=\rho \tag{4a}$$

or

$$\sigma=\rho^E \tag{4b}$$

The innovative sphericality measure (4a) has been successfully applied to classify the stability of the isomers of C_{50} and C_{66} fullerene [13, 8]. Formulae (4a, 4b) show some interesting differences.

In our tests in fact, ρ^E works better than ρ in providing, for fullerenes, a better correlation between *isomer sphericality* and *energy*, this diversity being probably related to the fact that the stability of hollow molecules is very sensible to the presence on their surface of regions with large deformations in respect to the ideal spherical profile.

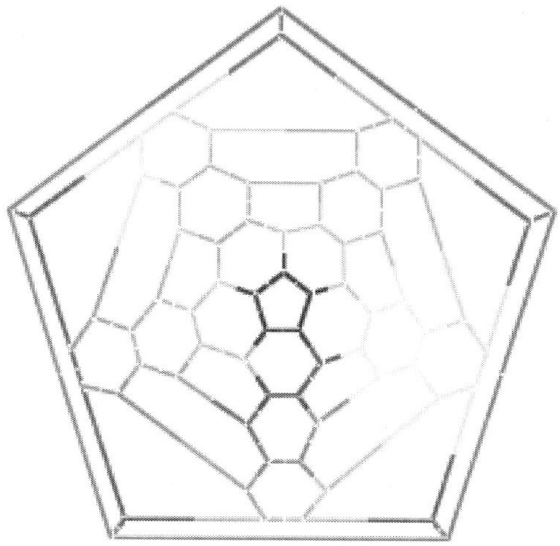

Figure 1. The Fullerene $E_n = C_{50+10n}$, $n=2$.

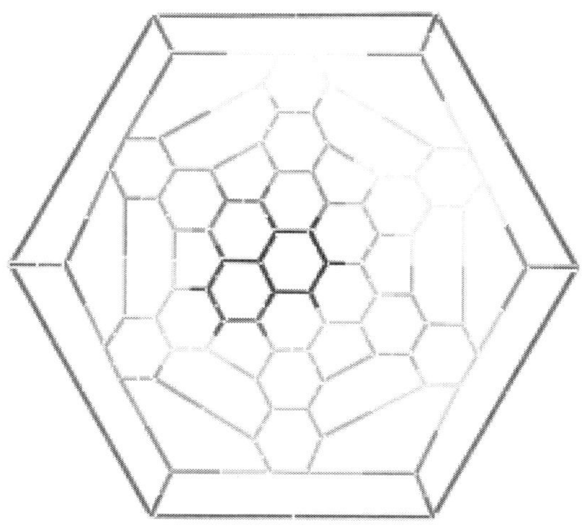

Figure 2. The Fullerene $F_n = C_{60+12n}$, $n=2$.

In the present numerical studies the definition (4b) of molecular *sphericality* will be normally assumed.

- The aim of the present work is then to summarize our recent findings about the correlation between molecular *Hückel energy* and *topological sphericality* of four classes E_n, F_n, G_n and H_n of fullerenes depicted in Figures 1-4 as planar Schlegel graphs and as *3D* structures in Figures 5-8.

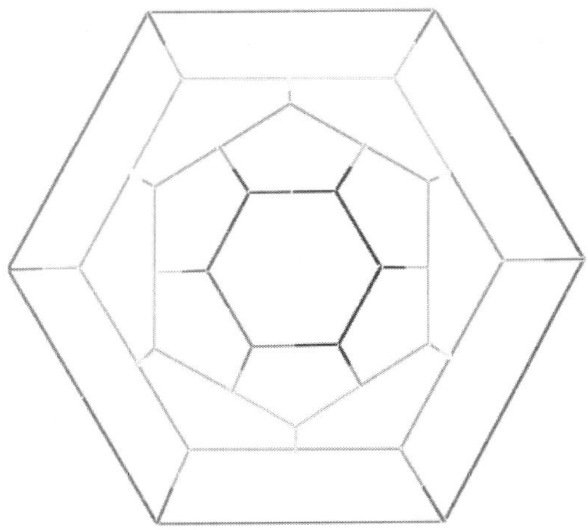

Figure 3. The Fullerene $G_n = C_{12n}$, $n = 3$.

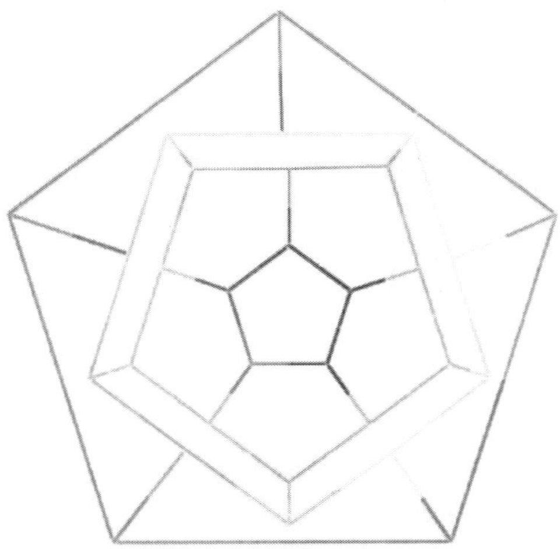

Figure 4. The Fullerene $H_n = C_{10n}$, $n = 3$.

Fullerene families $E_n = C_{50+10n}$ and $F_n = C_{60+12n}$ with $n \geq 1$, only include IPR fullerenes (e.g., cages respecting the Isolated Pentagon Rule that excludes fused pentagons); in particular $E_{n=1}$ corresponds to the icosahedral buckyball C_{60}-I_h, the most stable fullerene, for which the direct calculation shows that $\rho = \rho^E = 1$. The third and fourth classes of fullerenes however consist of non–IPR fullerenes only with exactly $12n$ and $10n$ carbon atoms, respectively, such that $n \geq 2$ is a positive integer.

All these structures may be schematically described a *tubular fullerenes* capped by four different types of *symmetrical poles* (whose structures are evident in the planar graphs in Figures 1-4) bridged by a proper number of intermediate hexagons. It is worth noticing that we are then devoting our attention to the study of the stability of an infinite set of fullerenes characterized by having – in the limit of large n – a *linear structures* growing proportionally with the number of hexagonal face f_6 along a given direction.

For these molecules, closed expressions for the topological descriptors (4a,4b) will be derived in the following sections.

3. Computation Details

In [14] Graver presented the catalog of all fullerenes with ten or more symmetries and in [15, 16], some of the present authors computed the symmetries and some distance based topological indices of four infinite classes of fullerenes.

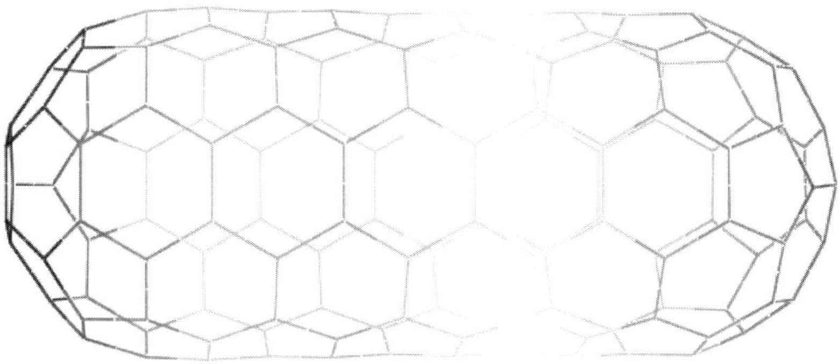

Figure 5. *3D* view of E_n, $n = 9$.

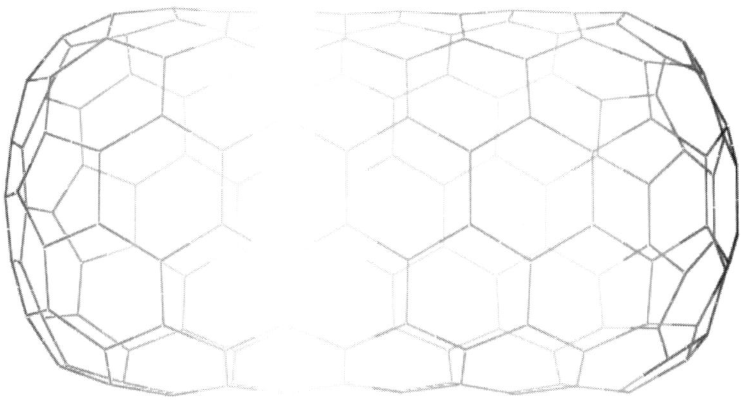

Figure 6. *3D* view of F_n, $n = 9$.

As previously mentioned, E_n, F_n classes are made of IPR fullerenes with $50 + 10n$ and $60 + 12n$ carbon atoms, respectively. Here, n is a natural number.

On the contrary, G_n and H_n sets of fullerenes are all non–IPR cages with $12n$ and $10n$ carbon atoms with $n \geq 2$. We refer to [17] for more information on symmetry and topology of fullerenes.

With scope of computing the topological sphericality, we first introduce some algebraic notions which are crucial throughout this paper. The cyclic group of order n is denoted by Z_n and the symmetry group of a regular n–gon is denoted by D_{2n}. The group D_{2n} is called the dihedral group of order $2n$. If G and H are two groups then the Cartesian product $G \times H$ has a group structure with natural operation in each component.

If Γ is a graph then the set of all permutations f on Γ such that f and f^{-1} preserve adjacency is called an automorphism of Γ. The set of all such mappings is denoted by $Aut(\Gamma)$. It is easy to see that $Aut(\Gamma)$ has a group structure under composition of functions.

The set $\bar{x} = \{\alpha(x) \mid \alpha \in Aut(\Gamma)\}$, is called an *orbit* in $Aut(\Gamma)$. The graph Γ is called *vertex transitive* if there exists x such that $\bar{x} = V(\Gamma)$.

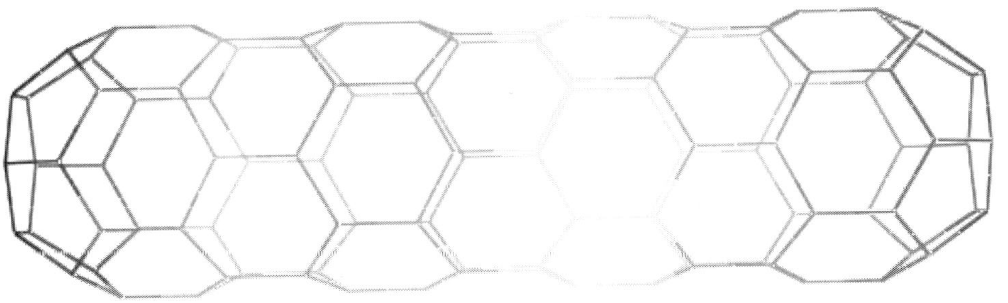

Figure 7. *3D* view of G_n, $n = 9$.

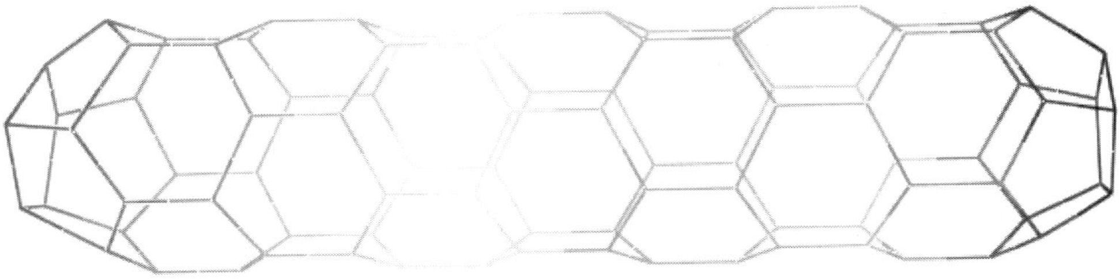

Figure 8. *3D* view of H_n, $n = 9$.

The first property of the topological efficiency indices may be now easily derived:

- If Γ is a vertex transitive chemical graph then for each i, j, $1 \leq i, j \leq n$, we have $w_i = w_j$ and then $\rho(\Gamma) = \rho^E(\Gamma) = 1$.

This is the case includes C_{60}-I_h and other highly symmetric chemical structures like benzene ring. Figures 9 and 10 report instead two fullerenes C_{40} and C_{240} containing 40 and 240 carbon atoms having topological efficiency (2) $\rho=\rho^E=1$ and three orbits under natural action of symmetry group. For an arbitrary labeling of the C_{40} carbon atoms these orbits include respectively 4 graph vertices (atom 1, 19, 21, 40), 12 atoms (2, 3, 38, 27, 8, 32, 24, 18, 20, 39, 35, 13) and 24 atoms (4, 16, 26, 5, 29, 17, 6, 10, 22, 14, 25, 36, 15, 28, 7, 11, 23, 33, 30, 9, 31, 12, 34, 37); topologically, this fullerene shows therefore a (1:3:6) C^{13} NMR theoretical signature.

In a similar way, C_{240} –I_h atoms are grouped in 3 distinct sets of symmetry-equivalent atoms (e.g., 3 orbits) with populations 60, 60, 120 respectively, producing a distinctive (1:1:2) resonance pattern.

This result shows that these two fullerenes are not vertex transitive. On the other hand, the C_{40} has symmetry group T_d and C_{240} has point symmetry group I_h. This shows that there are fullerenes which are not vertex transitive with point group symmetry $\neq I_h$ such that topological efficiency is one. On the other hand, assume that Le_C_{240} is the leapfrog of C_{240}, Figure 11. Then Le_C_{240} has a point group symmetry I_h, $\rho(Le\_C_{240}) = 1.02110078$ and $\rho^E(Le\_C_{240}) = 1.064189$. So, it is possible to find fullerenes with point group symmetry I_h with topological efficiency (and sphericality) larger than one. Current findings integrate previous [18] results which correctly classified C_{240} as the smallest I_h fullerene with non-symmetry equivalent atoms and maximal sphericality $\sigma=\rho=\rho^E=1$.

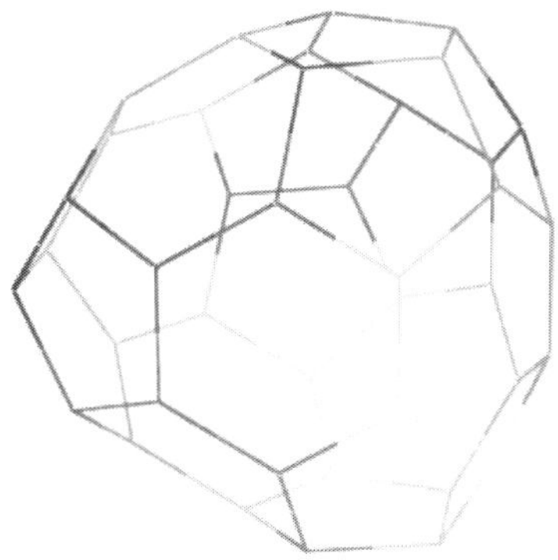

Figure 9. The C_{40}-T_d fullerene.

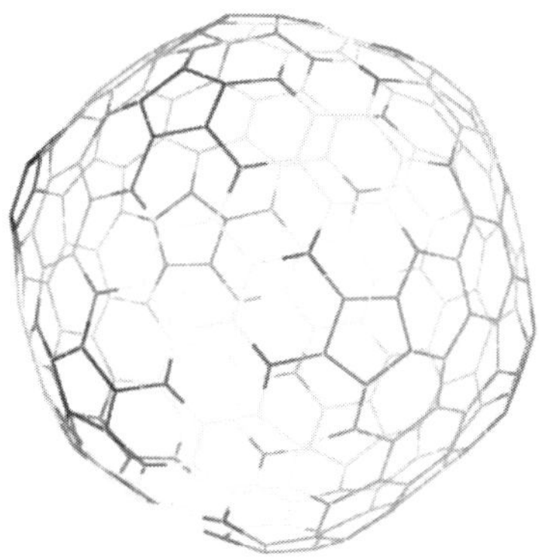

Figure 10. The C_{240}-I_h fullerene.

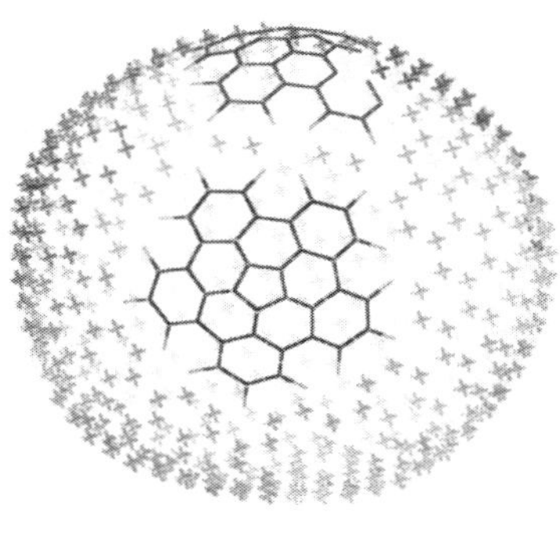

Figure 11. The Le_C_{240} -I_h leapfrog fullerene.

More generally, it has been proven [17] that $Aut(E_n)$ corresponds to D_{20} the dihedral group of order 20:

$$Aut(E_n) = D_{20}, \text{ with } n \geq 1 \tag{5a}$$

whereas for the automorphism groups of the remaining classes F_n $(n \geq 1)$ and G_n, H_n fullerenes $(n \geq 2)$ the following equations hold:

$$Aut(F_n) = \begin{cases} D_{24} & 2 \nmid n \\ Z_2 \times Z_2 \times S_3 & 2 \mid n \end{cases}; \tag{5b}$$

$$Aut(G_n) = \begin{cases} D_{24} & 2 \mid n \\ Z_2 \times Z_2 \times S_3 & 2 \nmid n \end{cases}; \tag{5c}$$

$$Aut(H_n) = \begin{cases} Z_2 \times A_5 & n = 2 \\ D_{20} & n > 2 \end{cases} \tag{5d}$$

Equations (5) provide an effective tool for evaluating topological sphericality of fullerene isomers – see the next section for details.

4. Main Results

In this section, the closed forms expressing the *topological efficiency* for the fullerene sequences $E_n = C_{50+10n}$, $F_n = C_{60+12n}$, $G_n = C_{12n}$ and $H_n = C_{10n}$ will be originally derived and, according to Equations (4a, 4b), applied to measure fullerene sphericality σ. This topological parameter is then used to find *the fast and efficient fitting* of the Hückel energies of these classes of C_n isomers.

4.1. Fullerene Series E_n

In this section, we derive the topological sphericality for $E_n = C_{50+10n}$ fullerenes. Having defined the group $G = Aut(E_n)$ the action of the automorphisms on the fullerene vertices is considered. G has the basal property to split in $(n/2 + 4)$ orbits the $(50+10n)$ atoms of a E_n isomer, each orbit having length (population) either 10 or 20 [19].

G has exactly 2 orbits of length 10, if n is odd. For even n, orbits of length 10 are 3. By adopting a suitable atom numbering [19], these orbits are as follows: $\{1, ..., 5, 46+10n, ..., 50+10n\}$ and $\{6,...,10, 41 + 10n,...,45 + 10n\}$.

Moreover, for even n the third orbit of length 10 contains $\{5n + 21, ..., 5n + 30\}$ atoms. On the other hand, the remaining orbits, of length 20, in C_{50+10n} fullerenes follow the general rule:

{11, …, 20, 31 + 10n, …, 40 + 10n}, {21, …, 30, 21 + 10n, …, 30 + 10n},

{31, …, 40, 11 + 10n, …, 20 + 10n}, {41, …, 50, 1 + 10n, …, 10 + 10n},

and analogous. As a general rule, easy to demonstrate, if v_r and v_s belong to the same orbit then $w_r = w_s$. Define $R = \{1, 6, 5n + 21, 11, 21, 31, \ldots, \}$. Then one may also see that for each r, $1 \le r \le 50 + 10n$, there is an element $j \in R$ such that $w_j = w_i$. *This implies that the minimum and maximum values of w_i will occur for atoms of the set R.* After drawing fullerenes in the E_n series by HyperChem [20], the application of TopoCluj software by Diudea and Nagy [21] allows to compute w_i's for each nodes of a given molecule. The calculations show that $Min_{E_2}\{w_i\}=347$, $Min_{E_3}\{w_i\}=420$, $Min_{E_4}\{w_i\}=493$, $Min_{E_5}\{w_i\}=570$. For the other n-sized $E_n = C_{50+10n}$ fullerenes, invariants $\underline{w}$ and $\overline{w}$ follow these polynomials curves:

$$Min\{w_i\} = \begin{cases} 5/2n^2 + 50n + 257 & n \text{ is even and } n \ge 6 \\ 5/2n^2 + 50n + 519/2 & n \text{ is odd and } n \ge 7 \end{cases}$$

$$Max\{w_i\} = \begin{cases} 5n^2 + 67n + 210 & n \text{ is even} \\ 5n^2 + 67n + 206 & n \text{ is odd} \end{cases}$$

Therefore, the topological efficiency indices exhibit the closed forms:

$$\rho(E_n) = \begin{cases} \frac{2}{3} \times \frac{10n^3 + 225n^2 + 1682n + 3063}{(5+n)(5n^2 + 100n + 519)} & n \text{ is odd and } n \ge 7 \\ \frac{2}{3} \times \frac{10n^3 + 225n^2 + 1682n + 3081}{(5+n)(5n^2 + 100n + 514)} & n \text{ is even and } n \ge 6 \end{cases}$$

$$\rho^E(E_n) = \begin{cases} \frac{2(5n^2 + 67n + 210)}{5n^2 + 100n + 514} & n \text{ is even and } n \ge 6 \\ \frac{2(5n^2 + 67n + 206)}{5n^2 + 100n + 519} & n \text{ is odd and } n \ge 7 \end{cases}$$

Table 1. Values of ρ and ρ^E for $E_n = C_{50+10n}$ fullerenes with n=2,3,4,5

n	E_n	ρ	ρ^E
2	C_{70}	1.018938	1.048991
3	C_{80}	1.032143	1.07619
4	C_{90}	1.055443	1.131846
5	C_{100}	1.072982	1.168421

Some special cases are recorded in Table 1. This proves that for $E_{n=}$ C_{50+10n} tubular fullerenes in the limit of large n:

$$\text{Lim}_{n\to\infty}\ \rho(E_n) = 4/3 \text{ and } \text{Lim}_{n\to\infty}\ \rho^E(E_n) = 2 \qquad (6)$$

It is worth noticing that the upper limit $\rho(E_n)<4/3$ in (6a) appears to be coherent with the topological efficiency value initially derived [7, 8] for infinite 1D nanostructures like polymers and carbon nanotubes. On the other hand, also the Wiener index of E_n graphs respects the n^3-dependence usually shown by polymers [12, 8]:

$$\mathbf{W(E_n)} = \begin{cases} \frac{50}{3}n^3 + 375n^2 + \frac{8410}{3}n + 5105 & n \text{ is odd} \\ \frac{50}{3}n^3 + 375n^2 + \frac{8410}{3}n + 5135 & n \text{ is even} \end{cases}$$

Present considerations about dimensionality remain valid for the other 3 classes of tubular fullerenes described below.

Table 2. The Hückel Energy (HE) for $E_{n=}$ C_{50+10n} fullerenes, $2 \le n \le 23$

n	$\|V(E_n)\|$	HE	$\frac{HE(E_n)}{HE(E_{n-1})}$
2	70	108.81312	
3	80	124.37634	1.143027
4	90	140.3847	1.128709
5	100	156.08364	1.111828
6	110	171.66224	1.099809
7	120	187.59942	1.09284
8	130	203.30214	1.083703
9	140	218.91228	1.076783
10	150	234.80318	1.07259
11	160	250.50838	1.066887
12	170	266.1399	1.062399
13	180	281.99822	1.059586
14	190	297.70356	1.055693
15	200	313.3493	1.052555
16	210	329.1819	1.050527
17	220	344.885	1.047703
18	230	360.53982	1.045391
19	240	376.351284	1.043855
20	250	392.05274	1.04172
21	260	407.70686	1.039929
22	270	423.5063	1.038752
23	280	439.20106	1.037059

Table 2 gives the Hückel Energy (HE) for $E_{n=} C_{50+10n}$ fullerenes, $2 \le n \le 23$ and Figure 12 shows the curve of the Hückel energy represented as a second degree polynomial of the topological sphericality $\sigma=\rho^E$ which is expressed by the following a parabolic curve $y = 871.97x^2 - 1787.7x + 1048.3$ being $x = \sigma$ and y=HE. Table 2 is computed by the TopoCluj software of Diudea and his team [15]. The level of this approximation appears to be quite good ($R^2 = 0.9966$) and it will be applied to the other fullerene families.

4.2. The Fullerene Series F_n

We now consider the case of $F_n = C_{60+12n}$. The symmetry group of this class of fullerenes can be computed as follows:

$$\mathrm{Aut}(F_n) = \begin{cases} D_{24} & n \text{ is odd} \\ Z_2 \times Z_2 \times S_3 & n \text{ is even} \end{cases}$$

Our calculations with this series of fullerenes show that $Min_{F_1} w_i = 354$, $Min_{F_2} w_i = 462$, $Min_{F_3} w_i = 570$, $Min_{F_4} w_i = 672$, $Min_{F_5} w_i = 776$, $Min_{F_6} w_i = 880$, $Min_{F_7} w_i = 990$, $Max_{F_1} w_i = 376$ and $Max_{F_2} w_i = 472$. For other values of n, we have:

$$Max\, w_i = \begin{cases} 6n^2 + 84n + 278 & n \text{ is even and } n \ge 4 \\ 6n^2 + 84n + 274 & n \text{ is odd and } n \ge 3 \end{cases}$$

$$Min\, w_i = \begin{cases} 3n^2 + 66n + 380 & n \text{ is even and } n \ge 8 \\ 3n^2 + 66n + 383 & n \text{ is odd and } n \ge 9 \end{cases}$$

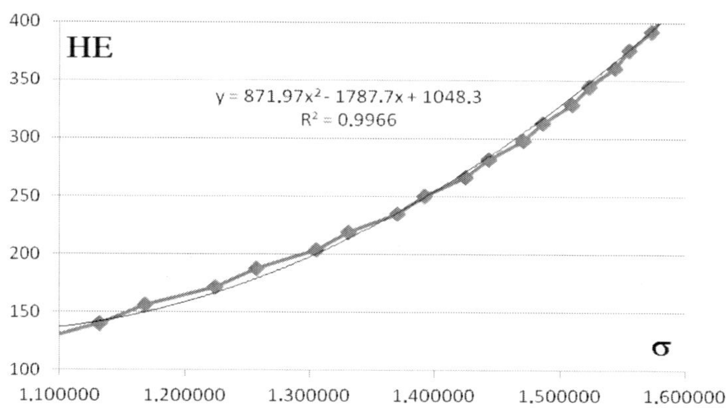

Figure 12. The Hückel Energy (HE) of E_n as a parabolic function of sphericality $\sigma=\rho^E$.

From above calculations, one can see that,

$$\rho(F_n)=\begin{cases}\dfrac{4n^3+96n^2+772n+1306}{3n^3+81n^2+713n+1915} & n \text{ is odd and } n\geq 9\\[2ex] \dfrac{4n^3+96n^2+772n+1316}{3n^3+81n^2+710n+1900} & n \text{ is even and } n\geq 8\end{cases}$$

$$\rho^E(F_n)=\begin{cases}\dfrac{6n^2+84n+278}{3n^2+66n+380} & n \text{ is even and } n\geq 8\\[2ex] \dfrac{6n^2+84n+274}{3n^2+66n+383} & n \text{ is odd and } n\geq 9\end{cases}$$

Hence, the asymptotic limits (6) are again valid. Special cases are reported in Table 3. By [17] $W(F_1)$ = 13284, $W(F_2)$ = 19716 being the general formulae:

$$\mathrm{W(F_n)}=\begin{cases}24n^3+576n^2+4632n+7836 & n \text{ is odd and } n\geq 3\\ 24n^3+576n^2+4632n+7896 & n \text{ is even and } n\geq 4\end{cases}$$

Based on the HE valued reported in Table 4, the computation of the best polynomial of $\sigma=\rho^E$ for approximating the energy of fullerene series F_n has been performed. Figure 13 reports the resulting diagram for the Hückel Energy dependence on topological sphericality for F_n fullerenes. Still the best polynomial approximating the energy of fullerene in the F_n series is a is the second-order polynomial $y = 668.62x^2 - 1073x + 562.83$, pointing out the good correlation ($R^2 = 0.9979$) between the two variables HE and σ.

4.3. The Fullerene Series G_n

We now consider the case of $G_n = C_{12n}$. In this case, the symmetry group of this class of fullerenes can be computed as follows:

$$\mathrm{Aut(G_n)}=\begin{cases}D_{24} & n \text{ is even}\\ Z_2\times Z_2\times S_3 & n \text{ is odd}\end{cases}$$

Table 3. The Special Cases of ρ and ρ^E for $F_n = C_{60+12n}$

n	F_n	ρ	ρ^E
2	C_{84}	1.016079	1.021645
3	C_{96}	1.007456	1.017544
4	C_{108}	1.024471	1.056548
5	C_{120}	1.039433	1.087629
6	C_{132}	1.060744	1.134091
7	C_{144}	1.076263	1.167677

Table 4. The Hückel Energy for F_n fullerenes with $1 \leq n \leq 19$

n	$\lvert V(E_n)\rvert$	HE	$\frac{HE(E_n)}{HE(E_{n-1})}$
1	72	112.06788	
2	84	130.74184	1.166630796
3	96	149.51864	1.143617376
4	108	168.69396	1.12824702
5	120	187.51864	1.111590717
6	132	206.26812	1.099987287
7	144	225.35188	1.092519193
8	156	244.20198	1.083647405
9	168	262.96704	1.076842374
10	180	282.0042	1.072393711
11	192	300.85636	1.066850636
12	204	319.63648	1.062422214
13	216	338.6368	1.059443528
14	228	357.48838	1.055669024
15	240	376.27226	1.05254403
16	252	395.2536	1.05044576
17	264	414.09532	1.047669952
18	276	432.88334	1.045371244
19	288	451.82238	1.04375091

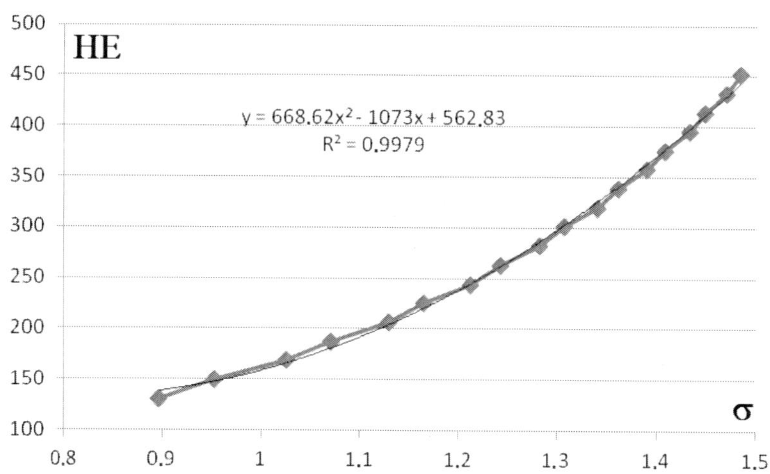

Figure 13. The Hückel Energy (HE) of F_n fullerenes as a parabolic function of sphericality $\sigma=\rho^E$.

Our calculations with this series of fullerenes show that $Min_{G_2} w_i = 66$, $Min_{G_3} w_i = 126$, $Min_{G_4} w_i = 186$, $Min_{G_5} w_i = 258$, $Min_{G_6} w_i = 334$, $Min_{G_7} w_i = 422$, $Min_{G_8} w_i = 517$, $Min_{G_9} w_i = 624$, $Min_{G_{10}} w_i = 739$, $Max_{G_2} w_i = 68$, $Max_{G_3} w_i = 128$ and $Max_{G_4} w_i = 210$.

For other values of n, we have:

$Max\, w_i = 12n^2 - 6n + 44,\ n \geq 5$

$Min\, w_i = 6n^2 + 140, \qquad n \geq 11$

From above calculations, one can see that,

$$\rho(G_n) = \frac{4n^3 + 69n - 136}{n(3n^2 + 70)}, \quad n \geq 11$$

$$\rho^E(G_n) = \frac{6n^2 - 3n + 22}{3n^2 + 70}, \quad n \geq 11$$

preserving the limits (6), being the special cases listed in Table 5.

As for the previous cases, special values were computed [17] $W(G_2) = 804$, $W(G_3) = 2292$, $W(G_4) = 4788$, $W(G_5) = 8514$ together with the following elegant expression for the Wiener index of $G_n = C_{12n}$ fullerenes:

$W(G_n) = 48n^3 + 828n - 1632,\ n \geq 6$

The best polynomial approximation for fullerene series G_n is represented in Figure 14 showing a good parabolic fit ($R^2 = 0.9799$) between HE and $\sigma=\rho^E$; this interpolation performs obviously better ($R^2 = 0.9982$) when only $n \geq 11$ and HE ≥ 205.9122 molecules are considered.

3.4. The Fullerene Series H_n

The case of $H_n = C_{10n}$ fullerenes admits a symmetry group which can be computed as follows:

$$\mathrm{Aut}(H_n) = \begin{cases} Z_2 \times A_5 & n = 2 \\ D_{20} & n > 2 \end{cases}$$

Table 5. The Special Cases of ρ and ρ^E for $G_n = C_{12n}$, $2 \leq n \leq 10$

n	G_n	ρ	ρ^E
2	C_{24}	1.015152	1.030303
3	C_{36}	1.010582	1.015873
4	C_{48}	1.072581	1.129032
5	C_{60}	1.1	1.217054
6	C_{72}	1.139721	1.317365
7	C_{84}	1.163846	1.398104
8	C_{96}	1.191489	1.477756
9	C_{108}	1.211182	1.541667
10	C_{120}	1.232476	1.602165

Table 6. The Hückel Energy G_n, $2 \leq n \leq 20$

n	$\lvert V(G_n)\rvert$	HE	$\frac{HE(G_n)}{HE(G_{n-1})}$
2	24	35.41938	
3	36	54.41528	1.536314
4	48	73.41442	1.349151
5	60	92.34556	1.257867
6	72	111.2499	1.204713
7	84	130.1414	1.169811
8	96	149.0734	1.145473
9	108	168.0386	1.12722
10	120	186.9828	1.112738
11	132	205.9122	1.101236
12	144	224.8293	1.09187
13	156	243.7374	1.0841
14	168	262.6374	1.077542
15	180	281.5302	1.071935
16	192	300.4189	1.067093
17	204	319.2993	1.062847
18	216	338.1774	1.059123
19	228	357.0496	1.055806
20	240	375.9162	1.05284

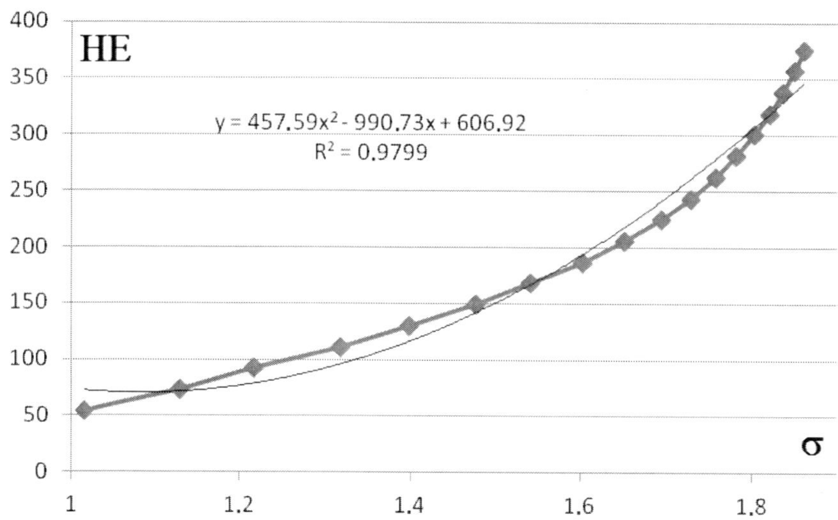

Figure 14. Hückel Energy (HE) of G_n fullerenes as a parabolic function of sphericality $\sigma=\rho^E$.

Our calculations with this series of fullerenes show that $Min_{H_2} w_i = 50$, $Min_{H_3} w_i = 93$, $Min_{H_4} w_i = 138$, $Min_{H_5} w_i = 193$, $Min_{H_6} w_i = 253$, $Min_{H_7} w_i = 323$, $Min_{H_8} w_i = 399$, $Max_{H_2} w_i = 50$ and $Max_{H_3} w_i = 99$.

For other values of n, we have:

$$Max\, w_i = 10n^2 - 5n + 26, n \geq 4$$

$$Min\, w_i = 5n^2 + 80, n \geq 9$$

From above calculations, one can see that,

$$\rho(H_n) = \frac{20n^3 + 235n - 402}{15n(n^2 + 16)}, n \geq 9$$

$$\rho^E(H_n) = \frac{10n^2 - 5n + 26}{5n^2 + 80}, n \geq 9$$

The special cases are listed in Table 7.

Previous relationships for ρ and ρ^E confirm the asymptotic trends $Lim_{n\to\infty}\rho(H_n) = 4/3$ and $Lim_{n\to\infty}\rho^E(H_n) = 2$ also for H_n infinite set. It is worth noticing that the condition $Lim_{n\to\infty}\rho(H_n) = 4/3$ leads to the following equivalent statement: given any natural number m, there is an infinite number of fullerenes with $4/3-\rho<10^{-m}$.

Previous computation [12, 22] show $W(H_2) = 500$, $W(H_3) = 1435$, $W(H_4) = 3035$ and the general curve:

$$W(H_n) = \frac{100}{3}n^3 + \frac{1175}{3}n - 670,\ n \geq 5.$$

We end the present detailed investigations by computing the best polynomial interpolating the Hückel Energy with the topological sphericality of fullerene series H_n. Our calculation is based on HE values (Table 8) and the formula for ρ^E (H_n). In Figure 15, the resulting parabolic fitting with $R^2 = 0.984$ is depicted. As for the previous case of G_n fullerenes, when the ρ^E curve is considered in the convergence region $n \geq 9$ or $HE \geq 138.737$ the correlation between the Hückel Energy and the sphericality $\sigma=\rho^E$ immediately increases, reaching the $R^2 = 0.9985$ level.

Table 7. The Special Cases of ρ and ρ^E for $H_n = C_{10n}$ fullerenes

n	H_n	ρ	ρ^E
2	C_{20}	1	1
3	C_{30}	1.028674	1.064516
4	C_{40}	1.099638	1.202899
5	C_{50}	1.13057	1.300518
6	C_{60}	1.16996	1.407115
7	C_{70}	1.194604	1.489164
8	C_{80}	1.223684	1.568922

Table 8. The Hückel Energy H_n, $2 \leq n \leq 16$

n	$\|V(H_n)\|$	HE	$\frac{HE(H_n)}{HE(H_{n-1})}$
2	20	29.41642	
3	30	44.87968	1.525668
4	40	60.40008	1.345822
5	50	75.98424	1.258016
6	60	91.61724	1.20574
7	70	107.2848	1.171011
8	80	122.9856	1.146346
9	90	138.737	1.128075
10	100	154.486	1.113517
11	110	170.2342	1.10194
12	120	185.9831	1.092513
13	130	201.7302	1.08467
14	140	217.4772	1.07806
15	150	233.2223	1.072399
16	160	248.9681	1.067514

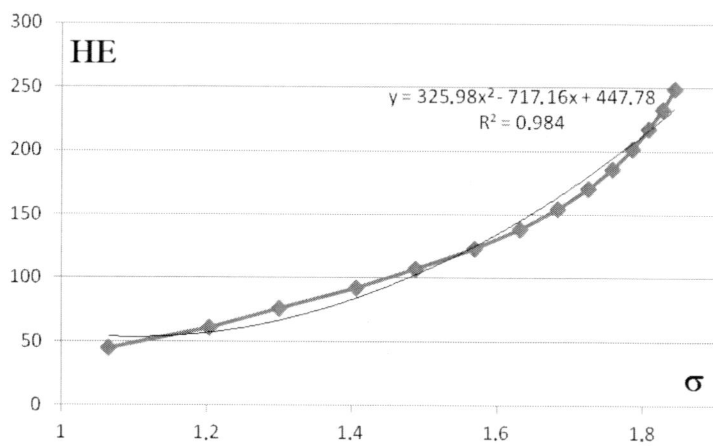

Figure 15. Hückel Energy (HE) of H_n fullerenes as a parabolic function of sphericality $\sigma=\rho^E$.

Conclusion

In this chapter the topological sphericality and Hückel energies of four series of tubular fullerenes are compared. These molecules grow along one axis and they are therefore describable as 1D nanostructures. In the case of large n, the discovered upper limit $\rho \to 4/3$ reproduces the usual condition valid for open 1D infinite molecular graphs [23].

The originally reported parabolic correlation between the molecular invariants HE and σ, represents a further demonstration of the fact that the electronic properties of sp^2 carbon systems are deeply rooted in the topology of their atomic networks, indicating the relevance of the *all-neighbors-interaction-picture* which is tipical of the topological long-range invariants and it is customary for parabolic and Hückel analysis, here generalized to the class of molecules (fullerenes) according with Primas' paradigmatic analysis aim [24]. Present fruitful results will be investigated on other classes of fullerene growing on an expanding spherical 2D surface, on the long range interaction while replacing the custom Coulomb interaction by topological potential with equivalent influence, yet carrying the quantum nature of nano-bondings via recently employed bondons [25].

ACKNOWLEDGMENTS

The first and second authors are partially supported by the university of Kashan under grant number 159020/74.

REFERENCES

[1] H. W. Kroto, J. R. Heath, S. C. O'Brien, R. F. Curl and R. E. Smalley, C60: Buckminsterfullerene, *Nature* 318 (1985) 162–163.
[2] P. W. Fowler and D. E. Manolopoulos, *An Atlas of Fullerenes*, Oxford Univ. Press, Oxford, 1995.
[3] M. Deza, P. W. Fowler, A. Rassat and K. M. Rogers, Fullerenes as Tilings of Surfaces, *J. Chem. Inf. Comput. Sci.* 2000, 40, 550-558.
[4] W. Myrvold, B. Bultena, S. Daugherty, B. Debroni, S. Girn, M. Minchenko, J. Woodcock, P. W. Fowler, FuiGui: A graphical user interface for investigating conjectures about fullerenes, *MATCH Commun. Math. Comput. Chem.* 58 (2007) 403 – 422.
[5] P. Schwerdtfeger, L. Wirz and J. Avery, Program fullerene – a software package for constructing and analyzing structures of regular fullerenes, version 4.4, *J. Comput. Chem.* 34 (2013) 1508 – 1526.
[6] D. Vukičević, F. Cataldo, O. Ori and A. Graovac, Topological efficiency of C_{66} fullerene, *Chem. Phys. Letters* 501 (2011) 442 –445.
[7] O. Ori, F. Cataldo and A. Graovac, Fullerenes, Topological ranking of C_{28} fullerenes reactivity, *Fullerenes, Nanotubes Carbon Nanostructures* 17 (2009) 308 – 323.
[8] J. Sedlar, D. Vukičević, F. Cataldo, O. Ori and A. Graovac, Compression ratio of Wiener index in 2D – rectangular and polygonal lattices, *Ars Mathematica Contemporanea* 7 (2014) 1 – 12.
[9] O. Ori, F. Cataldo, S. Iglesias–Groth and A. Graovac, Topological modeling of $C_{60}H_{36}$ hydrides, In: *Fulleranes: The Hydrogenated Fullerenes*, F. Cataldo, S. Iglesias–Groth (Eds.), Springer–Varlag, 2010; pp. 251 – 272.
[10] Z. Slanina, M.-C. Chao, S.-L. Lee and I. Gutman, On applicability of the Wiener index to estimate relative stabilities of the higher-fullerene IPR isomers, *J. Serb. Chem. Soc.*, 1997, 62(3), 211-217.
[11] J. R. Dias, *Molecular orbital calculation using chemical graph theory*. Springer, Berlin, 1993.

[12] D. Vukičević, *private communication*, Split University 2010.

[13] A. Graovac, A. R. Ashrafi, O. Ori, Topological efficiency approach to fullerene stability – Case study with C_{50} In: S. C. Basak, G. Restrepo, J. L. Villaveces (Eds.), *Advances in Mathematical Chemistry*, Bentham, Sharajah, 2014, in press.

[14] J. E. Graver, Catalog of all fullerenes with ten or more symmetries, Graphs and discovery, 167–188, *DIMACS Ser. Discrete Math. Theoret. Comput. Sci.*, 69, Amer. Math. Soc., Providence, RI, 2005.

[15] S. Djafari, F. Koorepazan–Moftakhar and A. R. Ashrafi, Eccentric sequences of two infinite classes of fullerenes, *J. Comput. Theor. Nanosci.* 10 (2013) 2636 – 2638.

[16] F. Koorepazan–Moftakhar and A. R. Ashrafi, Symmetry and PI index of C_{60+12n} fullerenes, *J. Comput. Theor. Nanosci.* 10 (2013) 2484 – 2486.

[17] F. Koorepazan–Moftakhar and A. R. Ashrafi, Fullerenes: Topology and Symmetry, In: *Topics in Chemical Graph Theory*, I. Gutman (Ed.), University of Kragujevac and Faculty of Science Kragujevac, 2014; pp. 163–176.

[18] O. Ori, M. V. Putz, I. Gutman and P. Schwerdtfeger, Generalized Stone-Wales Transformations for Fullerene Graphs Derived from Berge's Switching Theorem, In: I. Gutman, B. Pokrić, D. Vukičević (Eds.), Ante Graovac – Life and Works, Univ. Kragujevac, Kragujevac, 2014, pp. 259–272.

[19] F. Koorepazan–Moftakhar, A. R. Ashrafi and Z. Mehranian, Symmetry and PI polynomials of C_{50+10n} fullerenes, *MATCH Commun. Math. Comput. Chem.* 71 (2014) 425 – 436.

[20] *HyperChem package Release 7.5 for Windows*, Hypercube Inc., Florida, US, 2002.

[21] M. D. Diudea, O. Ursu and L. Cs. Nagy, *TOPOCLUJ*, Babes–Bolyai University, Cluj, 2002.

[22] A. Graovac, O. Ori, M. Faghani and A. R. Ashrafi, Distance Property of Fullerenes, *Iranian Journal of Mathematical Chemistry*, Vol. 2, No. 1, September 2011, pp. 99 – 107.

[23] F. Koorepazan–Moftakhar, A. R. Ashrafi, O. Ori, and M. V. Putz, Topological Efficiency of Fullerene, *J. Comput. Theor. Nanosci.* (2014) 000-000.

[24] H. Primas, Chemistry, Quantum Mechanics, and Reductionism: perspectives in theoretical chemistry. In: *Lect. Notes Chem.* Volume 24, Springer Verlag, Berlin, 1981.

[25] M. V. Putz, O. Ori, Bondonic Effects in Group-IV Honeycomb Nanoribbons with Stone-Wales Topological Defects. *Molecules* 19(4) (2014) 4157-4188.

INDEX

A

F

G

H

I

J

K

L

M

N

O

P

Q

R

S

T

U

V

W

X

Y

Z